DESIGN OF EXPERIMENTS FOR AGRICULTURE AND THE NATURAL SCIENCES

Second Edition

T0239604

DESIGN OF EXPERIMENTS FOR AGRICULTURE AND THE NATURAL SCIENCES

Second Edition

A. Reza Hoshmand

Daniel Webster College
Nashua, New Hampshire, U.S.A.

CRC Press
Taylor & Francis Group
Boca Raton London New York

CRC Press is an imprint of the
Taylor & Francis Group, an **informa** business

A CHAPMAN & HALL BOOK

CRC Press
Taylor & Francis Group
6000 Broken Sound Parkway NW, Suite 300
Boca Raton, FL 33487-2742

First issued in paperback 2020

© 2006 by Taylor & Francis Group, LLC
CRC Press is an imprint of Taylor & Francis Group, an Informa business

No claim to original U.S. Government works

ISBN 13: 978-0-367-57788-9 (pbk)
ISBN 13: 978-1-58488-538-2 (hbk)

Visit the Taylor & Francis Web site at
http://www.taylorandfrancis.com

and the CRC Press Web site at
http://www.crcpress.com

To my wife Lisa and

my children Anthony and Andrea

Preface and Acknowledgment

This book is written for students and applied researchers of agricultural and natural sciences. It is intended to serve as a guide in the design and analysis of agricultural and natural science experiments. The emphasis of the second edition, as was in the first, is on understanding the logical underpinnings of design and analysis. It is hoped that the reader can extend the principles of design and analysis presented here to those which are not discussed in this book. In the presentation of the various designs, the reasons for selecting a particular design and its analysis are discussed. Practical examples from different areas of agriculture are given throughout, to show how practical issues of design and analysis are handled.

It is assumed that the reader has familiarity with statistical concepts such as the normal, binomial, chi-square, and t distributions. No previous exposure to analysis of variance is assumed. Since this is meant to be a basic experimental design book rather than a reference book of design and analysis, a number of advanced topics have not been addressed.

In organizing the book, I have elected in the first 3 chapters of the book to discuss the practical issues of design, assumptions of the analysis of variance, and the problems associated with the data in agricultural research. Chapters 4, 5, and 6 present the experimental design and analysis of 1-, 2-, and 3-factors. The concept of treatment means comparisons are discussed in Chapter 7. Chapter 8 presents the design and analysis of experiments over time. Regression and correlational analysis are presented in Chapter 9. A new chapter has been added in this edition. The content of Chapter 10 deals with the analysis of covariance and how it is applied to several designs.

An attempt has been made to improve on the readability of the book. First, a step-by-step process is used to explain the concept of design and analysis. Second, equations are numbered consecutively in each chapter as they are introduced. If the same equation is repeated at a later point within the chapter, the original equation number appears.

I wish to gratefully acknowledge the assistance provided by Bob Stern of CRC Press in the completion of this book. A special note of gratitude is due my wife Lisa, who not only provided support, but editorial assistance in the preparation of the manuscript. Finally, I wish to thank Dr. Guangxiong Fang of the Engineering, Math, and Science Division of Daniel Webster College for his assistance with the use of Minitab.

A. Reza Hoshmand

About the Author

A. Reza Hoshmand is the Associate Dean for Graduate Studies and Chair of Business and Management Division at Daniel Webster College. A native of Kabul, Afghanistan, he holds a Ph.D. in Agricultural and Resource Economics from the University of Maryland. Prior to his current position he was a Professor of Economics and Finance at Lesley University. Professor Hoshmand has taught at Harvard and Tufts University courses in Economics and Statistics. His research interests are in economic development, international trade, and quantitative methods.

Hoshmand has published numerous articles in academic journals, and his latest book, *Business and Economic Forecasting for the Information Age* was published by Greenwood Publishing in 2002. Hoshmand has also authored *Statistical Methods for Environmental and Agricultural Sciences* published by CRC Press in 1997. He continues to be involved in projects sponsored by the U.S. Agency for International Development.

Table of Contents

1

The Nature of Agricultural Research

1.1 Fundamental Concepts

Agriculturalists and other scientists in biological fields who are involved in research constantly face problems associated with planning, designing, and conducting experiments. Basic familiarity and understanding of statistical methods that deal with issues of concern would be helpful in many ways. Practitioners (researchers) who collect data and then look for a statistical technique that would provide valid results may find that there may not be solutions to a problem and that the problem could have been avoided in the first place by conducting a properly designed experiment. Obviously, it is important to keep in mind that we cannot draw valid conclusions from poorly planned experiments. Second, the time and cost involved in many experiments are considerable, and a poorly designed experiment increases such costs. For example, agronomists who carry out a fertilizer experiment know the time limitations on the experiment. They know that, in the temperature zone, seeds must be planted in the spring and harvested in the fall. The experimental plot must include all the components of a complete design. Otherwise, what is omitted from the experiment will have to be carried out in subsequent trials in the next cropping season or the next year. This additional time and expenditure could be minimized by a properly planned experiment that would produce valid results as efficiently as possible.

Good experimental designs are the product of the technical knowledge of one's field, an understanding of statistical techniques, and skill in designing experiments. Any research endeavor, be it correlational or experimental, may entail the following phases: conception, design, data collection, analysis, and dissemination. Statistical methodologies can be used to conduct better scientific experiments if they are incorporated into the entire scientific process, i.e., from recognition of the problem to experiment design, data analysis, and interpretation. The intent of this book is to provide practitioners with the necessary guidelines and techniques for designing experiments, as well as the statistical methodology to analyze the results of experiments in an efficient way.

Agricultural experiments usually entail comparisons of crop or animal varieties. When planning agricultural experiments, we must keep in mind that large uncontrolled variations are common occurrences. For example, a crop scientist who plants the same variety of a crop in a field may find variations in yield that are due to periodic variations across a field or to some other factor that the experimenter has no control over. Throughout this book we are concerned with the methodologies used in designing experiments that will separate, with confidence and accuracy, varietal differences of crops and animals from uncontrolled variations.

It is essential that you become familiar with the terminology used in this book as we discuss the concept of experimentation. Agricultural experimentations are conducted in response to questions raised by researchers who are interested either in comparing the effects of several conditions on some phenomena or in discovering the unknown effects of a particular process. An experiment facilitates the study of such phenomena under controlled conditions. Therefore, the creation of controlled conditions is the most essential characteristic of experimentation. It has been said that wisdom consists not so much in knowing the right answers, as in knowing the right questions to ask (Gill, 1980). Hence, how we formulate our questions and hypotheses are critical to the experimental procedure that will follow.

Once we have established a hypothesis, we look at a method to objectively test the validity of that hypothesis. Our final results and conclusions depend to a large extent on the manner in which the experiment was statistically designed and how the data were collected. Agricultural researchers know that it is difficult to avoid differences in yield of the same crop variety planted in two adjacent fields, or the weight gain in two animals fed the same ration, because of uncontrolled variations. Such differences in yield resulting from experimental units treated alike are called *experimental error*. The intent of an optimal design is to provide a mechanism for estimation and control of experimental error in the field experiments conducted. Practitioners should keep in mind that it is very difficult to account for all the sources of natural variation even when an entire population of factors of interest is under study. The problem is further aggravated when we depend on a sample that is, at best, an approximation of the population or the characteristic of interest. Given this dilemma, what practitioners can hope for is a predictive model that minimizes experimental error. Such models are based on theoretic knowledge, empirical validity, and an understanding of experimental material. Before we discuss the various designs, and any methodology for the estimation and control of experimental error, a quick review of other commonly encountered terms is in order.

When discussing experimentation, you will encounter the term *experimental unit*. An experimental unit is an entity that receives a treatment. For example, for a horticulturist it may be a plot of land or a batch of seed, for an animal scientist it may be a group of pigs or sheep, and for an agricultural engineer it may be a manufactured item. Thus, an experimental unit may be looked upon as a small subdivision of the experimental material, which receives the treatment.

In choosing an experimental unit, the researcher must pay special attention to the size of the unit, the representative nature of the unit, and how independent the units are of one another. In terms of the size of the unit, technical and cost factors play a major role. The essential point to determine is how many units are needed to attain a specified precision in the most economical way.

As far as the representative nature of the unit is concerned, it is important that the conditions of the experiment be as close as possible to those of the actual study subject. For example, if an animal scientist is interested in a feeding experiment in which the results are to apply to Holstein dairy cows, then ideally the sample of cows in the experiment should be selected from a population of Holstein dairy cows.

The independence of experimental units from one another is also an important element in a properly conducted experiment. Independence of units implies that the researcher must ensure that the treatment applied to one unit has no effect on the observation obtained in another. Furthermore, the occurrence of unusually high or low observations in one unit should have absolutely no effect on what may be observed in another unit.

Replication refers to the repetition of the basic experiment or treatment. Thus, an animal scientist who treats ten animals with a particular antibiotic has ten replicates. Researchers replicate their experiments to control for experimental error. It is also used to estimate the error variance and to increase the precision of estimates. It is important to keep in mind that increased replication of experimental units is desirable for increasing the accuracy of estimates of means and other functions of the measured variable (Gill, 1980; Das and Giri, 1986). However, the cost of increased replication plays an important role in the number of replicates a researcher is able to afford.

As the experimenter makes a decision to use a particular design, the choice of the number of replicates must also be made. It has been suggested that present-day design practices utilize existing information to the fullest, and replication is sometimes not needed at all (Anderson and McLean, 1974). In some cases, such as factorial fractional replication (discussed in Chapter 6, Section 6.4), only a part of all the treatment combinations is used in an experiment.

No matter how many replicates are used, the important point to keep in mind is whether the experiment will provide valid results in terms of estimates of effects, and whether there are enough degrees of freedom for error to adequately test for various effects.

Randomization is simply an objective method of random allocation of the experimental material or allocation of treatment to experimental units. We also use randomization to make certain that the order in which individual trials of experiments are performed is determined randomly. The advantages of randomization are: (1) it allows for protection against systematic error caused by subjective assignment of treatments; that is, each treatment will have an equal chance of being assigned to an experimental unit; (2) it helps in "averaging out" the effects of uncontrolled conditions or extraneous factors that persist over long or repeated experiments; and (3) it validates the statistical assumption that states that observations (or errors) are independently distributed random variables. In sum, we could say that randomization is the cornerstone of statistical theory in the design of experiments. Fisher (1956) stated to ensure the error estimate will be a good and valid one, we must randomize experimental treatments among experimental units. For further reading on this topic, Fisher (1956), Ogawa (1974), Gill (1980), Gomez and Gomez (1984), Mead (1988), Mead, Curnow, and Hasted (1993), Montgomery, (2000), and Samuels and Witmer (2003) may be consulted.

We shall refer to the experimental variables as *factors*, or we may define a controllable condition in an experiment as a *factor*. For example, a fertilizer, a new feed ration, and a fungicide are all considered factors. Factors may be quantitative or qualitative, and may take a finite number of values and types. Quantitative factors are those described by numerical values on some scale. The strength of a drug dosage (such as milligrams of a sulfa drug), the rate of application of a fertilizer, and temperature are examples of quantitative factors. Qualitative factors, such as type of protein in a diet, sex of an animal, age, or the genetic makeup of a plant or animal are factors that can be distinguished from each other, but not on a numerical scale.

Different factors are included in experimental designs for different reasons. The decision of including or excluding a factor, the levels of each factor in the experiment, and the criterion for selecting such factors are the responsibility of the experimenter, who may be influenced by theoretical considerations. What is essential to remember is that factors should be related to one another in simple ways. The relation of one factor to another implies how the levels of that factor are combined with the levels of another factor. When choosing factors for any experiment, the researcher should ask the following questions:

1. What treatments in the experiment should be related directly to the objectives of the study?
2. Does the experimental technique adopted require the use of additional factors?
3. Can the experimental unit be divided naturally into groups such that the main treatment effects vary for different groups?
4. What additional factors should one include in the experiment to interact with the main factors and shed light on factors of direct interest?
5. How desirable is it to deliberately choose experimental units of different types?

A *treatment* in a broad sense refers to a controllable quantitative or qualitative factor imposed at certain levels by the experimenter. For an agronomist, several fertilizer concentrations applied to a particular crop or a variety of a crop is a treatment. Similarly, an animal scientist regards several concentrations of a drug given to an animal species as a treatment. In agribusiness, we may consider the impact of an advertising strategy on sales as a treatment. To an agricultural engineer, different levels of irrigation may constitute a treatment. It should be kept in mind that a treatment is not a condition of the experiment unless the design consists only of treatment factors. This is because a *condition*, by definition, requires one level from every design factor for its specification (Lee, 1975). When there is only one criterion for classifying the imposed factor or condition, the levels of a factor may also be referred to as treatments.

Frequently, researchers include a treatment factor in their design not because they are interested in the treatment factor *per se*, but because the factor might prove to have importance, or the researcher may wish to control the effects of unavoidable variations in the experiment. For example, if the time of day has an effect on taking blood samples from an animal even though the experimenter had no real interest in studying such an effect, "time of day" may be included in the design so that the effects would not enter in a haphazard way. Factors that function to systematically take into account the effects of extraneous variables are called *control factors*.

The *levels* of a factor refer to its presence or absence in an experiment. Another way of stating the level of a factor is the number of ways in which it is varied. So when a factor takes on different values, such as a high and a low dose of an antibiotic or two modes of an application such as drip or sprinkler irrigation, we are referring to these as the level of the factor (independent variable) of interest. Similarly, when a factor is varied in three ways, we refer to it as having three levels. In experiments in which there are two or more factors with two or more levels each, a treatment consists of a combination of one level of each factor. When all possible combinations of one level of each factor, along with an equal number of observations for each treatment, are included in an experiment, we have a complete factorial experiment with equal replications.

Blocking refers to a methodology that forms the units into homogeneous or preexperimental similar-subject groups. It is a method to reduce the effect of variations in the experimental material on the error of the treatment comparison (Dodge, 1985). For example, animal scientists may decide to group animals based on age, sex, breed, or other factor that they may believe has an influence on the characteristic being measured. Effective blocking removes a considerable measure of variation from experimental error. A researcher must keep in mind that the selection of the source of variability to be used as the basis for blocking, as well as block shape and orientation, is crucial for effective blocking technique (Gomez and Gomez, 1984). Experimenters include a blocking factor in their design to investigate observational differences among blocks and to increase the power of the design to detect treatment effects.

The last three concepts, namely *replication, randomization*, and *blocking*, are quintessential parts of experimental designs. Each of these concepts is explained fully as different designs are introduced in subsequent chapters.

Good design is important for good research (results). The following examples point out the necessity for good designs to yield good research. First, a nutrition specialist in a developing country is interested in determining whether mother's milk is better than powdered milk for children under age one. The nutritionist has compared the growth of children in village A, who are all on mother's milk, with the children in village B, who use powdered milk. Obviously, such a comparison ignores the health of the mothers, the sanitary conditions of the villages, and other factors that may have contributed to the differences observed without any connection to the advantages of mother's milk or powdered milk on the children. A proper design would require that both mother's milk and powdered milk be alternatively used in both villages, or some other methodology to make certain that the differences observed are attributable to the type of milk consumed and not to some uncontrollable factor. Second, a crop scientist who is comparing two varieties of maize, for instance, would not assign one variety to a location where such factors as sun, shade, unidirectional fertility gradient, and uneven distribution of water would either favor or handicap it over the other. If such a design were to be adopted, the researcher would have difficulty in determining whether the apparent difference in yields was due to varietal differences or resulted from these factors. Such examples illustrate the types of poorly designed experiments that are to be avoided.

1.2   Research by Practitioners

Researchers in different disciplines within the food and fibers fields are commonly concerned about problems and their solutions that have implications for research, production, nutrition, handling, processing, marketing, and distribution, to name a few. Many of the experiments in agriculture tend to be comparative and correlational in nature. That is, an agronomist or an animal scientist would compare a number of varieties of some crops or breeds of animals, respectively, under different treatment conditions, to determine whether the yield or weight gain of one variety or breed is superior to the other, and if so, by how much. The determination of the superiority of a crop variety or an animal breed is based on the statistical significance attached to the results. In Chapter 4 through Chapter 7, various designs are discussed in which tests are performed to determine varietal differences. Correlational studies elaborate on the degree of association between variables, such as how the yield of a crop is affected by the application of fertilizer and the amount of rainfall. Correlational analysis (discussed in Chapter 9) allows the researcher to study the association between variables, alone or in combination.

The objectives of a study set in motion the plan to conduct an experiment, which in turn requires an experimental design that fits the objectives of the experiment. Experimenters have at their disposal a variety of experimental designs and techniques to obtain answers to research questions. The choice of the design determines the parameters of the experiment. This then allows the experimenter to make a number of technical decisions, such as the size of the experiment, the manner in which treatments are allotted to experimental units, and the grouping of the experimental units, etc., using in each case the fundamental concepts discussed in the previous section. As there are many designs (from the very simple to the highly complex) available to researchers, the critical point to consider is which design will be suited for a particular experiment and how efficient the design is in providing answers to research questions, given limited resources.

Once a design is selected, the choice and the level of treatments in the experiment, the number of replications to be used, and how the analysis is to be performed are all critical planning elements. Improper attention to these factors will have major implications for how the experiment is conducted and analyzed. To illustrate the importance of analysis in the design of experiments, consider an experiment in which an animal scientist is comparing a new sulfa drug (*B*) with an established drug (*A*), the performance of which has already been tested. The animal scientist argues that the emphasis should be placed on the new drug and feels it is not necessary to reexamine drug *A* as fully. Thus, in comparing drugs *A* and *B*, the investigator subjects 19 of the 20 pigs in the experiment to the new drug and only 1 to the established drug. In performing the analysis, the difference between the mean response to the drugs is tested, that is, $(\bar{x}_A - \bar{x}_B)$, and the standard error of the difference computed. The standard error of the difference for the samples of 1 and 19 is

$$\sigma\sqrt{\left(\frac{1}{1} + \frac{1}{19}\right)} = 1.023\sigma$$

To make the point clear, consider the situation in which the experimenter appropriately designs the experiment, testing both drugs under identical conditions. This means that the sample of 20 pigs was equally divided between drugs *A* and *B*. The standard error for this design is

$$\sigma\sqrt{\left(\frac{1}{10} + \frac{1}{10}\right)} = 0.447\sigma$$

Notice the difference in the standard errors under the two conditions. The smaller standard error of the second design, a highly desirable characteristic of any good experiment, improves the precision of the experiment by a factor of more than two. Obviously, not considering statistical analysis can also lead to poorly designed experiments.

In this section a few selected examples are given to illustrate the common types of comparisons made in agriculture and how practitioners design experiments to obtain answers to comparison questions.

Example 1.2.1

A crop scientist is interested in determining whether varieties of rice respond differently to a fertilizer treatment. In this particular instance the research question is simply about varietal differences. For such a comparative study, often the researcher performs a field trial in which the experimental area is divided into plots and different varieties are assigned, one to each plot. The yield is then measured or estimated for each plot, and varieties are thus compared. In making a comparison such as this, the researcher has to be cautious with respect to several factors that contribute to differences in yield. Yield differences due to the fertility gradient of the soil, systematic trends, or difference in space and time may be significant. Research has shown that yields may vary from plot to plot by as much as 20 to 30% from their mean (Cox, 1992). Such a variation is considerable when the difference of a small percentage such as 5 to 10% may have practical implications for the adoption of a variety.

From a design point of view, an experiment such as this would have to be arranged in such a way that the varietal differences are accurately separated from the uncontrolled variations mentioned earlier. As the aim of an experiment such as this is to find out which variety is superior, rather than an absolute determination of the yield per acre, it would be economical to make a direct comparison of varieties, using separate experiments for each variety, and then compare the mean yield of each variety under representative conditions. However, if the researcher is interested in the absolute determination of the yield per acre for each variety, the design would be different from that conducted by the crop scientist.

Example 1.2.2

In another experiment, a researcher is interested in determining whether hybrids or plant densities interact with the effect of 15-in. row spacing on grain yield and stalk breakage of corn grown under nonirrigated conditions. To get answers to his questions, the researcher conducts a $2 \times 2 \times 4$ factorial experiment replicated three times in a randomized complete block design arranged in a split-split-plot layout (this is discussed more fully in Chapter 6).

You will note that in this experiment there are three factors, namely, two corn hybrids (factor 1), two row spacings (factor 2), and four target plant densities (factor 3). As the emphasis in this experiment is on the interactions, the layout of the experiment would be such that hybrids are allocated to the main plots, row spacings to subplots, and plant densities to sub-subplots. In this experiment there are three levels of precision, the main plot factor receiving the lowest degree of precision and the sub-subplot receiving the highest degree of precision. The design chosen for this experiment correctly places emphasis on the interactions rather than on the main effects. Stated differently, the present design sacrifices precision on the main effects in order to provide higher precision on interactions.

Example 1.2.3

When plant breeders are interested in selecting qualitative traits, nursery tests can be conducted very easily. However, with complex traits such as yield, which is influenced by environment, the study should be conducted in several localities. Practitioners can use various techniques on many crops to evaluate genotype stability over a range of environmental conditions. The most widely used technique is regression analysis (Chapter 9 in this book), which was proposed by Yates and Cochran (1938), amplified by Finley and Wilkinson (1963), and further refined and adopted by other practitioners (Eberhart and Russell, 1966; Shukla, 1972; Ntare and Aken'Ova, 1985).

Other techniques such as genotype grouping, which does not use regression analysis, have also been used in yield-stability studies (Francis and Kannenberg, 1978).

Example 1.2.4

A group of crop scientists (Peterman, Sears, and Kanemasu, 1985) was interested in determining genotypic differences in the rate and duration of spikelet initiation in ten winter wheat cultivars. To study the differences among cultivars, the experiment was conducted on ten cultivars with varying phenotypes. The cultivars were planted on two dates with four replications each. The scientists were interested in making three comparisons regarding these cultivars. Specifically, they compared the rate and duration of spikelet initiation among benchmark cultivars, between semidwarf and standard-height wheats, and between semidwarf pure lines and hybrids.

The experimental design for this study was a randomized complete block (discussed in Chapter 4). Analysis of variance and least significant difference (Chapter 7, Section 7.2.1) was used to determine significant differences. The study also included a regression analysis (Chapter 9) to evaluate the correlation of interest.

Example 1.2.5

In a study to evaluate soybean yield response to sprinkler irrigation (center pivot) management strategies, specifically those scheduled by growth stage vs. those indicated by soil moisture depletion in different row widths, the scientists were interested in identifying the yield component mechanisms that respond to sprinkler irrigation strategies and in evaluating the effect of sprinkler irrigation on reproductive organ abscission (Elmore, Eisenhauer, Specht, and Williams, 1988).

This experiment was conducted over a 3-yr period in a research station in Nebraska. The scientists used a split-split-plot randomized complete block experimental design (see Chapter 6, Section 6.2) with four replications. The main plots were irrigation treatments and included a nonirrigated check, irrigation scheduled by soil moisture depletion, irrigation commenced no earlier than flowering, and irrigation commenced no earlier than pod elongation. The researchers limited the application of water to 1.5 in./week in order to simulate a minimum-capacity center pivot system. The subplots consisted of 10- and 30-in. row widths, and were 30 ft long. The sub-subplots were planted with three cultivars of soybean. To further minimize error, the harvest areas within the subplots were limited to 20 ft in length and consisted of the center 2 of the four rows that were 30 in. wide, and the center 6 of the ten rows that were 10 in. wide.

Standard analyses of variance for a split-split-plot design were used. The single degree of freedom comparisons (see Chapter 7, Section 7.3.1) were made for 1982 vs. 1983, 1984, and 1983 vs. 1984. The least significant difference (LSD) was used for mean comparisons.

Example 1.2.6

In determining the yield difference due to the application of P and K on kale, a horticulturist conducts a randomized complete block design experiment with four replications and four treatments. She is particularly interested in the following questions: (1) Is there a difference in yield between the fields receiving P and those receiving K? (2) Are there yield differences between the fields receiving P and K separately and those receiving P and K together? (3) Is there a response to fertilization at all? In response to these questions, the horticulturist designs the following set of treatments: (1) control, (2) P added to the plots, (3) K added to the plots, and (4) both P and K added to the plots.

This experiment is exactly the same as the one given in Chapter 7, Section 7.3.1. The reader is referred to this section of the book for the approach used to obtain answers to the researcher's questions.

Example 1.2.7

Agronomists have hypothesized that the successful establishment of legumes in tall fescue (*Festuca arrundinacea* Schreb.) pasture is limited by the amount of light that can be transmitted through the grass canopy. To ascertain the proportion of incident light transmitted to the ground through canopies of various biomasses, the agronomists conducted a study to quantify the relationship between leaf area index (LAI), canopy biomasses, and light penetration into the canopy of an established tall-fescue pasture (Trott, Moore, Lechtenberg, and Johnson, 1988).

Regression analysis was used in this particular study, as it was an objective of the study to determine the relationship between canopy biomass and leaf area.

Example 1.2.8

Animal researchers were interested in determining if the quantity of beef produced per acre could be increased by applying N fertilizer at 0, 100, and 200 lb/acre to tall fescue-ladino clover pastures or by feeding a supplemental grain-protein mix (creep feed) to calves from mid-November to mid-April (Morrow et al., 1988).

The experimental design used for this study was a randomized complete block with two replications, where treatments were 0, 100, and 200 lb of elemental nitrogen per acre, and creep feed vs. no creep feed for the calves for a total of twelve pastures (experimental units).

The researchers used six test cows and one or two substitute cows. The test cows were randomly assigned to each of the twelve pastures. Each cow remained on her assigned pasture system for the entire year and received no supplemental feed. Substitute cows were grazed with the test cows and were used for replacement if a test cow lost a calf or failed to breed back during the cycle. Self-feeders were used for creep feed treatments.

In performing the analysis of variance, the sources of variation tested were on pasture replications (2), sex of calf (2), creep feeding (2), years or environment (4), levels of nitrogen fertilization (3), and all the interactions. The researchers used the LSD test for mean comparisons.

The previous examples are only a small sample of the variety of research questions that are addressed by researchers in the various fields of agriculture. In subsequent chapters of this book, you will be introduced to a number of research designs that are often used by practitioners in agriculture.

References and Suggested Readings

Anderson, V.L. and McLean, R.A. 1974. *Design of Experiments: A Realistic Approach.* New York: Marcel Dekker.

Cochran, W.G. 1954. The combination of estimates from different experiments. *Biometrics* 10: 101–129.

Cochran, W.G. and Cox, G.M. 1957. *Experimental Designs.* 2nd ed. New York: John Wiley & Sons.

Cox, D.R. 1954. The design of an experiment in which certain treatment arrangements are inadmissible. *Biometrika* 41: 287.

Cox, D.R. 1992. *Planning of Experiments.* New York: John Wiley & Sons.

Das, M.N. and Giri, N.C. 1986. *Design and Analysis of Experiments.* 2nd ed. New York: John Wiley & Sons.

Dodge, Y. 1985. *Analysis of Experiments with Missing Data.* New York: John Wiley & Sons.

Eberhart, S.A. and Russell, W.A. 1966. Stability parameters for comparing varieties. *Crop Sci.* 6: 36–40.

Elmore, R.W., Eisenhauer, D.E., Specht, J.E., and Williams, J.H. 1988. Soybean yield and yield component response to limited-capacity sprinkler irrigation systems. *J. Prod. Agric.* July–September, 1(3): 196–201.

Finley, K.W. and Wilkinson, G.N. 1963. The analysis of adaptation in plant breeding programme. *Aust. J. Agric. Res.* 14: 742–754.

Fisher, R.A. 1956. *Statistical Methods for Research Workers.* 12th ed. Edinburg: Oliver and Boyd.

Francis, T.R. and Kannenberg, L.W. 1978. Yield stability studies in short season maize. I. A descriptive method of grouping genotypes. *Can. J. Sci.* 58: 1029–1034.

Gill, J.L. 1980. *Design and Analysis of Experiments in the Animal and Medical Sciences.* Ames, IA: Iowa State University Press.

Gomez, K.A. and Gomez, A.A. 1984. *Statistical Procedures for Agricultural Research.* 2nd ed. New York: John Wiley & Sons.

John, P.W.M. 1998. *Statistical Design and Analysis of Experiments.* New York: Macmillan.

Lee, W. 1975. *Experimental Design and Analysis.* San Francisco: W.H. Freeman.

Mead, R. 1988. *The Design of Experiments: Statistical Principles for Practical Applications.* chap. 9. Cambridge: Cambridge University Press.

Mead, R., Curnow, R.N., and Hasted, A.M. 1993. *Statistical Methods in Agriculture and Experimental Biology.* 2nd ed. London: Chapman & Hall.

Montgomery, D. 2000. *Design and Analysis of Experiments.* 5th ed. New York: John Wiley & Sons.

Morrow, R.E., Stricker, J.A., Garner, G.B., Jacobs, V.E., and Hines, W.G., 1988. Cow-calf production on tall fescue-ladino clover pastures with and without nitrogen fertilization or creep feeding: fall calves. *J. Prod. Agric.* (2): 145–148.

Ntare, B.R. and Aken'Ova, M. 1985. Yield stability in segregating populations of cowpea. *Crop Sci.* March–April, 25(2): 208–211.

Ogawa, J. 1974. *Statistical Theory of the Analysis of Experimental Designs.* New York: Marcel Dekker.

Peterman, C.J., Sears, R.G., and Kanemasu, E.T. 1985. Rate and duration of spikelet initiation in 10 winter wheat cultivars. *Crop Sci.* March–April, 25(2): 221–225.

Samuels, M.L. and Witmer, J.A. 2003. *Statistics for the Life Sciences.* 3rd ed. New Jersey: Pearson Education.

Schork, M.A. and Remington, R.D. 2000. *Statistics with Applications to the Biological and Health Sciences.* 3rd ed. New Jersey: Pearson Education.

Shukla, G.K. 1972. Some statistical aspects of partitioning genotype environment components of variability. *Heredity* 29: 237–245.

Trott, J.O., Moore, K.J., Lechtenberg, V.L., and Johnson, K.D. 1988. Light penetration through tall fescue in relation to canopy biomass. *J. Prod. Agric.* April–June, 1(2): 137–140.

Yates, F. and Cochran, W.G. 1938. The analysis of groups of experiments. *J. Agric. Sci.* 28: 556–580.

2

Key Assumptions of Experimental Designs

2.1 Introduction

In designing an experiment, the researcher's main interest is in creating controlled conditions to easily measure those characteristics that are of interest to him or her. That is, the experiment is designed in such a way as to maintain those factors uniform that are not part of the treatment. While doing so, the researcher must keep in mind that the design must satisfy the assumptions required for proper interpretation of data. Failure to meet the assumptions affects not only the significance level but also the sensitivity of the F-test and the t-tests. The assumptions underlying most of the designs in this book are that: (1) the effects of blocks, treatments, and error are additive (this implies no interaction), (2) the observations have a normal distribution (experimental errors are normally distributed), (3) the observations are distributed independently (experimental errors are independent), and (4) the variance of the observations is constant, which means homogeneity of variance. This implies that the treatment effects are constant and that experimental errors have common variance. In Appendix K, the reader can use MINITAB to evaluate whether any of these assumptions are violated. We must keep in mind that in certain conditions, not all of these assumptions are met. For example, when data are expressed as percentages (such as the percentage of plants infected with a disease or the percentage of germinated seed in a plot), the observations have a binomial distribution and, hence, the variance is not a constant. Similarly, when we are dealing with count data (such as the number of rare insects in a particular field or the number of infested plants in a greenhouse), we have a Poisson distribution, in which the variance is equal to the mean (more about the assumptions and their violations in the sections to follow).

2.2 Assumptions of the Analysis of Variance (ANOVA) and Their Violations

Additivity: This assumption states that the effects of two factors are additive if the effect of one factor remains constant over all levels of the other factor. This means that each factor influences the dependent variable solely through its impact. For example, if treatment and blocks are the two factors of interest, this assumption implies that the treatment effect remains constant for all blocks and that the block effect remains constant for all treatments. One interpretation of additivity is that the blocks and the treatment effects do not interact; that is, all population interaction effects are zero.

We can also express the basic principle of additivity in the following manner: Denote the treatments as $T_1, T_2, ..., T_n$. It is then assumed that the observation obtained on any unit following a particular treatment such as T_1 differs from T_2 by a constant $a_1 - a_2$. As you would expect, there are constants $a_1, a_2, ..., a_n$ for each treatment. The object of most of the experimental designs is to estimate the $a_1 - a_2$ differences. This difference is called the true *treatment effect*. If, for example, different varieties of a crop (*i*) are subjected to different treatments (*j*), then the true or expected yield of the *i*th variety when subjected to the *j*th treatment, under a certain growing condition, is μ_{ij}. Thus, additivity implies that under the general experimental conditions of the test, the true mean yield of one variety is greater (or less) than the true mean yield of another variety by an amount, an additive constant, that is the same for each of the treatments concerned. Conversely, the true mean yield with one treatment is greater (or less) than the true mean yield with another treatment, by an amount that does not depend on the variety concerned.

The mathematical model for the different experimental designs is called a *linear additive model*, which may be written as:

$$X_i = \mu + t_i + e_i \qquad \text{(Completely randomized design) (2.1)}$$

$$X_{ij} = \mu + t_i + b_j + e_{ij} \qquad \text{(Randomized complete block design) (2.2)}$$

In Equation 2.1, the value of the experimental unit (X_i) is made up of the general mean (μ) plus the treatment effect (t_i) plus an error term (e_i). Similarly, in Equation 2.2, the value of the experimental unit (X_{ij}) is made up of the general mean (μ) plus a treatment effect (t_i) plus a block effect (b_j) plus an error term (e_{ij}). This model implies that a treatment effect is the same for all blocks and that the block effect is the same for all treatments. What is crucial to remember is that in both models the terms are added, and hence the term *additivity*.

There are instances in agricultural experimentation in which the additivity assumption is not met. Let us take Equation 2.2 as an example of how the additivity assumption may be violated. The data may only rarely be so obliging as to conform to Equation 2.2. More often than not, the variability observed in such an equation is also a function of the interaction between the treatment effect and the block effect. Such a condition requires a revision of the model to include an interaction term. This revised model may be written as:

$$X_{ij} = \mu + t_i + b_j + (tb)_{ij} + e_{ij} \qquad (2.3)$$

where $(tb)_{ij}$ is the interaction effect of the ith treatment and the jth block. Another circumstance that leads to the violation of the additivity assumption is when the treatment effect increases yield by a percentage or proportion. This is referred to as the *multiplicative treatment effect*. This type of condition is often observed when experiments are conducted to measure the number of insects per plot or the number of egg masses per unit area. When faced with a multiplicative model, it is appropriate to work with the logarithms of the original observations. Because $\log (xy) = \log x + \log y$, by transforming the original data into logarithms we are converting the multiplicative effect to the additive effect. The following hypothetical example is given to illustrate the concept of additivity and when this assumption is violated.

Example 2.2.1
In a randomized complete block design with two treatments and two blocks, a scientist found the data shown in Table 2.2.1. Do the data satisfy the additivity assumption?

SOLUTION
You will notice in this example that the treatment effect remains constant over all levels of the blocking factor. Similarly, the block effect remains constant for all treatments. Given this condition, we would say that the effects of treatment and block are additive.

On the other hand, when the treatment effect is not constant over blocks and block effect is not constant over treatments, the factors are said to have

TABLE 2.2.1

Additive Effects of Treatment and Block Using Hypothetical Data

Treatment	Block		Block Effect (I − II)
	I	II	
A	160	120	40
B	140	100	40
Treatment effect (A − B)	20	20	—

TABLE 2.2.2

Multiplicative Effects of Treatment and Block Using Hypothetical
Data

Treatment	Block		Block Effect	
	I	II	(I − II)	(I − II)100/II
A	100	50	50	100
B	80	40	40	100
Treatment effect (A − B)	20	10	—	—
(A − B)100/B	25	25	—	—

a multiplicative rather than an additive effect. Table 2.2.2 shows a set of
hypothetical data with multiplicative effects of treatment and block.

You will note that the treatment effect for blocks I and II are 20 and 10,
respectively, whereas the block effect is 50 for treatment A and 40 for treatment B. When the treatment and block effects are expressed in percentage
terms, as shown in the last column of Table 2.2.2, you will observe that the
treatment effect is 100% in both blocks, and the block effect is 25% for both
treatments. What is important to remember is that a variable may interact
with the treatments employed to the extent that the combined effects of
treatments and error may be more multiplicative than additive. Therefore,
it is desirable to detect nuisance variables in the early stages of experimentation. This means making supplementary observations where appropriate
and assigning the treatments to the experimental units in such a way as to
detect variations.

The assumption of additivity is also violated when one or more aberrant
observations (outliers) impact the data. It is crucial to determine whether
the outlier is just an extreme value from the defined population or if it should
be classified with another population. Outliers should not be excluded arbi-
trarily as their exclusion may seriously bias estimates of the treatment effect
and underestimate the experimental error. It is the responsibility of the
experimenter to use common sense in dealing with outliers. The experimenter may choose from a variety of statistical methods devised by Dixon
and Massey (1957), Anscombe and Tukey (1963), Grubbs and Beck (1972),
and Samuels and Witmer (2003) to determine whether an outlier is excluded
(or included) in the analysis.

Normality: This assumption states that errors are normally distributed.
Certain types of data do not satisfy this assumption. The validity of such an
assumption depends upon the measure chosen. For example, count data
such as the number of rare insects found in a soil sample, the number of
lesions per leaf, or the number of particular impurities in milk samples
usually follow a Poisson distribution. As the data in the form of counts are
discrete, they do not follow the normal distribution, which assumes a con-
tinuous variable. Percentage scores, such as that of seeds germinated in a
greenhouse or the percentage of plants infected with a disease, have a binomial distribution and, hence, the variance, which depends on the unknown

percentage, is not a constant. The normality assumption is crucial in probability statements (where decisions are based on tests of hypotheses) as well as in the reliability of estimates (where confidence intervals are used). In most agricultural experiments, two statistical facts provide support to the normality assumption: first, the central limit theorem that states that as the sample size n becomes sufficiently large, the sampling distribution of the mean tends toward a normal distribution; second, the F-test of the hypothesis of treatment effects is known to be robust, i.e., the probabilities of errors of type I and type II are not affected severely by moderate departures from normality (Pearson, 1931; Donaldson, 1968; Tiku, 1971; Gill, 1978).

There are certain situations in which a researcher, by looking at the data, may suspect departures from the normality assumption. For example, it is difficult to assume that when the mean values for the different treatments differ by a factor of three or more that there will not be greater variability. A researcher should also be cautious when there are substantial differences between blocks. It is difficult to assume that treatment differences can remain constant over blocks.

To evaluate the significance of the departures of the obtained distribution from the assumed normal distribution, we can analyze the residuals, $e_{ij} = X_{ij} - \bar{X}_i$, of the sample. Depending on the size of the sample, we can use various tests. The test of goodness of fit that utilizes the χ^2 distribution provides a good approximation when the sample size is large ($n = 100$). The same test could be used for smaller samples; however, with some loss of accuracy. Others have suggested that the Kolmogorov–Smirnov test is a better approximation for small samples (Siegel, 1956; Sokal and Rohlf, 1969).

In cases in which nonnormality is suspected, one may transform the data to ensure near normality of distribution for the transformed data. The most common transformations used are square root, inverse square root, log, or reciprocal. These transformations are discussed later in Section 2.4. It should be recognized that although transformation improves the accuracy of the probability statements, interpretation of the results is greatly handicapped by the transformation.

Independent errors: This assumption states that normally distributed errors are independent of each other. In other words, the error of an observation is not correlated with that of another. The validity of this assumption is ensured when proper randomization is applied in any designed experiment. There are instances, however, in which the design requires a systematic assignment of treatments to the experimental units. In such a design (systematic design), the assumption of independence of error is violated. Figure 2.2.1 shows a systematic design and the location of the problem with respect to the violation of the assumption of independence of errors.

You will note in Figure 2.2.1 that treatment A and treatment B are adjacent to each other in all replications, and treatment B and treatment E are always separated by two blocks. Furthermore, it should be noted that even when the design is not systematic, crop yields on neighboring plots tend to be positively correlated. Therefore, treatments that are close to each other tend

Replication I	A	B	C	D	E	F
Replication II	F	A	B	C	D	E
Replication III	E	F	A	B	C	D
Replication IV	D	E	F	A	B	C
Replication V	C	D	E	F	A	B
Replication VI	B	C	D	E	F	A

FIGURE 2.2.1
Layout of a systematic design involving six treatments (A, B, C, D, E, and F) and six replications.

to have similar errors, more so than those that are farther apart. Thus, in a systematic design such as that shown in Figure 2.2.1, the assumption of independence of errors is violated. As a precaution, to ensure independence of experimental errors, it is highly desirable to properly randomize the experimental layout. In the chapters to follow, we will discuss in detail the randomization and layout of experiments for each design.

Homogeneous variance: One of the most frequently occurring and serious violations of the assumptions of analysis of variance (ANOVA) is the presence of heterogeneous variance among treatment groups. This means that the experimental error variance is not constant over all observations. Another way of stating the assumption of homogeneous variance is that the separate variance estimates provided by samples are estimates of the same population variance. When error variance is heterogeneous, the *F*- and *t*-tests tend to give too many significant results. This disturbance is usually only moderate if every treatment has the same number of replications (Scheffé, 1959). The validity of the homogeneity of variance depends on: (1) how careful the experimenter is in administering experimental treatments in a consistent manner to all subjects, (2) the type of measure taken, and (3) the treatment levels employed. Undoubtedly, the uniform application of treatments to subjects will tend to stabilize variances from group to group. However, despite appropriate experimental methodology, we can expect heterogeneity of variances. Heterogeneous variance is also associated with data whose distribution is not normal. For example, when we have a Poisson distribution in which the mean and the variance are equal ($\mu = \sigma^2$), they are either homogeneous or heterogeneous over treatments. Poisson distributions occur when the observations consist of counts. Once we recognize the nature of the functional relationship, data can be transformed on a scale on which the error variance is more nearly constant. Bartlett (1947) found several transformation methods such as the log, square root, and inverse sines extremely helpful. Such transformations are also useful in cases in which treatment and environmental effects are not additive, as was mentioned earlier.

Several methods are available for detecting departures from homogeneity of variance. These will be discussed in the following section.

2.3 Measures to Detect Failures of the Assumptions

Failure to take into account major departures of the data from the assumptions mentioned in the previous section have serious implications for the conclusions of any analysis. Minor departures, on the other hand, do not greatly disturb the conclusions reached from the standard analysis (Cochran and Cox, 1957; Snedecor and Cochran, 1980; Montgomery, 2000). Just as increased replications and proper randomization are essential to minimizing experimental error, so is the recognition of the departures of the data from the assumptions for the conclusions of an experiment. The reader is referred to Scheffé (1959) for an excellent review of the consequences of various types of failures of the assumptions.

In the previous section we discussed the assumptions and some of the theoretical conditions that lead to the violation of such assumptions. Now we will present specific examples of how to detect the failures of the assumptions and show procedures to transform the data to conform to the assumptions.

The assumption of additivity, which states that the block effects are approximately the same for all treatments, can be examined using *Tukey's test*. The test requires that we compute the sum of squares (SS) for nonadditivity in the following manner:

$$SS \text{ nonadditivity } = \frac{N^2}{\left(\sum d_i^2\right)\left(\sum d_j^2\right)} \qquad (2.4)$$

where

N = each cell of the table of raw data multiplied by the corresponding treatment and block effect and the sum of all the products, or

$$N = \sum w_i d_i = \sum \sum X_{ij} d_i d_j$$

$$\sum d_i^2 = \text{sum of the treatment effect squared}$$

$$\sum d_j^2 = \text{sum of the block effect squared}$$

Equation 2.4 is basically the contribution of nonadditivity with one degree of freedom to the error sum of squares. This value is then tested against the remainder of the residual sum of squares to determine whether the hypothesis of additivity is tenable. This test can be applied to any two-way classification, such as the randomized complete block design experiments, in which the data are classified by treatments and blocks. Example 2.3.1 shows how this test is performed.

Example 2.3.1

An agricultural biologist attempting to determine insect infestation in a field has collected the following data over four periods in four traps as shown in Table 2.3.1. To find out whether the data meet the additivity assumption, we perform Tukey's test.

SOLUTION

Before we perform Tukey's test, the data are analyzed using the method discussed in Chapter 4, Subsection 4.2.2. We need this analysis so that we can show the partitioning of the error sum of squares into its components. The result of the analysis is presented in Table 2.3.2.

Now we are ready to perform Tukey's test. The steps in computing the intermediate values for performing the test are as follows:

Step 1: Compute the treatment effect (d_i) and the period effect (d_j) as shown in Table 2.3.1. Remember that both effects add up to exactly 0.

TABLE 2.3.1

Insects Caught over Four Different Periods in Four Traps in an Experiment

Period	Trap 1	Trap 2	Trap 3	Trap 4	Mean $\bar{X}_{i.}$	Treatment Effect $di = \left(\bar{X}_{i.} - \bar{X}_{..}\right)$	$W_i = \sum x_{ij} d_j$
1	8	10	18	32	17.00	−5.75	400.00
2	12	16	26	46	25.00	2.25	560.00
3	16	20	26	42	26.00	3.25	423.00
4	10	22	20	40	23.00	0.25	144.50
Mean $(\bar{X}_{.j})$	11.50	17.00	22.50	40.00	—	—	—
$d_j = \left(\bar{X}_{.j} - \bar{X}_{..}\right)$	−11.25	−5.75	−0.25	17.25	—	0.00	—
$\bar{X}_{..}$	—	—	—	—	22.75	—	—

TABLE 2.3.2

ANOVA of the Data Presented in Table 2.3.1

Source of Variation	Degrees of Freedom	Sum of Squares	Mean Square	F
Periods	3	195	65.0	7.39**
Traps	3	1829	609.7	69.28**
Error	9	79	8.8	—
Total	15	—	—	—

** = Significant at the 1% level.

Step 2: Compute w_i for each period, using the following formula. The computed values are shown in Table 2.3.1.

$$w_i = \sum X_{ij} d_j \tag{2.5}$$

Thus, for our example we have:

$$w_1 = 8(-11.25) + 10(-5.75) + 18(-0.25) + 32(17.25) = 400.00$$

$$w_4 = 10(-11.25) + 22(-5.75) + 20(-0.25) + 40(17.25) = 446$$

Step 3: Compute N, which is the sum of the product of w_i and d_i. Another way of stating N is that it is the sum of the product of each cell of Table 2.3.1 multiplied by the corresponding "treatment" and "period" effects. Thus, we have:

$$N = \sum w_i d_i = \sum \sum X_{ij} d_i d_j \tag{2.6}$$

For our example, the value of N is:

$$N = 400(-5.75) + 560(2.25) + 423(3.25) + 446(0.25) = 446.25$$

Step 4: Compute the sum of the value of the treatment and period effects as follows:

$$\sum d_i^2 = (-5.75)^2 + (2.25)^2 + (3.25)^2 + (0.25)^2 = 48.75$$

$$\sum d_j^2 = (-11.25)^2 + (-5.75)^2 + (-0.25)^2 + (17.25)^2 = 457.25$$

Step 5: Perform the test using Equation 2.4 as follows:

$$SS \text{ nonadditivity} = \frac{(446.25)^2}{(48.75)(457.25)} = 8.93$$

This is the contribution of the nonadditivity to the error sum of squares with one degree of freedom. This value is tested against the residual, as shown in Table 2.3.3. The test result indicates strong evidence that the assumption of additivity is correct. Thus, the researcher is assured that the data gathered do meet at least the additivity assumption.

To test the assumption of normality, we have to look carefully at the error terms associated with each observation to determine whether they are randomly distributed or not. Recall from Equation 2.2 that any observation (X_{ij}) of a two-way table in a randomized complete block design is simply made up of the general mean (μ_j), the treatment effect (t_i), the block effect (b_j), and the error term (e_{ij}). To find out the error term for any cell of a two-way table, we can rewrite Equation 2.2 as follows:

$$e_{ij} = X_{ij} - \mu - t_i - b_j \tag{2.7}$$

Before we can determine the error term using Equation 2.7, we must find out the value of the treatment and block effects. Because the treatment and block effects are the difference between the treatment and block mean from the general mean, they can be written as:

$$\text{Treatment effect } (t_i) = \overline{X}_{i.} - \mu \tag{2.8}$$

$$\text{Block effect } (b_j) = \overline{X}_{.j} - \mu \tag{2.9}$$

Substituting Equation 2.8 and Equation 2.9 into Equation 2.7, we get

$$e_{ij} = X_{ij} - \mu - (\overline{X}_{i.} - \mu) - (\overline{X}_{.j} - \mu) \tag{2.10}$$

Using Equation 2.10 we are now able to determine the value of the error term for each cell of a two-way table.

TABLE 2.3.3

ANOVA Table and the Test of Additivity

Source of Variation	Degrees of Freedom	Sum of Squares	Mean Square	F
Error (Period × Trap)	9	79.00	8.8	—
Nonadditivity	1	8.93	8.93	<1
Residual	8	72.83	9.10	—

TABLE 2.3.4

Error Components of the Insects-Trapped Experiment

Period	Trap 1	2	3	4	Total
1	2.25	−1.25	1.25	−2.25	0.0
2	−1.75	−3.25	1.25	3.75	0.0
3	1.25	−0.25	0.25	−1.25	0.0
4	−1.75	4.75	−2.75	−0.25	0.0
Total	0.0	0.0	0.0	0.0	—

Let us use the data given in Table 2.3.1 to test the normality assumption for this example. Using Equation 2.10, we will remove the general mean, the block effect, and the treatment effect from each cell of the data. For example, to find the error term for e_{11}, we would have

$$e_{11} = 8 - 22.75 - (17 - 22.75) - (11.50 - 22.75)$$

$$= -14.75 - 17 + 22.75 - 11.50 + 22.75$$

$$= 2.25$$

The error terms for each of the cells are computed similarly and are presented in Table 2.3.4.

Looking at the error terms, we note that the distribution does not appear to be normal, as there are two modal classes (see Figure 2.3.1). This test tells

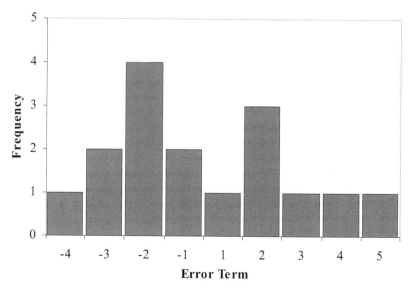

FIGURE 2.3.1
Frequency distribution of the error terms.

us that the normality assumption is untenable in the present case. To remedy this problem, transformation of the data is required. In the next section we will discuss different transformation techniques.

To test the presence of heterogeneous variances and whether the variances are functionally related to the mean, a simple procedure is to use a scatter diagram. The steps to follow are as follows:

Step 1: Compute the mean and the variance for each treatment across all replications.

Step 2: Using the mean value and the variance, plot a scatter diagram. It should be noted that one can also use the mean and the range in plotting the scatter diagram.

Step 3: Upon examining the scatter diagram, identify the relationship, if any, between the mean and the variance. Three possible outcomes may be observed: (1) homogeneous variance; (2) heterogeneous variance, when there is a functional relationship between the variance and the mean; and (3) heterogeneous variance, when there is no functional relationship between the variance and the mean.

Figure 2.3.2 illustrates the presence and absence of homogeneity of variance. In Figure 2.3.2(a), we note that the relationship between the mean and variance is such that the variances of the treatments do not vary significantly from one another at different means and are, thus, considered homogeneous. However, in Figure 2.3.2(b) and Figure 2.3.2(c), the variances are heterogeneous as a result of the proportionality of the variance to the mean in the case of Figure 2.3.2(b) and an absence of functional relationship between the variance and mean in the case of Figure 2.3.2(c). The scatter diagram can be used as a signal for using the tests of the homogeneity of variance. There are several tests that could be applied to verify the homogeneity of variance. The reader may find the review of these methods given by Anderson and McLean (1974) helpful. In this text, the following two tests are suggested for use by practitioners.

Bartlett (1937) introduced a homogeneity-of-variance test that involves computing a statistic whose sampling distribution is closely approximated

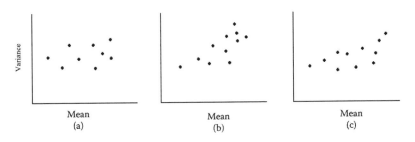

FIGURE 2.3.2
Illustration of homogeneous and heterogeneous variance: (a) homogeneous variance, (b) heterogeneous variance in which variance is proportional to the mean, and (c) heterogeneous variance in which variance and the mean are not functionally related.

by the chi-square distribution with $k - 1$ degrees of freedom. The test has become a well-established measure. However, it should be kept in mind that the test is a bit sensitive to nonnormality, especially if the tails of the distribution are too long. When this occurs, the test tends to show significance too often.

The test criterion, when there is $k > 2$ independent estimate of variance s_i^2, and all have the same number of degrees of freedom v, using logarithms to base e is:

$$\chi^2 = \frac{M}{C} \tag{2.11}$$

where

$$M = v\left(k \ln \bar{s}^2 - \sum \ln s_i^2\right) \tag{2.12}$$

$$C = 1 + \frac{k+1}{3(k)(v)} \tag{2.13}$$

$$\bar{s}^2 = \frac{\sum s_i^2}{k} = \text{pooled variance} \tag{2.14}$$

v = degrees of freedom per variance
k = number of treatments or samples

To see how Bartlett's test is applied, let us use the following example.

Example 2.3.2

In a randomized experiment with four replications, the number of weeds found per plot after the fields were sprayed with different rates of a general herbicide are given in Table 2.3.5. Verify the homogeneity of variance for the data using Bartlett's test.

SOLUTION

Step 1: Compute the variance for each of the treatments as shown in Table 2.3.5.

Step 2: Compute the logarithm of each of the variances, the total for the variances, and the total for the logarithm of the variances as shown in Table 2.3.5.

Step 3: Compute the estimate of the pooled variances where $k = 6$:

TABLE 2.3.5

Number of Weeds per Plot following Different Herbicide Treatments

Treatment	Rep. I	Rep. II	Rep. III	Rep. IV	Total	Mean	s_i^2	$\ln s_i^2$
A	15	12	17	20	64	16.00	11.33	2.427
B	12	5	7	4	28	7.00	12.66	2.538
C	11	7	8	6	32	8.00	4.66	1.539
D	18	10	11	9	48	12.00	16.66	2.813
E	16	18	20	14	68	17.00	6.66	1.896
Control	20	28	24	28	100	25.00	14.66	2.685
Total	—	—	—	—	—	—	66.30	13.898

$$\bar{s}^2 = \frac{\sum s_i^2}{k} = \frac{66.63}{6} = 11.10$$

Step 4: Compute M, given that $v = 3$:

$$M = (3)[6(2.407) - 13.747]$$

$$= 2.085$$

Step 5: Compute C, the correction factor as:

$$C = 1 + \frac{(6+1)}{3(6)(3)} = 1.13$$

Step 6: Compute the χ^2 as:

$$\chi^2 = \frac{2.085}{1.13} = 1.85$$

Step 7: Compare the computed χ^2 with the table value at five degrees of freedom (one less than the number of treatments) given in Appendix A. You will note that $\chi^2_{.05,5} = 11.07$. Thus, we cannot reject the null hypothesis and conclude that the variances are homogeneous.

When the samples are of unequal size, the test statistic is similar to Equation 2.11, except that computing M and C are as follows:

$$M = \left(\sum v_i\right) \ln \bar{s}^2 - \sum v_i \ln s_i^2 \qquad (2.15)$$

$$C = 1 + \frac{1}{3(k-1)}\left(\sum \frac{1}{v_i} - \frac{1}{\sum v_i}\right) \tag{2.16}$$

where

$$\bar{s}^2 = \frac{\sum v_i s_i^2}{\sum v_i} \tag{2.17}$$

v = degrees of freedom per variance
k = number of treatments or samples

Hartley (1950) proposed another measure for testing the homogeneity of variance. This test is based on the ratio of the largest to the smallest within-groups variance. This is known as the *test of homogeneity of variance*. If this ratio is nonsignificant, variances are said to be homogeneous. On the other hand, if the ratio is significant, the variances are said to be heterogeneous. The *F*-test for the homogeneity of variance is defined as:

$$F = \frac{s_1^2}{s_2^2} \tag{2.18}$$

where

s_1^2 = the larger sample variance

s_2^2 = the smaller sample variance

2.4   Data Transformation

Having discussed the various approaches that could be applied to determine whether the assumptions of the ANOVA are violated, we are now ready to review data transformation as a means to remedy the violation of the various assumptions. Let us take the violation of the assumption of homogeneity of variance to show how data transformation improves the situation.

In Figure 2.3.2(b) and Figure 2.3.2(c), the functional relationship between the mean and the variance shows the presence of heterogeneous variances.

Once the nature of the functional relationship between the mean and the variance is determined, the original data can be transformed to a scale on which the error variance is more nearly constant. In agricultural research, we often perform one of the following three commonly used transformations.

Square-root transformation: The square-root transformation is applied when the means and the variances are proportional for each treatment. That is, $\sigma_j^2 = k\mu_j$. Transformation of the data allows the variance to be nearly independent of the mean.

Researchers often encounter conditions in which the mean and variance are proportional to each other. This situation is not unusual when the data are in the form of frequency counts (for example, the number of weeds found in a plot) or when the dependent variable is the number of correct or "yes" responses. In such circumstances, the ANOVA is performed on X' rather than X (the original data), where

$$X' = \sqrt{X} \tag{2.19}$$

Bartlett (1936) suggests that if most of the values of X are small (i.e., less than 10 and, especially, some with a value of 0), homogeneity of variance is likely to be produced by using the following transformation:

$$X' = \sqrt{X + 0.5} \tag{2.20}$$

Freeman and Tukey (1950) have suggested that for the Poisson distribution (counts of rare events), it is best to use transformation $\sqrt{X} + \sqrt{X + 1}$ for better results.

We will use the data from Example 2.3.2 to show how it could be transformed before further analyses are performed on the data.

SOLUTION

Step 1: Transform the data, using the square-root transformation. Observe that a large number of values in the data set are greater than 10 and there are no observations with a value of 0. Hence, the $X' = \sqrt{X}$ transformation. The transformed data X' is shown in Table 2.4.1.

Step 2: Observe that both the range and the estimates of the variance of the data as shown in Table 2.4.1 indicate that the transformed data are now on a scale on which the error variance is more nearly constant.

Step 3: Perform the ANOVA on the transformed data. Researchers sometimes prefer to report the results in the original scale, as it is recognized that it is more easily understood on this scale. Hence, the means are reconverted to the original scale by squaring.

Logarithmic transformation: Researchers often face situations in which they suspect that the standard deviation (not variances) of the data is proportional

TABLE 2.4.1

Data from Table 2.3.5 Transformed with a Square-Root Transformation

Treatment	Transformed Data							
	Rep. I	Rep. II	Rep. III	Rep. IV	Total	Mean	Range	S^2
A	3.87	3.46	4.12	4.47	15.92	3.98	1.01	0.18
B	3.46	2.23	2.65	2.00	10.34	2.59	1.46	0.41
C	3.31	2.65	2.00	2.45	10.41	2.60	1.31	0.29
D	4.24	3.16	3.31	3.00	13.71	3.43	1.24	0.30
E	4.00	4.24	4.47	3.74	16.45	4.11	0.73	0.09
Control	4.47	5.29	4.90	5.29	19.95	4.99	0.82	0.15

to the mean or in which the treatment and replication (block) effects are multiplicative. Under such conditions, logarithmic transformation of the data is appropriate. Once transformed, the data will not show any apparent relationship between the mean and the standard deviation. Furthermore, this transformation changes the multiplicative effect into an additive one, thus meeting the additivity assumption. In the following text, with the use of some examples, we will transform the data to illustrate the effectiveness of the logarithmic transformation.

Typically, data that require transformation are whole numbers and may cover a wide range of values. The most common situation that gives rise to a logarithmic transformation is associated with the data on growth: for example, changes in the weight of animals caused by diets, the number of insects per plot, or the number of germinated seeds per greenhouse. As we apply this transformation, we should keep in mind two data situations in which logarithmic transformation is problematic. First, negative values cannot be transformed. Second, if the data set contains values of 0, its logarithm is minus infinity. If the number of zeros in a data set is large, it is best to use some other transformation. However, a small number of zeros in a data set can easily be handled by adding 1 to each data point before the data are transformed. You may use logarithms of any base, but it is easiest to use the common logarithm (to the base 10). To avoid negative logarithms, it is best to multiply the data set with a constant. Such manipulation of the data is legitimate, as it has no effect on subsequent analyses.

To show how logarithmic transformation changes the multiplicative effect into an additive effect so as to meet the additivity assumption, let us use the hypothetical data given in Table 2.4.2. To transform the data, simply take the logarithm of each data point as shown in Table 2.4.2. You will note in Table 2.4.2 that the treatment effect is 0.0969 in both blocks and 0.3010 for both treatments. Thus, with the transformation, we have converted the multiplicative effect into an additive effect.

Logarithmic transformation can also be applied to data in which the heterogeneity of variance is suspected. Using the data in Table 2.3.5, we can determine the relationship that exists between the mean and the variance or

TABLE 2.4.2

Logarithmic Transformation of the Multiplicative
Effects of the Hypothetical Data from Table 2.2.2

Treatment	Block I	II	Block Effect
A	2.0000	1.6989	0.3010
B	1.9031	1.6020	0.3010
Treatment effect (A − B)	0.0969	0.0969	—

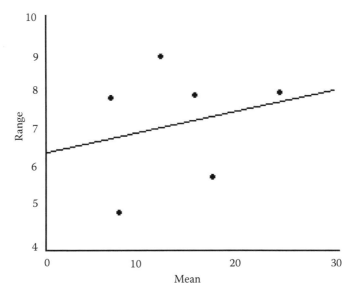

FIGURE 2.4.1
Relationship between treatment mean and the range of the data of Table 2.3.5.

the range. For this purpose, we plot the value of the mean against the range
as shown in Figure 2.4.1.

Figure 2.4.1 shows that there is a linear relationship between the mean and
the range, implying that the range increases proportionately with the mean.
To remedy this problem, we perform a logarithmic transformation of the
data as shown in Table 2.4.3. Because there are several data points with
values of less than 10, it is appropriate to use log $(X + 1)$ rather than log X,
where X is the original data.

Note that in Table 2.4.3 there seems to be a much more consistent pattern
about the transformed data, though there is still some random variation. We
now can perform Bartlett's test of homogeneity of variance as before to see
if the transformed data meet this assumption. As you may recall from our
earlier discussions, whenever we take the logarithms of any value less than
1, we have the problem of negative logarithms. To remedy this, we need to
multiply each number by any constant we choose. This will not alter the

TABLE 2.4.3

Transformed Data from Table 2.3.5 Using a Log $(X + 1)$ Scale
Transformation

Treatment	Transformed Data				Total	Mean
	Rep. I	Rep. II	Rep. III	Rep. IV		
A	1.2041	1.1139	1.2552	1.3222	4.8954	1.2238
B	1.1139	0.7782	0.9031	0.6989	3.4941	0.8735
C	1.0792	0.9031	0.9542	0.8451	3.7816	0.9454
D	1.2788	1.0414	1.0792	1.0000	4.3994	1.0998
E	1.2304	1.2788	1.3222	1.1761	5.0075	1.2518
Control	1.3222	1.4624	1.3979	1.4624	5.6449	1.4112

TABLE 2.4.4

Number of Weeds per Plot following Different Herbicide
Treatments

Treatment	Mean	s_1^2	Coded s_1^2	Log Coded s_1^2
A	1.2238	0.0020	20.00	2.99
B	0.8735	0.0169	169.00	5.13
C	0.9454	0.0089	89.00	4.89
D	1.0998	0.0003	3.00	1.09
E	1.2518	0.0034	34.00	3.53
Control	1.4112	0.0192	192.00	5.26
Total	6.8055	—	507.00	22.89
Mean	—	—	84.50	—
Log of Mean	—	—	4.44	—

Bartlett's test we perform. In the present example, the values of all the
variances computed are less than 1. Thus, they are multiplied by 10,000 and
are recorded as the coded variance in Table 2.4.4.

Bartlett's test of homogeneity of variance, as given before, is:

$$\chi^2 = \frac{M}{C} \tag{2.11}$$

For the present example, the values of M and C are:

$$M = (3)[6(4.44) - 22.89]$$

$$= 11.25$$

$$C = 1 + \frac{6+1}{3(6)(3)} = 1.13$$

Thus, χ^2 is computed as follows:

$$\chi^2 = \frac{11.25}{1.13} = 9.96$$

Compare the computed χ^2 with the table value at five degrees of freedom (one less than the number of treatments) given in Appendix A. From Appendix A you will note that $\chi^2_{.055} = 11.07$. Thus, we cannot reject the null hypothesis and therefore conclude that the variances are homogeneous.

We are now able to perform the ANOVA on the transformed data. The results are shown in Table 2.4.5. The results indicate that both the replication and the treatment effects are significant at the 1% level.

Arc sine or angular transformation: This transformation is most appropriate when researchers encounter data in the form of counts or the binomial proportions (p). In the binomial distribution, in contrast to the normal distribution, the mean and the variance are related in such a way that the variances tend to be small at the two ends of the range of values and large in the middle. This means that small variances are associated with values close to 0 and 100%, whereas large variances are expected around the middle or 50%. Another way of stating this is that with binomial data ("0" or "1" response), the variance is proportional to [mean (1 − mean)], and the appropriate transformation is arc sine $\sqrt{p}$. In the angular scale, the percentages near 0 or 1 are spread out so as to increase their variance. If all the error variance is binomial, the error variance in the angular scale is about $821/n$ (Snedecor and Cochran, 1980). It has been shown empirically that the variances are fairly stable in the range of $p = .30$ and $p = .70$. Thus, it is not necessary to transform the data for this range of values. Arc sine transformation is most appropriate for the data if the range of percentages is greater than 40.

In using the arc sine or angular transformation, it has been suggested by Bartlett (1947) that with $n < 50$ (where n is the number upon which the percentage data are based), we should count a 0% proportion as $1/(4n)$ and a 100% proportion as $(n − 1/4)/n$ before transforming to angles. This will improve the equality of the variance in the angles. It should be kept in mind, however, that transformation will not improve the inequalities in variance if there are differing values of n. Under such conditions, it is appropriate to use a weighted analysis in the angular scale.

TABLE 2.4.5

ANOVA of the Transformed Data

Source of Variation	Degrees of Freedom	Sum of Squares	Mean Square	F
Replication	3	0.0554	0.1845	16.62**
Treatment	5	0.8136	0.1627	14.65**
Error	15	0.1668	0.0111	—
Total	23	—	—	—

** = Significant at the 1% level.

TABLE 2.4.6

Percentages of Germinated Seeds in a Greenhouse

Treatment	Rep. I	Rep. II	Rep. III	Rep. IV	Total	Mean
A	42.2	36.4	43.2	52.5	174.3	43.6
B	24.5	29.7	26.2	29.0	109.4	27.4
C	0.0	0.0	0.0	0.0	0.0	0.0
D	44.0	70.1	81.2	69.3	264.6	66.2
E	96.0	98.0	95.4	100.0	389.4	97.4
F	20.0	28.1	24.9	28.4	101.4	25.3
G	13.5	14.8	20.1	12.1	60.5	15.1
H	0.0	0.0	0.0	0.0	0.0	0.0
I	55.6	62.8	64.6	58.9	241.9	60.5
J	33.2	39.7	42.1	37.5	152.5	38.1

TABLE 2.4.7

Transformed Data using the Arc Sine Transformation

Treatment	Rep. I	Rep. II	Rep. III	Rep. IV	Total	Mean
A	40.5	37.1	41.1	46.4	165.1	41.3
B	29.7	33.0	30.8	32.6	126.1	31.5
C	0.6	0.6	0.6	0.6	2.4	0.6
D	41.6	56.9	64.3	56.4	219.2	54.8
E	78.5	81.9	77.6	5.7	243.7	60.9
F	26.6	32.0	29.9	32.2	120.7	30.2
G	21.6	22.6	26.6	20.4	91.2	22.8
H	0.6	0.6	0.6	0.6	2.4	0.6
I	48.2	52.4	53.5	50.1	204.2	51.1
J	35.2	39.1	40.5	37.8	152.6	38.2

To illustrate the application of the arc sine transformation, let us assume that in a completely randomized design with ten treatments and four replications and $n = 25$ seeds, an ornamental horticulturist found the percentages of seeds germinated in a greenhouse, as shown in Table 2.4.6.

You will note in Table 2.4.6 that the percentages of seeds that germinated range from 0 to 100%. Given this range of values, it is appropriate that we use the arc sine transformation. Remember from our earlier discussion that all 0 and 100% values need to be replaced by $1/(4n)$ and $(n - 1/4)/n$, respectively. Thus, for treatments C, E, and H we have:

Treatment C = $[1/4(25)] = 0.01$

Treatment E = $[(25 - 1/4)/25] = 0.99$

Treatment H = $[1/4(25)] = 0.01$

The transformed data, using Appendix B, are given in Table 2.4.7.

We will now perform the ANOVA of the angles, following the procedures detailed in Chapter 4, Subsection 4.2.2. The results are shown in Table 2.4.8. Note that the error mean square in Table 2.4.8 is 150.11. This is greater than

TABLE 2.4.8

ANOVA of the Angles

Source of Variation	Degrees of Freedom	Sum of Squares	Mean Square	F
Replication	3	420.77	140.26	0.93
Treatment	9	15,556.81	1,728.53	11.52**
Error	27	4,052.96	150.11	—
Total	39	—	—	—

** = Significant at the 1% level.

$821/n = 821/25 = 32.8$, implying that some variation in excess of the binomial may be present.

In summary, the intent of this chapter was to present the basic assumptions of experimental designs and to discuss the implications of the violation of these assumptions. Several approaches are available for detecting failures of the assumptions, and data transformation was suggested as a means to remedy the violation of the various assumptions.

References and Suggested Readings

Anderson, V.L. and McLean, R.A. 1974. *Design of Experiments: A Realistic Approach.* New York: Marcel Dekker.

Anscombe, F.J. and Tukey, J.W. 1963. The examination and analysis of residuals. *Technometrics* 5: 141–160.

Bartlett, M.S. 1936. Square-root transformation in analysis of variance. *J.R. Stat. Soc. Suppl.* 3: 68–78.

Bartlett, M.S. 1937. Some examples of statistical methods of research in agriculture and applied biology. *J.R. Stat. Soc. Suppl.* 4: 137–183.

Bartlett, M.S. 1947. The use of transformations. *Biometrics* 3: 39–52.

Cochran, W.G. and Cox, G.M. 1957. *Experimental Designs.* 2nd ed. New York: John Wiley & Sons.

Dixon, W.J. and Massey, F.S. 1957. *Introduction to Statistical Analysis.* 2nd ed. New York: McGraw-Hill.

Donaldson, T.S. 1968. Robustness of the F-test to errors of both kinds and the correlation between the numerator and denominator of the F-ratio. *J. Am. Stat. Assoc.* 63: 660–667.

Freeman, M.F. and Tukey, J.W. 1950. Transformations related to the angular and the square root. *Ann. Math. Stat.* 21: 607–611.

Gill, J.L. 1978. *Design and Analysis of Experiments: In the Animal and Medical Sciences.* Vol. 1. Ames, IA: Iowa State University Press. pp. 154.

Grubbs, F.E. and Beck, G. 1972. Extension of sample sizes and percentage points for significance tests of outlying observations. *Technometrics* 14: 847–854.

Hartley, H.O. 1950. The maximum F-ratio as a short-cut test for heterogeneity of variance. *Biometrika* 37: 308–312.

Montgomery, D. 2000. *Design and Analysis of Experiments.* 5th ed. New York: Wiley.

Pearson, E.S. 1931. The analysis of variance in cases of non-normal variation. *Biometrika* 23: 114–133.

Samuels, M.L. and Witmer, J.A. 2003. *Statistics for the Life Sciences*. 3rd ed. New Jersey: Pearson Education.

Scheffé, H. 1959. *The Analysis of Variance*. New York: John Wiley & Sons.

Siegel, S. 1956. *Non-Parametric Statistics*. New York: McGraw-Hill.

Snedecor, G. and Cochran, W.G. 1980. *Statistical Methods*. 7th ed. Ames, IA: Iowa State University Press.

Sokal, R.R. and Rohlf, F.J. 1969. *Biometry*. San Francisco: W.H. Freeman.

Tiku, M.L. 1971. Power function of the F-test under non-normal situations. *J. Am. Stat. Assoc.* 66: 913–916.

Tukey, J.W. 1949. One degree of freedom for nonadditivity. *Biometrics* 5: 232–242.

Turan, M. and Angin, I. 2004. Organic chelate assisted phytoextraction of B, Cd, Mo and Pb from contaminated soils using two agricultural crop species. *Acta Agric. Scand. Sect. B, Soil Plant Sci.* 54: 221–231.

Exercises

1. Using the data given below, perform Tukey's test to determine if the data meet the assumption of additivity.

Treatment	Block			
	I	II	III	IV
A	20	32	22	24
B	18	20	23	25
C	19	21	28	27
D	24	24	26	28

2. To determine the impact of two newly developed fungicides on roses, a horticulturist used a randomized block experiment. Each block contained six plots, and each fungicide was used on two of the plots within each block. Two weeks after the spray, 50 leaves were examined on each plot, and the number of leaves showing rust problems was recorded as shown below. Determine whether any transformation is required before an analysis can be performed.

Block	Fungicide					
	F1		F2		Control	
I	6	8	7	9	25	18
II	5	6	8	5	18	19
III	3	4	7	8	14	26

3. Using a randomized complete block design, an animal scientist performed an experiment in which the impact of linseed cake meal on

weight gain (in pounds) of several groups of animals was observed. Perform a logarithmic transformation and analysis on the data.

Treatment (Species)	Block			
	I	II	III	IV
Pig — Linseed Cake	95	102	110	96
Pig — Control	82	85	80	75
Sheep — Linseed Cake	112	130	114	129
Sheep — Control	90	86	89	85
Chicken — Linseed Cake	3.0	2.9	2.6	2.8
Chicken — Control	1.5	1.9	2.0	1.2

4. In a randomized experiment with three replications, the number of insects found per plot after the fields were sprayed with different rates of a general insecticide are shown in the table. Verify the homogeneity of variance for the data using Bartlett's test.

Number of Insects Found Per Plot following Different Insecticide Treatments

Treatment	Rep. I	Rep. II	Rep. III
A	10	12	7
B	9	15	10
C	5	16	12
D (Control)	15	22	19

5. Slow germination limits establishment of perennial warm season forage grasses, and temperature is a major factor influencing it. A crop scientist conducted a completely randomized design experiment in which the following percentages of seeds germinated were recorded.

Influence of Temperature on Germination Percentage of Seeds of Warm-Season Forage Grasses

Species	Temperature (°C)					
	9	12	15	20	25	30
Native big bluestem	6	32	36	30	27	20
Rountree big bluestem	36	58	68	64	59	49
Caucasian bluestem	0	1	8	10	12	14
Blackwell switchgrass	4	14	18	8	9	12
Osage	30	52	56	54	44	32

(a) What type of transformation should be used?
(b) Perform an analysis of variance on the data.

3

Designs for Reducing Error

3.1 Introduction

Our main focus in this chapter is on designs that reduce experimental error. The intent of an optimal design is to provide a mechanism for the estimation and control of experimental error in the many field experiments that are conducted. The success in reducing errors depends to a large extent on using general knowledge of the experimental material and appropriate groupings to detect true treatment effects.

Generally, there are two sources of experimental error to which a researcher must pay attention. First, the inherent variability in experimental units introduces errors into the experiment. Second, lack of uniformity or failure to standardize experimental technique contributes to experimental error. Lack of attention to either contributes to difficulties in assessing and interpreting results. Several strategies are available to reduce error and enhance the accuracy of the experiment. These approaches can be broadly grouped into (1) increasing the size of the experiment, (2) refining the experimental conditions, and (3) reducing variability in the experimental material. Each of these strategies, along with their implications on experimental error, are discussed in the following text.

3.1.1 Increasing the Size of the Experiment

As a means of increasing the accuracy of the experiment, the size of the experiment can be increased either by increasing the number of replications or by adding more treatments to the experiment.

As was mentioned in Chapter 1, *replication* refers to the repetition of the basic experiment or treatment. By having several replications in an experiment, the experimenter is attempting to reduce the error associated with the differences between the average results of the treatments concerned. It is also used to estimate the error variance that is a function of the difference among observations from an experimental unit receiving the same treatment. The error variance is steadily reduced in a randomized experiment as more replications are added. The rate of reduction of such error is predictable from statistical theory. For example, if σ^2 is the error variance per unit and there are r replications, the error variance of the difference between the means for two treatments is $\sqrt{2\sigma^2/r}$. This result remains valid as long as increased replication does not translate into the use of less homogeneous experimental material and less careful technique. Although more replication is helpful in improving the precision of the estimate, the cost associated with additional replication serves as a limiting factor in its use. Additionally, the sensitivity of statistical methods depends on the number of replications. A discussion of how to determine the number of replications is given next.

Determination of number of replications: When comparing two treatments, it is essential that the experiment be large enough to ensure that if there is a true difference between the treatments, the experiment will obtain a significant result. Researchers have suggested that for crop and vegetable investigations, an experiment with four to eight replications is sufficient to provide a reasonable degree of precision (Little and Hills, 1978). Similarly, Cochran and Cox (1964) have suggested a convenient method for estimating the number of replications required to detect a specified difference. We can also use a test of significance to determine the number of replications. It was mentioned earlier that the error variance is used as a measure of precision or accuracy of an experiment. The precision desired in an experiment could be set either by specifying the size of the true difference or by specifying the width of the confidence interval. In conducting an experiment, the researcher may be interested in knowing if two treatments differ in their effects by less than a certain quantity, say, d. Should a greater difference than d be observed, the results are significant. The significance of the difference between two treatments (d) is tested by performing a t-test, as shown below.

$$t = \frac{\bar{x}_i - \bar{x}_j}{\sqrt{2s^2/r}} \tag{3.1}$$

where

$\bar{x}_i, \bar{x}_j$ = arithmetic means of treatments (i) and (j)

s^2 = variance

r = replications

Given that d is the difference between $\bar{x}_i$ and $\bar{x}_j$, we can substitute the absolute value of d in Equation 3.1 and from it determine the number of replications needed for an experiment as follows:

$$t_o = \frac{|d|}{\sqrt{2s^2/r}} \tag{3.2}$$

where

t_o = critical value of t from the t table

s^2 = variance

Equation 3.2 serves as the basis for the calculation of the number of replications needed so that a significant difference between two treatments can be observed. Rewriting Equation 3.2, we get the following:

$$r = \frac{2t_o^2 s^2}{d^2} \tag{3.3}$$

Equation 3.3 uses the measure of error variance for a set of experimental units, along with the level of significance, to determine the number of replications needed for the results to be significant.

Another way to improve the precision of an experiment is to carefully select the treatments for the experiment. For example, if a horticulturist is interested in studying the effect of a fungicide on roses, it is more useful to the experimenter to determine how the experimental unit (roses) responds to increasing doses of the selected fungicide, rather than to see if there is a significant difference between two successive doses. This allows the experimenter to conduct tests of significance that are more sensitive than assessment based on mere observation of the differences between adjacent means in an array.

3.1.2   Refining the Experimental Conditions

Once the researcher has defined the problem and has selected a design that efficiently tests his or her hypothesis, it is essential to look at some of the critical elements in conducting an experiment. Some of these elements were touched upon in Chapter 1. Other refinements in the experimental conditions are the uniform application of the treatments and selection of an appropriate measure for the treatment effects. Researchers can also improve precision by taking precautions that will prevent gross errors (unusually large or unusually small values that may have been mistakenly recorded) and by controlling the external influences.

Poor attention to these elements will negate the superiority of the design and contribute to the likely increase in experimental error.

3.1.3   Reducing Variability in the Experimental Material

To avoid problems associated with the variability of the experimental material, uniform material is frequently prepared for the experiment. For example, in some agricultural experiments, researchers use inbred lines of plants and animals to ensure uniform experimental material. Although appropriate in some conditions, the use of the inbred lines has been questioned by Biggers and Claringbold (1954). They have argued that inbred lines are not more homogeneous than randomly bred material and have suggested the use of F_1 hybrids between the inbred lines as being more appropriate than the inbred lines themselves.

Researchers can also use a sample that is chosen for its homogeneity from a large batch of experimental material. For example, in an animal experiment, the researcher may choose the initial weight as the criterion. Thus, the two animals with the highest weight are grouped in a pair and the two with the next highest weight in the second group, and this is continued until all the animals (experimental material) have been paired in this fashion.

Undoubtedly, such pairing of animals and plants reduces the variability of the experimental material. However, the substantial increase in precision sometimes may be at the cost of getting conclusions that are not representative of a wider class of experimental units.

3.2   Approaches to Eliminating Uncontrolled Variations

Reducing the effect of uncontrolled variation on the errors in treatment comparisons is an essential element of a good design. Agricultural scientists, in conducting experiments on varietal differences, for example, are concerned with arranging the experiment in such a way that they can, with confidence and accuracy, separate the uncontrolled variations from the varietal differences.

In the chapters to come, you will be introduced fully to the different designs that allow the experimenter to separate the uncontrolled variation from the treatment effects under a variety of circumstances. Here we are interested in giving an overview of the different designs, from the simplest to the most complex, and how they can be used appropriately to remove the effect of the uncontrolled variation from the experiment.

In the most simple case of comparing two treatments, the researcher is interested in evaluating the differences between treatments rather than the treatments themselves. Thus, to be able to accurately evaluate the differences between treatments, the experimenter must select pairs of similar experimental units. For example, a horticulturist may choose a pair of roses that have the same genetic makeup, and an animal scientist may choose two male piglets from the same litter.

Another way of selecting a pair of observations is by means of *self-pairing*. This involves making observations on the same experimental unit on two different occasions. No matter which pairing method is used, the researcher is interested in having a number of pairs of experimental units; the two units in each pair are expected to provide, as nearly as possible, identical observations in the absence of treatment differences.

As was suggested in Subsection 3.1.3, there are many ways of pairing experimental units so that the experimenter is assured of uniform material. Proper randomization of the paired experimental units also provides added protection from the uncontrolled variations in an experiment.

When there are more than two treatments to be compared, each treatment must be applied to several different units. The reason for this is obvious. If each treatment is applied to only a single unit, the experimenter will have difficulty distinguishing between the difference that might have been caused by the different treatments and that due to the inherent difference between the units.

Researchers, when faced with *t* alternative treatments, can group the units into sets of *t*, the units in each set being expected to give, as nearly as possible, the same observation if the treatments are equivalent in their effect. Each set of the *t* units is called a *block*. An experimental design in which the block differences are removed from the error, and the only recognizable difference between units is the treatment, is called a *randomized block design*. In agricultural research there are many ways of blocking the experimental units, and we shall return to this in Chapter 4, Chapter 5, and Chapter 6.

The statistical model for a randomized block design is:

$$X_{ij} = \mu + \tau_i + \beta_j + \varepsilon_{ij} \left\{ \begin{matrix} i = 1, 2, \dots, a \\ j = 1, 2, \dots, b \end{matrix} \right\} \tag{3.4}$$

where

μ = the overall mean

τ_i = the *i*th treatment effect

β_j = the *j*th block effect

ε_{ij} = the random error term

In agricultural experiments, the general principle for grouping plots into blocks is to make sure that the plots chosen minimize the uncontrolled variation from plot to plot within blocks. This is usually achieved by arranging the plots within a block in a compact square area. Again, proper randomization within blocks will allow the experimenter to statistically assess uncontrolled variation between plots.

Even though blocking provides an added mechanism to remove uncontrolled variation from the results, experimenters should be cautious about the erratic variations that might result from the order in which plots are cultivated and harvested. For example, when harvesting takes more than a day, it is recommended that the plots in one block be harvested on the same day. This allows the constant differences between days to become identified with block differences and to not contribute to the error of the experiment. Furthermore, care must be taken in the collection of data and in the tests to be used in evaluating the data.

Researchers also use covariance analysis, as discussed in Chapter 10, as a mechanism in reducing error. The purpose of the covariance analysis is to allow the researcher to use supplementary data to reduce experimental errors by eliminating the effects of variations within a block. We refer to this approach as an *indirect* or *statistical control* to increase the precision of an experiment.

3.3   Error Elimination by Several Groupings of Units

So far we have discussed how the single grouping of units into blocks can be used to reduce error in an experiment. There are several other types of design that utilize the blocking principle. The Latin square, the combined Latin squares, and the Greco-Latin squares are such designs. Each of these designs allows for the removal of the known sources of variability from the experimental error. Hence, depending on the objective of the study, the researcher has at his or her disposal any one of these designs. In Subsection 3.3.1, a brief discussion of the use of the Latin square design is given, simply to provide the reader with an overview of the contribution of this design in reducing error. For a detailed discussion of the Latin square, the reader is referred to Chapter 4 (Subsection 4.2.3). In Subsection 3.3.2 and Subsection 3.3.3, we discuss the use of the combined Latin and Greco-Latin squares, respectively.

3.3.1   Latin Square

This design is used to eliminate the effect of two known sources of uncontrolled variation. By using row and column blocking, the design allows the experimenter to remove two sources of variability from the experimental error. A conditional requirement of the design is that the treatments be randomly assigned to the rows and columns, such that each treatment occurs only once in each row and column.

To satisfy the requirement that each treatment occurs once in each row and column, the number of replications and treatments must be equal. The implication of this requirement is that when there are a large number of treatments in the experiment, there must be an equally large number of replications. Technically and economically, this places a limitation on the use of this design. The statistical model for a Latin square is:

$$X_{ijk} = \mu + \alpha_i + \tau_j + \beta_k + \varepsilon_{ijk} \begin{cases} i = 1,2,\dots,n \\ j = 1,2,\dots,n \\ k = 1,2,\dots,n \end{cases} \tag{3.5}$$

where

X_{ijk} = the observation of the *i*th row and *k*th column for the *j*th treatment

μ = the overall mean

α_i = the *i*th row effect

τ_j = the *j*th treatment effect

β_k = the *k*th column effect

ε_{ijk} = the random error

The model is completely additive; that is, there is no interaction between rows, columns, and treatments.

Researchers should also recognize that when using a Latin square design, especially when there are only three or four treatments in an experiment, the degrees of freedom associated with error $(t-1)(t-2)$ are not adequate to estimate the σ^2. For example, when the number of treatments is three, the degrees of freedom for error are only two. This is not sufficient for an adequate estimate of σ^2. In such experimental conditions it is necessary to use more than one square in the experiment.

Although the Latin square design is a good design for many types of field experiments, it is suggested that it should not be used for experiments with fewer than four treatments or more than eight (Kempthorn, 1952). The Latin square is most valuable to those experimenters who use only restrictions on randomization (not true factors of interest) for the rows and columns. Furthermore, the design is helpful in analyzing data where there are no interactions between factors. It should, however, be kept in mind that agricultural researchers have used the Latin square design to screen for major effects even when there has existed a small amount of interaction.

In Appendix C, a list of standard Latin squares are provided. Different Latin squares can be constructed using the randomization process. The randomization of a 3 × 3 standard Latin square for example, is accomplished by first randomizing the rows and then the columns. For a 4 × 4 Latin square, we first select at random one of the four tabulated squares from Appendix C and then randomize the rows and columns, respectively. For the higher-order squares, the rows, columns, and treatments are all randomized independently.

To illustrate the randomization process for a 4 × 4 Latin square, we can use a table of permutations or some other random selection procedure, such as drawing pieces of paper from a bowl in which four pieces of paper, with a number from 1 to 4 written on each, are placed. Suppose that the first number drawn is "2." We draw a second piece of paper from the bowl; suppose that this number is "1." We continue this process until we have drawn all four numbers. This process is repeated two more times so that we have three sets of four numbers each. The first set of numbers is used to select a Latin square from Appendix C. The other two sets are used for randomizing the rows and columns of the selected 4 × 4 Latin square. Suppose the three sets of numbers drawn are: 2, 1, 4, 3; 2, 3, 1, 4; and 1, 4, 3, 2. Using the first set (2, 1, 4, 3) we start with the second 4 × 4 tabulated square from Appendix C. The selected Latin square is:

A	B	C	D
B	C	D	A
C	D	A	B
D	A	B	C

Now we use the second set of numbers (2, 3, 1, 4) to randomize the rows of the above 4 × 4 Latin square. This is shown below:

Original Column Number

2	B	C	D	A
3	C	D	A	B
1	A	B	C	D
4	D	A	B	C

The third set of numbers (1, 4, 3, 2) is used to randomize the columns as illustrated:

Original Column Number

1	4	3	2
B	A	D	C
C	B	A	D
A	D	C	B
D	C	B	A

The same procedure could also be applied easily to larger squares. For a discussion of the theoretical basis of the randomization, the reader is referred to Yates (1933).

In the next section we will discuss the use of the multiple squares, which is used when there are not sufficient treatments for an adequate estimation of the σ^2. The multiple squares are an extension of the Latin square.

3.3.2 Combined Latin Square

The problem of inadequate degrees of freedom associated with error arises when the number of treatments in the experiment is not enough. The researcher can deal with this problem either by extending one of the sides of the square in space or time or by using a separate multiple Latin square.

Table 3.3.2.1(*a*) shows the extension of the rows of a 4 × 4 Latin square. This form of a multiple Latin square implies that the two squares (columns 1–4 and columns 5–8) have common row effects. When it is assumed that the row effects are consistent over several sets of columns, a less stringent requirement for the occurrence of each treatment in rows is allowed. This means that each treatment appears once in each column and twice or more in each row. This is not the case for separate multiple Latin squares.

As the rows (or columns) of a Latin square are extended, the resultant design can be looked at as a Latin rectangle rather than a square. This is not a problem, as the analysis of a Latin rectangle follows the model of a single Latin square design, in which the sum of squares for rows, columns, treatment, and error are computed for the analysis.

The completely separate multiple square, as shown in Table 3.3.2.1(*b*), indicates that there is no relation between the rows (or columns) in the different squares. In either form of the Latin square, the number of treatments

TABLE 3.3.2.1

Layout of Multiple Latin Squares with Common Row Effects (*a*) and Completely Separate Squares Shown in (*b*)

a

Row	Column 1	2	3	4	5	6	7	8
1	A	B	C	D	A	B	C	D
2	B	C	D	A	B	C	D	A
3	C	D	A	B	C	D	A	B
4	D	A	B	C	D	A	B	C

b

Row	Column 1	2	3	4	5	6	7	8
1	A	B	C	D				
2	B	C	D	A				
3	C	D	A	B				
4	D	A	B	C				
5					A	B	C	D
6					B	C	D	A
7					C	D	A	B
8					D	A	B	C

TABLE 3.3.2.2

The ANOVA Table for a Latin Rectangle

Source of Variation	Degrees of Freedom	Sum of Squares
Rows	1	$\dfrac{R_1^2 + R_2^2 + R_3^2 + R_4^2}{8} - \dfrac{G^2}{32}$
Columns	7	$\dfrac{C_1^2 + C_2^2 + \ldots + C_8^2}{4} - \dfrac{G^2}{32}$
Treatments	3	$\dfrac{T_1^2 + T_2^2 + T_3^2 + T_4^2}{4} - \dfrac{G^2}{32}$
Error	20	By subtraction
Total	31	$\sum \left(x_{ijkl}^2 \right) - \dfrac{G^2}{32}$

increases, thus allowing the researcher to estimate the σ^2 with adequate precision.

The difference in practice between the designs is best observed by considering what types of systematic variation are eliminated from the error by the two designs.

What makes the two forms of the design different from one another is whether the row differences (or the column differences) might be expected to be similar for the different squares. For example, a horticulturist uses a leaf as an experimental unit, and plants and the leaf position are chosen as the two blocking criteria. If two groups of plants are used for the two squares, it would seem reasonable to assume that the differences in leaf position should be similar for plants in both groups. This example shows how one of the blocking criteria is consistent over squares.

Based on the assumption that the row effects are the same for both the original squares and for all columns, the form of the analysis differs. For example, the analysis for a design such as the one depicted in Table 3.3.2.1(*a*), in which the width of the rows of a Latin square is extended, is presented in Table 3.3.2.2.

In Table 3.3.2.2 the notation used for the various sums of squares are:

$R_1 \ldots R_4$ = row totals

G = grand total

$C_1 \ldots C_8$ = column totals

$T_1 \ldots T_4$ = treatment totals

X_{ijkl} = a particular observatio

TABLE 3.3.2.3

ANOVA for Two Independent Latin Squares

Source of Variation	Degrees of Freedom	Sum of Squares (SS)
Between squares	1	$\dfrac{G_1^2 + G_2^2}{16} - \dfrac{G^2}{32}$
Rows	9	(Row SS)$_1$ + (Row SS)$_2$
Columns	9	(Column SS)$_1$ + (Column SS)$_2$
Treatments	3	$\dfrac{T_1^2 + T_2^2 + T_3^2 + T_4^2}{4} - \dfrac{G^2}{32}$
Error	9	By subtraction
Total	31	$\displaystyle\sum \left(x_{ijkl}^2 \right) - \dfrac{G^2}{32}$

The form of the analysis of variance (ANOVA) for two independent squares such as the one shown in Table 3.3.2.1(*b*) differs slightly. For this design, the sum of squares for rows and columns is calculated separately for each square. Additionally, we calculate a sum of squares for the overall difference between squares. Table 3.3.2.3 shows the ANOVA table for this design.

The notations for the various sums of squares given in Table 3.3.2.3 are similar to those given in Table 3.3.2.2, except that G_1 and G_2 represent the totals for square 1 and square 2, and subscripts 1 and 2 for the rows and columns refer to the rows and columns of each square, respectively.

3.3.3 Greco-Latin Square

This is an associated design of the Latin square that allows for one more restriction on randomization. Recall that in the case of the randomized block design, experimental units were grouped in one way. When the Latin square design is used, the experimental units are simultaneously grouped in two ways. If it is desired to group the units in three or more ways, use is made of the Greco-Latin square.

In this design, the treatments are grouped into replicates in three different ways, such that the effects of the three sources of variation are equalized for all treatments. This means that we are able to systematically control three sources of variability by blocking in three directions. The additional grouping that is made possible by this design is usually represented by Greek letters. The sum of squares due to the Greek letter factor is computed directly from

TABLE 3.3.3.1

A 4 × 4 Greco-Latin Square

Row	Column			
	1	2	3	4
1	Aα	Bβ	Cγ	Dδ
2	Bδ	Aγ	Dβ	Cα
3	Cβ	Dα	Aδ	Bγ
4	Dγ	Cδ	Bα	Aβ

the Greek letter totals, and the experimental error is further reduced by this amount.

To see how this third grouping is incorporated into the design, consider an $n \times n$ Latin square and superimpose on it a second $n \times n$ Latin square in which the treatments are denoted by Greek letters. When the second Latin square is superimposed on the first, the resultant Latin square is called a Greco-Latin square if each Greek letter appears once, and only once, with each Latin letter. Table 3.3.3.1 shows a Greco-Latin square with four treatments.

Experimenters use the Greco-Latin square to investigate four factors. Factors could be assigned to the rows, columns, Latin letters, and Greek letters, each at n levels with n^2 runs. The statistical model for a Greco-Latin square is given as:

$$X_{ijkl} = \mu + \theta_i + \tau_j + \omega_k + \rho_l + \varepsilon_{ijkl} \left\{ \begin{array}{l} i = 1,2 \ldots , n \\ j = 1,2 \ldots , n \\ k = 1,2 \ldots , n \\ l = 1,2 \ldots , n \end{array} \right\} \tag{3.6}$$

where

X_{ijkl} = the observation in row i and column l for Latin letter j and Greek letter k

θ_i = the effect of the ith row

τ_j = the effect of the Latin letter treatment j

ρ_l = the effect of column l

ε_{ijkl} = the random error

The statistical analysis of the results of the Greco-Latin square experiment is a straightforward extension of the Latin square. Table 3.3.3.2 is the ANOVA table for the Greco-Latin square. The sum of squares for the various sources

TABLE 3.3.3.2

ANOVA Table for a Greco-Latin Square

Source of Variation	Sum of Squares (SS)	Degrees of Freedom
Latin letter treatments	$\left(\sum\limits_{j=1}^{n} \dfrac{T_L^2}{n}\right) - C$	$n-1$
Greek letter treatments	$\left(\sum\limits_{K=1}^{n} \dfrac{T_G^2}{n}\right) - C$	$n-1$
Rows	$\left(\sum\limits_{i=1}^{n} \dfrac{R_L^2}{n}\right) - C$	$n-1$
Columns	$\left(\sum\limits_{l=1}^{n} \dfrac{C_l^2}{n}\right) - C$	$n-1$
Error	Error SS computed by subtraction	$(n-3)(n-1)$
Total	$\sum\limits_{i}\sum\limits_{j}\sum\limits_{k}\sum\limits_{l} X^2 - C$	n^2-1

of variation is computed in the same manner as for the Latin square given in Chapter 4, Subsection 4.2.3.

In computing the various sum of squares in Table 3.3.3.2, the term C is the correction factor, computed as:

$$C = \frac{G^2}{N} \tag{3.7}$$

where G = grand total and N = total number of observations.

In Appendix C several Greco-Latin squares are presented. The designs have been constructed for all numbers of treatments from 3 to 12, except 6 and 10 (Cochran and Cox, 1957). It is generally thought that no $n \times n$ Greco-Latin square exists whenever the size n leaves a remainder of 2 when divided by 4 (Cox, 1958).

The purpose of this chapter was to give the reader an overview of the various approaches to reducing error. Error can be reduced by increasing the size of the experiment, refining the experimental technique, and reducing variability in the experimental material. The uncontrolled variation in an experiment can be reduced further by the design used in an experiment. The selection of the design is based on the objective of the experiment and other factors such as technical and cost issues. In trying to minimize experimental error, several ways of grouping experimental units were discussed. The Latin square and its associated designs were reviewed to show how researchers can use these designs to reduce error. In the subsequent chapters of this book, each design is further detailed.

References and Suggested Readings

Biggers, J.D. and Claringbold, P.J. 1954. Why use inbred lines? *Nature* 174, 596.

Cochran, W.G. and Cox G.M. 1957. *Experimental Designs*. New York: John Wiley & Sons.

Cox, D.R. 1958. *Planning of Experiments*. New York: John Wiley & Sons.

Kempthorn, O. 1952. *The Design and Analysis of Experiments*, New York: John Wiley & Sons.

Little, T.L. and Hills, F.J. 1978. *Agricultural Experimentation: Design and Analysis*. New York: John Wiley & Sons.

Yates, F. 1933. The formation of Latin squares for use in field experiments. *Emp. J. Exp. Agric.* 1: 235–244.

4

Single-Factor Experimental Designs

4.1 Introduction

Many agricultural experiments involve the use of a single-factor experimental design. In this type of design, a single factor varies while all other factors are held constant. For example, when an agronomist is interested in finding whether one variety is superior to another, he or she will use a single-factor experiment in which the single variable factor is the variety, and the treatments or factor levels are different variables. In such an experiment, all cultural practices — water, fertilizer application, pest control, and other management activities — remain the same.

Single-factor experiments can be grouped under two distinct experimental designs. The first design involves a small number of treatments and is called *complete block design*. As the name implies, complete block designs are characterized by blocks, each of which contains at least one complete set of treatments. The second grouping of designs is *incomplete block designs*. These designs contain a large number of treatments and are also characterized by blocks. However, each block contains only a fraction of the number of treatments.

In this book you will be introduced to three types of complete block designs (*completely randomized design* [CRD], *randomized complete block design* [RCB], and *Latin square design*) and two incomplete block designs (*lattice design* and *partially balanced lattice design*). As we discuss each of these designs, we will introduce you to the advantages of using the design and the randomization procedures, plot layout, and analysis of variance (ANOVA) for the particular design.

4.2 Complete Block Designs

4.2.1 Completely Randomized Design (CRD)

This design is the simplest ANOVA design from the standpoint of assignment of experimental units to treatment or treatment combinations. In this design, treatments are allotted to the experimental units entirely at random or by chance. More specifically, if a treatment is to be applied to eight units, each unit has the same chance or probability of receiving the treatment and, consequently, there exists no systematic source of error. In a CRD, the units are taken as a single group. Hence, as far as possible, the units forming the group should be homogeneous. This characteristic of the design suggests that it should be used in which environmental effects are easily controlled. Agricultural field experiments do not lend themselves to such homogeneous conditions. Therefore, this design is recommended mostly for laboratory and greenhouse experiments in which environmental effects are easily controlled. Additionally, a CRD is a logical choice when performing a first-stage pilot experiment in which experimental units and conditions are homogeneous.

The advantage of using the CRD is its simplicity. By using it, an experimenter may avoid making certain dubious assumptions. The inherent simplicity of the design extends to all aspects of the design, namely, the experimental layout, the model underlying the data analysis, and the computations involved in such analysis.

As far as the layout of the experiment is concerned, CRD only requires that treatments be assigned randomly to the experimental units. Any number of treatments and replicates may be used. In contrast, stratified designs require the division of treatments into strata or levels on the basis of some additional measure, followed by random assignment to the experimental units or plots. Other designs, as will be discussed later, involve yet other complications in laying out the experiment.

The model underlying the CRD involves fewer assumptions than that of any other experimental design. As a consequence, the derivations of parameter estimates are simpler in this case than in any other situation. Furthermore, fewer assumptions mean fewer violations of such assumptions. Hence, less can go wrong as far as the statistical inference is concerned.

Because the CRD involves fewer terms, computation of such designs are simple. Furthermore, related analyses such as the analysis of covariance and the estimation of the missing data will also generally be simpler. Statistical analysis is generally easy to perform, whether the number of replicates is equal or not. See Appendix K on the use of MINITAB for performing statistical analyses for a CRD.

The major disadvantage in using the CRD is its lack of accuracy, or its inefficiency. The error variance will usually be large in comparison to that of other designs. This outcome is because randomization is not restricted in any way, to ensure that the units that receive one treatment are similar to those that receive another treatment. This disadvantage is partly offset by the fact that no design yields as many degrees of freedom for the error variance as does the CRD, assuming some fixed amount of data. The greater the degrees of freedom, the higher is the power of a test. When experiments require a large number of experimental units, CRDs cannot ensure precision of the estimates of treatment effects.

1	2	3	4	5
6	7	8	9	10
11	12	13	14	15

FIGURE 4.2.1.1
Layout of an experimental plot.

Randomization: There are several methods for the random allocation of treatments to the experimental units. You will be introduced to two such procedures, described in a step-by-step fashion in the text that follows.

Example 4.2.1.1

An animal scientist has five treatments (A, B, C, D, and E), each replicated thrice. Determine the random layout of the experiment for the scientist.

Step 1: Determine the total number of experimental plots. This is simply the product of the number of treatments and replications:

$$n = t \times r$$

$$n = 5 \times 3$$

$$n = 15$$

Step 2: Assign each experimental plot a number. In the present case, the plots will be numbered from 1 to 15 consecutively, as shown in Figure 4.2.1.1.

Step 3: Assign each treatment to an experimental plot, using either the random digits technique (method A) or an alternative to random allocation (method B).

Method A — random digit technique: To illustrate the use of a table of random numbers, such as the one given in Appendix D, we present a partial table of random numbers in Table 4.2.1.1. You may choose to start randomly at any point in the table. For our example, we start at the left-hand corner of the table. The first number is 85967. One may select the first three digits, the middle three digits, or the last three digits of this number; any of these choices are appropriate. We have elected to choose the first 3 digits of the number and to read downward until 15 numbers have been selected.

TABLE 4.2.1.1

A Partial Table of Random Numbers from Appendix D

85967	73152	14511
07483	51453	11649
96283	01898	61414
49174	12074	98551
97366	39941	21,25
90474	41469	16812
28599	64109	09497
25254	16210	89717
28785	02760	24359
84725	86576	86944
41059	66,56	47679
67434	41045	82830
72766	68816	37643
92079	46784	66125
29187	40350	62533
74220	17612	65522
03786	02407	06098
75085	55558	15520
09161	33015	19155

The 15 numbers selected from the first column of Table 4.2.1.1 are:

1.	859	6.	904	11.	410
2.	074	7.	285	12.	674
3.	962	8.	252	13.	727
4.	491	9.	287	14.	920
5.	973	10.	847	15.	291

Once the numbers are selected, rank the numbers in either ascending or descending order. In this example, we have ranked the numbers in ascending order:

Random Number	Rank	Random Number	Rank	Random Number	Rank
859	11	904	12	410	6
074	1	285	3	674	8
962	14	252	2	727	9
491	7	287	4	920	13
973	15	847	10	291	5

Divide the 15 ranks into 5 groups, each consisting of 3 numbers, as follows:

Group Number	Rank in the Group		
1	11	1	14
2	7	15	12
3	3	2	4
4	10	6	8
5	9	13	5

Now we are ready to assign the treatments to the experimental plots by using the group number as the treatment number and the ranks in each group as the plot numbers. In our example, the first group is assigned to treatment A and the last group (group 5) receives treatment E. Plots that will receive treatment A are numbers 11, 1, and 14. Similarly, treatment B is applied to plots 7, 15, and 12. The remaining treatments are applied in the same manner to the rest of the experimental units or plots. Finally, we will have randomly assigned all treatments to the experimental plots, as shown in Figure 4.2.1.2.

Method B — alternative to random allocation: If a table of random numbers is not available, treatments can be allocated to the experimental units by drawing "lots" as described next.

Let the number of the ith treatment be written on r_j pieces of papers ($j = 1, 2, ... , k$). All the pieces of paper are then folded individually to protect the identity of the numbers written on them. They are placed in a container and mixed thoroughly. One piece at a time is drawn from the container without replacement. The treatment that is drawn in the jth draw is allotted to the jth plot ($j = 1, 2, ... , k$). In the present example there should be

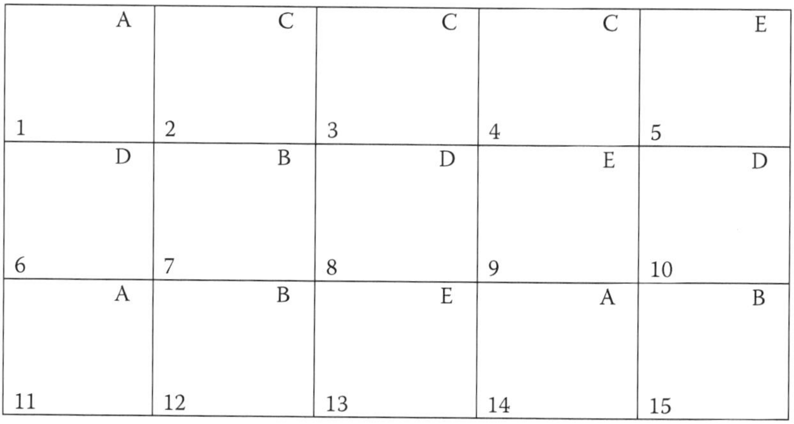

A	C	C	C	E
1	2	3	4	5
D	B	D	E	D
6	7	8	9	10
A	B	E	A	B
11	12	13	14	15

FIGURE 4.2.1.2
A completely randomized layout for five treatments and three replications.

TABLE 4.2.1.2

General Format of a 1-Factor analysis of Variance Table

Source of Variation	Degree of Freedon	Sum of Squares	Mean Square	F
Treatment	$t - 1$	SST_r	$MST_r = SST_r/t - 1$	MST_r/MSE
Error	$n - t$	SSE	$MSE = SSE/n - t$	
Total	$n - 1$	SST		

15 pieces of paper, 3 each with treatments A, B, C, D, and E written on them. Other procedures are also used to randomly assign treatments to experimental units.

ANOVA: In a CRD there are two sources of variation that one encounters in an experiment. The first is the treatment variation, and the second is the variation due to experimental error. The treatment variation or treatment sum of squares ($SSTr$) is due to variability between sample results, whereas variability within sample results is measured by the sum of squares error (SSE). It is the relative size of these two sources of variation that determines whether the differences observed among treatments is real or due to chance.

Table 4.2.1.2 gives the general format of a one-factor ANOVA table. As was mentioned earlier, computation of the ANOVA for experiments with equal or unequal replication is simple. The following examples illustrate how the ANOVA is applied to experiments with equal as well as unequal replications.

Equal replications: In this type of experiment, the numbers of replications are equal. The methods of calculating each of the sum of squares and performing the ANOVA are given in the following example.

Example 4.2.1.2

An animal scientist, in performing a study of vitamin supplementation, randomly assigned five heifers to each of eight treatment groups. The following weight gains (kg) were recorded from five replicates:

Treatment	Weight Gain (kg)				
	Rep. I	Rep. II	Rep. III	Rep. IV	Rep. V
A	4.5	5.2	6.2	3.9	4.9
B	5.6	4.7	4.3	4.4	6.1
C	6.4	6.7	6.8	6.1	6.9
D	5.2	5.0	6.8	3.6	5.6
E	4.0	4.9	4.3	4.8	4.2
F	7.1	6.5	6.2	6.8	6.1
G	6.1	4.9	4.2	3.9	6.8
H (Control)	4.6	4.0	4.9	3.8	4.2

TABLE 4.2.1.3

Weight Gain by Animals from a CRD with Eight Treatments (t) and Five Replications (r)

Treatment	Weight Gain (kg)					Treatment Total (*T*)	Treatment Mean
A	4.5	5.2	4.5	3.9	4.9	24.7	4.94
B	3.8	4.7	4.3	4.4	3.7	25.1	5.02
C	5.4	4.7	4.6	3.7	3.9	32.9	6.58
D	5.2	5.0	4.8	3.6	5.6	26.2	5.24
E	4.0	4.9	4.3	4.8	4.2	22.2	4.44
F	4.1	4.5	5.2	4.8	5.1	32.7	6.54
G	5.1	4.9	4.2	3.9	3.8	25.9	5.18
H (Control)	4.6	5.2	4.9	3.8	4.2	21.5	4.30
Grand Total (*G*)	—	—	—	—	—	211.2	—
Grand Mean	—	—	—	—	—	—	5.28

Perform the ANOVA to determine if there are differences between the treatments.

SOLUTION

Step 1: Calculate the treatment totals (*T*), their respective means, and the grand total (*G*) as shown in Table 4.2.1.3.

Step 2: Determine the degree of freedom (*d.f.*) for each source of variation as shown below:

$$\text{Total } d.f. = (r)(t) - 1 = (5)(8) - 1 = 39$$

$$\text{Treatment } d.f. = t - 1 = 8 - 1 = 7$$

$$\text{Error } d.f. = t(r - 1) = 8(5 - 1) = 32$$

Step 3: Calculate the correction factor (*C*) and the various sums of squares (*SS*) using the following equations:

$$C = \frac{G^2}{N} \qquad (4.1)$$

where

C = correction factor

G = grand total

N = the total number of experimental plots [(*r*)(*t*)]

X_i = an observation in the *i*th plot

T_i = treatment total for the *i*th plot

r_i = the number of replication of the *i*th treatment

$$\text{Total } SS = \sum X_i^2 - C \qquad (4.2)$$

$$C = \frac{G^2}{N} = \frac{(211.2)^2}{40} = 1,115.14$$

$$\text{Total } SS = \left[(4.5)^2 + (5.2)^2 + \quad + (3.8)^2 + (4.2)^2 \right] - 1,115.14$$

$$= 1,160.26 - 1,115.14$$

$$= 45.12$$

$$\text{Treatment } SS = \frac{\sum T_i^2}{r} - C \qquad (4.3)$$

$$= \frac{(24.7)^2 + (25.1)^2 + \quad + (25.9)^2 + (21.5)^2}{5} - 1,115.14$$

$$= 1,140.83 - 1,115.14$$

$$= 25.69$$

$$\text{Error } SS = \text{Total } SS - \text{Treatment } SS \qquad (4.4)$$

$$= 45.12 - 25.69$$

$$= 19.43$$

Step 4: Calculate the mean square for the treatment (treatment *MS*) and error (error *MS*) variations by dividing each *SS* by its respective *d.f.*'s.

$$\text{Treatment } MS = \frac{\text{Treatment } SS}{t - 1} \qquad (4.5)$$

$$= \frac{25.69}{8 - 1}$$

$$= 3.67$$

$$\text{Error } MS = \frac{\text{Error } SS}{t(r - 1)} \qquad (4.6)$$

$$= \frac{19.43}{(8)(4)}$$

$$= 0.61$$

Step 5: Calculate the F value using the treatment MS and error MS as:

$$F = \frac{\text{Treatment } MS}{\text{Error } MS} \tag{4.7}$$

$$= \frac{3.67}{0.61}$$

$$= 6.02$$

The ANOVA table for the present example is shown in Table 4.2.1.4.

Step 6: Compare the computed F given in Table 4.2.1.4 with the tabular F value given in Appendix E. For our example, the tabular F (for 7 *d.f.*'s for treatment and 32 *d.f.*'s for the error source of variation) is 2.32 for the 5% level of significance and 3.25 for the 1% level of significance.

Step 7: Make a decision on the significance of the difference among treatments by applying the following rules:

1. A treatment difference is said to be *highly significant* if the computed F value is greater than the tabular F at the 1% level of significance. Conventionally, two asterisks are placed on the computed F value in the ANOVA.

2. A treatment difference is said to be *significant* if the computed F value is greater than the tabular F at the 5% level of significance but smaller than or equal to the tabular F value at the 1% level of significance. Conventionally, one asterisk is placed on the computed F value in the ANOVA.

3. A treatment difference is said to be *nonsignificant* if the computed F value is smaller than or equal to the tabular F at the 5% level of significance. Conventionally, "*ns*" is placed on the computed F value in the ANOVA.

What you observe in the present example is an F value that is highly significant. This means that the test indicates the existence of some differ-

TABLE 4.2.1.4

ANOVA of Animal Weight Gain Data in a CRD

Source of Variation	Degree of Freedom	Sum of Squares	Mean Square	F
Treatment	7	25.69	3.67	6.02**
Error	32	19.43	0.61	
Total	39	45.12		

Note: Coefficient of variation (cv) = 14.8%; computation of cv is presented in Step 7; ** = highly significant or significant at 1% level.

ences among the treatments, but it does not specify which treatments differ from one another.

You should keep in mind that if we had observed a nonsignificant test result, the outcome would not mean that all treatments are the same. What it would indicate is the failure of the experiment to detect any observable differences among the treatments. Failure to detect treatment differences could be a result of very small treatment differences, a very large experimental error, or both. When nonsignificant results are obtained from an experiment, it is helpful to take a look at the size of the experimental error and the numerical differences among treatment means. These will serve as a guide to the researcher in carrying out further trials. If both values are small, it implies that the treatment differences are so small that they cannot be detected. Under such a condition, no additional trials will be helpful. On the other hand, if both values are large, additional trials may reduce the experimental error and, possibly, the treatment differences detected.

As a way of determining the reliability of the experiment, the coefficient of variation (*cv*) for the experiment should be calculated. The *cv* indicates the degree of precision with which the treatments are compared. In actuality, *cv* expresses the experimental error as a percentage of the mean. Thus, the higher the value of the *cv*, the lower is the reliability of the experiment. Conventionally, the *cv* value is reported below the ANOVA table for an experiment.

The *cv* for the present example is calculated as shown below:

$$\text{Grand mean} = \frac{G}{n} \tag{4.8}$$

$$= \frac{211.2}{40}$$

$$= 5.28$$

$$cv = \frac{\sqrt{\text{Error } MS}}{\text{Grand mean}} \times 100 \tag{4.9}$$

$$= \frac{\sqrt{0.61}}{5.28} \times 100$$

$$= 14.8\%$$

As mentioned earlier, the higher the value of the coefficient, the less reliable the experiment. This *cv* indicates that there is reliability in the experiment. It should be noted that the *cv* varies with the type of experiment and the characteristics measured. Gomez and Gomez (1984, p. 17) have reported that the acceptable range of *cv* is 6 to 8% for variety trials, 10 to 12% for fertilizer trials, and 13 to 15% for insecticide and herbicide trials. Furthermore, they point out that in a field experiment *cv* for rice yield is about 10%, that for tiller number is about 20%, and for plant height *cv* is about 3%.

Unequal replications: In the preceding example we performed the one-factor ANOVA when there was an equal number of replications. Sometimes it is not possible to have an equal number of replications for all treatments. For instance, animal scientists may be performing experiments on different breeds of animals, in which the number of a particular breed may be limited to a few. In such situations we may not have an equal number of replications.

In another situation, a researcher may start with an experiment that has equal replications. However, for unforeseen reasons, elements of the experimental unit may be destroyed or lost in the process. If such conditions prevail, the steps followed in performing the ANOVA are similar to those for equal replications.

Example 4.2.1.3

In a recent study on maize hybrid's response to nitrogen fertilization in the northern "Corn Belt," agronomists obtained the following data. Determine if there is a difference in yield among the hybrids.

Hybrid	Yield (bu/ac)		
	Rep. I	Rep. II	Rep. III
P3747	133	154	175
P3732	125	151	161
Mo17 × A634	—	146	152
LH74 × LH 51	—	166	167
CP 18 × LH 54	122	—	159
Control	120	159	147

SOLUTION

Step 1: Calculate the treatment totals (*T*), their respective means, and the grand total (*G*) as shown in Table 4.2.1.5.

TABLE 4.2.1.5

Effect of Hybrid on Grain Yield from a CRD Experiment with Unequal Number of Replications

Treatment	Yield (bu/ac)			Treatment Total (*T*)	Treatment Mean
	Rep. I	Rep. II	Rep. III		
P3747	133	154	175	462	154.0
P3732	125	151	161	437	145.7
Mo17 × A634	—	146	152	298	149.0
LH74 × LH 51	—	166	167	333	166.5
CP 18 × LH 54	122	—	159	281	140.5
Control	120	159	147	426	142.0
Grand Total (*G*)				2237	
Grand Mean					149.6

Step 2: Determine the *d.f.* for each source of variation as shown:

Total *d.f.* = (*n* (4.4) − 1) = 15 − 1 = 14

Treatment *d.f.* = *t* − 1 = 6 − 1 = 5

Error *d.f.* = (*n* − 1) (*t* − 1) = 14 − 5 = 9

Step 3: Calculate the correction factor (C) and the various sums of squares using the following notations:

X_i = an observation in the *i*th plot

T_i = treatment total for the *i*th plot

G = grand total

r_i = the number of replication of the *i*th treatment

N = the total number of experimental plots [(*r*)(*t*)]

$$C = \frac{G^2}{N} = \frac{(2,237)^2}{18} = 278,009.4 \tag{4.1}$$

$$\text{Total } SS = \sum X_i^2 - C \tag{4.2}$$

$$\text{Total } SS = \left[(133)^2 + (154)^2 + \ldots + (122)^2 + (159)^2 \right] - 278,009.4$$

$$= 337,697 - 278,009.4$$

$$= 59,687.6$$

$$\text{Treatment } SS = \frac{\sum T_i^2}{r} - C \tag{4.3}$$

$$= \left[\frac{(462)^2}{3} + \frac{(437)^2}{3} + \frac{(298)^2}{2} + \ldots + \frac{(426)^2}{3} \right] - 278,009.4$$

$$= 292,996.5 - 278,009.4$$

$$= 14,987.1$$

$$\text{Error } SS = \text{Total } SS - \text{Treatment } SS \tag{4.4}$$

$$= 59,687.6 - 14,987.1$$

$$= 44,700.5$$

Step 4: Calculate the mean square for the treatment and error variations by dividing each sum of squares by their respective *d.f.*'s.

$$\text{Treatment } MS = \frac{\text{Treatment } SS}{t-1} \tag{4.5}$$

$$= \frac{14,987.1}{6-1}$$

$$= 2,997.42$$

$$\text{Error } MS = \frac{\text{Error } SS}{(n-1)(t-1)} \tag{4.6}$$

$$= \frac{44,700.5}{9}$$

$$= 4,966.72$$

Step 5: Calculate the F value, using the treatment and error mean squares as:

$$F = \frac{\text{Treatment } SS}{\text{Error } MS} \tag{4.7}$$

$$= \frac{2,997.42}{4,966.72}$$

$$= 0.604$$

The ANOVA table for the present example is shown in Table 4.2.1.6.

Step 6: Compare the computed F given in Table 4.2.1.6 with the tabular F value given in Appendix E. For our example, the tabular F (for five *d.f.*'s for treatment and nine *d.f.*'s for the error source of variation) is 3.48 for the 5% level of significance and 6.06 for the 1% level of significance.

TABLE 4.2.1.6

ANOVA of Effect of Hybrid on Grain Yield from a CRD Experiment with Unequal Number of Replications

Source of Variation	Degree of Freedom	Sum of Squares	Mean Square	F
Treatment	5	14,987.1	2,997.42	0.604[ns]
Error	9	44,700.5	4,966.72	
Total	14	59,687.6		

Note: $cv = 47.1\%$; [ns] = nonsignificant.

Step 7: Make a statistical decision. Given that the computed value of 0.493 is less than the tabular value of 3.48 at the 5% level, the treatment difference is said to be nonsignificant.

The *cv* for the present example is:

$$cv = \frac{\sqrt{\text{Error } MS}}{\text{Grand Mean}} \times 100 \qquad (4.9)$$

$$= \frac{\sqrt{4,966.72}}{149.6} \times 100$$

$$= 47.1\%$$

This high *cv* indicates that the reliability of the experiment is questionable. The researcher may wish to reexamine the way in which the treatments are compared.

4.2.2 Randomized Complete Block (RCB) Design

In agricultural research, the experimental unit, often being plots of land or animals, will by nature be different from place to place or animal to animal. Therefore, differences among treatment level means will reflect not only the variations that are attributable to the different treatments but also the variation that can be directly linked to individual differences existing prior to experimentation. It should be noted that some variation due to individual differences is inevitable. However, it is possible to isolate some of the variation so that it does not appear in the estimate of treatment effects and the experimental error. The RCB design is one such experimental design that accomplishes this task. The design is used extensively in agricultural research in which experimental material is divided into groups, such that each group constitutes a single trial or replication. The design requires that each block contain the same number of experimental units. Experimental units are assigned to *k* blocks so that the variability within each block is less than the variability among blocks. In this design the variation associated with an extraneous factor is isolated by means of a *blocking technique*, whereby blocks are formed from experimental units that are homogeneous. The experimental units in the blocks are known as *plots*. Homogeneity within each block is achieved through matching units to each block, or using litter-mates or subjects with similar heredities. Additionally, observation of each unit under all *k* treatment levels accomplishes homogeneity within each block. In agricultural experiments, we often use neighboring plots of land, animals from the same litter, set of similar trees, etc., for such grouping.

The major advantages of this design are its accuracy of results, flexibility of design, and ease of statistical analysis. Accuracy of results in comparison to the CRDs is due to further partitioning of the experimental units by blocking. In essence, blocking reduces the experimental error by accounting for the differences between blocks rather than within blocks. This increased precision is due to the fact that error variance, which is a function of comparisons within blocks, is smaller because of homogeneous blocks. The fact that the design allows for any number of treatments and replicates to be used provides more flexibility to the experimenter. Statistical analysis performed on this design is not difficult or complicated. Even when the experimenter is faced with missing data, techniques such as the one developed by Yates (1936b) can be applied for analysis. It should be mentioned, however, that if the gaps in the data are numerous, this design is less convenient than the CRD.

Before describing the randomization procedure for this design, it is relevant to discuss the blocking technique often used in agricultural experimentation.

Blocking technique: As mentioned earlier, the principal objective of blocking is to reduce experimental error. This is accomplished by accounting for a known source of variation among experimental units. To achieve an appropriate and effective blocking technique, the experimenter must identify the experimental unit's source of variability that can be used as the basis for blocking. In field experiments, soil characteristics and slope of the land are examples of sources of variation that can be used for blocking purposes. Once such a source of variability is identified, it can be used to serve as a block. Then a decision must be made on the shape and orientation of the block. Following the suggestions of Gomez and Gomez (1984), the following guidelines can be used to maximize variability among blocks:

1. Use narrow and long blocks when the field gradient is unidirectional. Additionally, block orientation should be such that their lengths are perpendicular to the direction of the gradient, as shown in Figure 4.2.2.1.

2. Use the stronger gradient when the fertility gradient runs in two directions in a field. That is, when there is a weak and a strong fertility gradient, the weaker gradient should be ignored. The block orientation should be perpendicular to the direction of the strong gradient.

3. Choose one of the following alternatives when two equally strong gradients are encountered in a field:

   • Use blocks that are as square as possible.

   • Long and narrow blocks with their lengths perpendicular to the direction of one gradient can be used. Covariance technique should be applied for the other gradient.

   • Latin square design with two-way blockings, one for each gradient, is quite helpful.

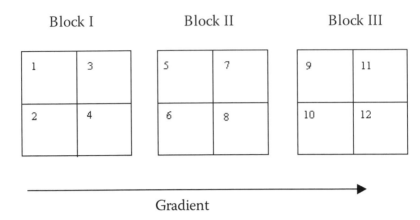

FIGURE 4.2.2.1
Layout of experimental plots when there is unidirectional field gradient.

4. Use square blocks when the pattern of variability is not predictable in a field.

Randomization: The randomization process for the RCB design entails numbering the treatments from 1 to k, in any order. Similarly, the units in each block are numbered from 1 to k. Then the k treatments are allotted at random to the k units in each block. The random allocation of treatments to the blocks may follow the procedures outlined in the CRD discussed in Subsection 4.2.1. Essentially, the random allocation of each treatment to units in each block reduces the error variance as, now, a portion of the variation is attributed to block differences. It should be kept in mind that if variation between blocks is not significantly large, grouping of the units as suggested in this design does not lead to any advantages; rather, some $d.f.$ of the error variance is lost, without a decrease in the error variance. Hence, under such conditions, it is preferable to use the CRDs. Another concern that a researcher must deal with in using this design is when the number of treatments is too large to allow for a homogeneous grouping of the units. For example, if the number of treatments is greater than ten, it is not suitable to use this design. In such cases, incomplete block designs that permit smaller groupings of homogeneous subjects may be preferable; or if it is possible to form homogeneous groups with larger number of units, this RCB design can be used.

The layout for an RCB design is shown in Figure 4.2.2.2, using a field experiment with four treatments (A, B, C, and D) and three replications. The steps followed in this layout are described in the following text.

Step 1: Following the blocking technique described earlier, divide the experimental area into r equal blocks. Each block represents a replication. In the present example, the experimental area is divided equally into three blocks or replications.

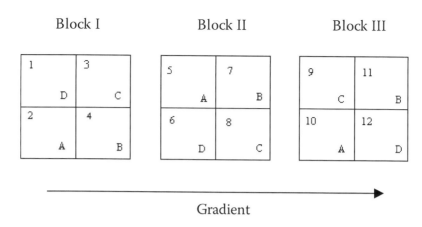

FIGURE 4.2.2.2
Randomized layout of an experimental area having three blocks and four treatments.

Step 2: Subdivide each block into *t* experimental plots, where each *t* represents a treatment. Number the plots in ascending order from 1 to *t* and then assign the treatments at random to the *t* plots. Random allocation may follow any of the earlier schemes described in Subsection 4.2.1. In the present example, we have four treatments. Therefore, each block is subdivided into four equal-sized plots. The plots are usually numbered consecutively from top to bottom and from left to right as shown in Figure 4.2.2.1.

Step 3: Using a table of random numbers such as the one given in Appendix D or Table 4.2.1.1, select four three-digit random numbers. For the present example, we start at the intersection of the ninth row and seventh column of Table 4.2.1.1. The first number selected is 276. We read downward from 276 to select the remainder of the numbers as shown below:

Random Number	Sequence
276	1
657	2
645	3
104	4

Step 4: Rank the numbers from smallest to the largest, as shown below:

Random Number	Sequence	Rank
276	1	2
657	2	4
645	3	3
104	4	1

Step 5: Assign the four treatments to the four plots by using the sequence of occurrence of the random numbers as the treatment numbers and the ranks as the plot numbers. Hence, treatment A will be assigned to plot 2, treatment B to plot 4, treatment C to plot 3, and treatment D to plot 1. The assignment of the treatments to the plots for blocks II and III follows the same steps as for block I. Figure 4.2.2.2 shows the randomized layout for the present example.

This design illustrates an important design principle. If a source of variability can be separated from the treatment effects and experimental error, the sensitivity or power of the resulting experiment may be increased. Variations that cannot be estimated remain a part of the uncontrolled sources of variability in the experiment and, thus, are automatically a part of the experimental error.

ANOVA: In this design the objective is to isolate the effect of not one but two variables of interest, namely, the treatment and the block (replication). Data gathered from the *t* treatments and *r* blocks represent a two-way classification with *tr* cells, each containing one observation. The data are orthogonal and, hence, the design is called an *orthogonal design*. The RCB design accounts for three sources of variability. Variability due to treatment, block (replication), and error. The following example provides the procedural steps in performing an ANOVA for this design.

Example 4.2.2.1

In a study to determine whether method of placement of fertilizer (P, K) would have an impact on corn yield, the following yield data were obtained by an agronomist. Perform an ANOVA.

Treatments	Grain Yield (bu/ac)		
	Rep. I	Rep. II	Rep. III
Control	147.0	130.1	142.2
2 × 2 in. band	159.4	167.3	150.5
Broadcast	158.9	166.2	159.1
Deep band	173.6	170.8	162.5
Disk applied	158.4	169.3	160.2
Sidedress	157.1	148.8	139.0

SOLUTION

Step 1: Calculate treatment totals (T), replication totals (R), their respective means, and the grand total (G), as shown in Table 4.2.2.1.

Step 2: Determine the *d.f.* for each source of variation as follows:

$$\text{Total } d.f. = (r)(t) - 1 = (3)(6) - 1 = 17$$

$$\text{Treatment } d.f. = t - 1 = 6 - 1 = 5$$

$$\text{Replication (Block) } d.f. = r - 1 = 3 - 1 = 2$$

$$\text{Error } d.f. = (t - 1)(r - 1) = (6 - 1)(3 - 1) = 10$$

TABLE 4.2.2.1

Corn Yield from Six Different Application Methods with Three Replications

Treatments	Grain Yield (bu/ac)			Treatment Totals (T)	Treatment Mean
	Rep. I	Rep. II	Rep. III		
Control	147.0	130.1	142.2	419.3	139.8
2 × 2 in. band	159.4	167.3	150.5	477.2	159.1
Broadcast	158.9	166.2	159.1	484.2	161.4
Deep band	173.6	170.8	162.5	506.9	168.9
Disk applied	158.4	169.3	160.2	487.9	162.6
Sidedress	157.1	148.8	139.0	444.9	148.3
Rep. Total (R)	954.4	952.5	913.5		
Grand Total (G)					2820.4
Grand Mean					156.7

Step 3: Calculate the correction factor (C) and the various sums of squares using the following notations:

X_i = an observation in the ith plot

T_i = treatment total for the ith plot

G = grand total

r_i = the number of replication of the ith treatment

N = the total number of experimental plots $[(r)(t)]$

$$C = \frac{G^2}{rt} = \frac{(2{,}820.4)^2}{(3)(6)} = 441{,}925.3 \qquad (4.10)$$

$$\text{Total } SS = \sum_{j=1}^{t}\sum_{i=1}^{r} X_{ij}^2 - C \qquad (4.11)$$

Total SS = $[(147.0)2 + (159.4)2 + \ldots + (139.0)2] - 441{,}925.3$

$\qquad$ = 444,263.99 − 441,925.3

$\qquad$ = 2,338.69

$$\text{Replication } SS = \frac{\sum R_i^2}{t} - C \tag{4.12}$$

$$= \frac{(954.4)^2 + (952.5)^2 + ... + (913.5)^2}{6} - 441,925.3$$

$$= 442,102.98 - 441,925.3$$

$$= 177.7$$

$$\text{Treatment } SS = \frac{\sum T_i^2}{r} - C \tag{4.3}$$

$$= \frac{(419.34)^2 + (477.2)^2 + ... + (444.9)^2}{3} - 441,925.3$$

$$= 443,637.33 - 441,925.3$$

$$= 1,712.00$$

$$\text{Error } SS = \text{Total } SS \text{ Replication } SS - \text{Treatment } SS \tag{4.13}$$

$$= 2,338.69 - 177.7 - 1,712.00$$

$$= 448.99$$

Step 4: Calculate the mean square for each source of variations by dividing the treatment, replication, and error sum of squares by their respective *d.f.*'s.

$$\text{Replication } MS = \frac{\text{Replication } SS}{r-1} \tag{4.14}$$

$$= \frac{177.7}{2}$$

$$= 88.89$$

$$\text{Treatment } MS = \frac{\text{Treatment } SS}{t-1} \tag{4.15}$$

$$= \frac{1712}{5}$$

$$= 342.4$$

$$\text{Error } MS = \frac{\text{Error } SS}{(r-1)(t-1)} \tag{4.16}$$

$$= \frac{448.99}{10}$$

$$= 44.89$$

Step 5: Calculate the *F* value using the treatment and error mean squares as:

$$F = \frac{\text{Treatment } SS}{\text{Error } MS} \tag{4.7}$$

$$= \frac{342.4}{44.89}$$

$$= 7.63$$

Step 6: Compare the computed *F* with the tabular *F* value given in Appendix E. For this example, the tabular *F* (for five *d.f.*'s for treatment and ten *d.f.*'s for the error source of variation) is 3.33 for the 5% level of significance and 5.64 for the 1% level of significance. The computed *F* value is greater than the tabular *F* at the 1% level of significance; it is therefore concluded that there is a significant difference among the six treatments.

To determine the degree of precision with which the treatments are compared, we can compute *cv* as follows:

$$cv = \frac{\sqrt{\text{Error } MS}}{\text{Grand Mean}} \times 100 \tag{4.9}$$

$$= \frac{\sqrt{44.89}}{156.7} \times 100$$

$$= 4.28\%$$

The ANOVA table for the present example is shown in Table 4.2.2.2.

Efficiency of RCB Design: In our earlier discussion, we mentioned that an advantage of using the block design over the CRD was the ability to further partition the experimental units into blocks. In essence, such partitioning reduces the experimental error by accounting for the differences between blocks rather than within blocks. This increased precision is due to the fact that error variance, which is a function of comparisons within blocks, is smaller because of homogeneous blocks. To determine the magnitude of the reduction in experimental error due to blocking, the relative

TABLE 4.2.2.2

ANOVA of Corn-Yield Data in an RCB Design

Source of Variation	Degrees of Freedom	Sum of Squares	Mean Square	F
Treatment	5	1,712.00	342.4	7.63**
Replication	2	177.7	88.85	
Error	10	448.99	44.89	
Total	17	2,338.69		

Note: cv = 4.28%; ** = significant at 1% level.

efficiency (*R.E.*) of an RCB design, relative to that of the CRD, is computed as follows:

$$R.E. = \frac{(r-1)s_B^2 + r\ (t-1)s_e^2}{(rt-1)s_e^2} \tag{4.17}$$

where

r = replications or blocks

t = treatments

s_B^2 = replication (block) mean square

s_e^2 = error mean square

It should be noted that when the error *d.f.*'s is less than 20, Fisher (1974) proposed an adjustment to account for the discrepancies in *d.f.* He suggests that the *R.E.* parameter be multiplied by an adjustment factor as shown below:

$$R.E. = \frac{(r-1)s_B^2 + r\ (t-1)s_e^2}{(rt-1)s_e^2} \cdot \frac{\left[(r-1)(t-1)+1\right]\left[t(r-1)+3\right]}{\left[(r-1)(t-1)+3\right]\left[t(r-1)+1\right]} \tag{4.18}$$

For the present example, the *R.E.* value is:

$$R.E. = \frac{(3-1)(88.85)+3\ (6-1)(44.89)}{(18-1)(44.89)} \cdot \frac{(11)(9)}{(13)(7)}$$

$$= (1.12)(1.08)$$

$$= 1.21$$

The result shows that by using the RCB design instead of the CRD, the efficiency of the experiment has improved by 21%. Thus, a block design has proved to be more efficient than a CRD.

4.2.3 Latin Square Designs

In the previous section we have seen how effective single grouping of the units into blocks was in reducing the experimental error. Such reduction in error is a major improvement over the CRDs. Error reduction was made possible by removing the effect of a single factor from the experiment. The Latin square design extends this further by allowing the removal of two sources of variation from the experimental error. As the name implies, the design has its origins in an ancient puzzle in which Latin letters were arranged in a square matrix, such that each letter appeared once, and only once, in each row and column. This design enables an experimenter to isolate two known sources of variation among experimental units. The levels of one source of variation are assigned to the rows, and the levels of the second source of variation are assigned to the columns of a Latin square; the t treatment levels are assigned to the t^2 cells of the square, such that each level appears once in each row and once in each column. Another way of expressing the basic criteria for this design is that the experimental material should be arranged as two independent (row and column) blocks such that the differences among row blocks and column blocks reflect the major sources of variation.

Latin square designs are appropriate for those experiments that meet the general assumptions of the ANOVA model as well as the following conditions:

1. The number of treatments must equal the number of categories for each of the two factors or sources of variation; that is, the number of replicates equals the number of treatments.

2. There should be an absence of interactions among the two sources of variation and the treatment. This means that there is no interaction between the treatment effect and either block effect.

3. Treatment levels are randomly assigned to the cells of the square, with the proviso that each treatment level must appear once in each column and once in each row.

The advantages of using the Latin square design are in requiring fewer subjects, removing systematic treatment biases through counterbalancing, and removing of error variance through two-way blocking. The basis for the need of fewer subjects stems from one of the design requirements, which states that the number of replications must equal the number of treatments. For example, an animal scientist, in conducting an experiment to evaluate the effects of four different rations on weight gain, may choose a design that takes into consideration other factors (such as the initial weight and age of animals) that might explain the weight-gain variability.

Thus, the animal scientist might block the sample subjects into four initial weight and four age categories. In such a situation, in which the scientist wants replications of the four treatments with two blocking factors, each also having four categories, $4 \times 4 \times 4 \times n = 64n$ subjects are required. In factorial experiments with large number of factors and numerous categories of each factor, the number of subjects required will indeed be very large. The Latin square design permits us to reduce the number of subjects required, yet allows assessment of the relative effects of various treatments when a two-directional blocking restriction is imposed on the experimental unit. In the case of the feed ration experiment, we will need only 16 instead of 64 subjects. The disadvantage of the design lies in the requirement that the number of treatments equal the number of replications. This is especially problematic when there are large numbers of treatments, hence requiring large numbers of replications. This has the added problem that as the square gets larger, the experimental error per unit also increases. On the other hand, smaller Latin squares do not have large *d.f.*'s for the estimation of error. For example, a 2×2 square has no *d.f.*'s, a 3×3 has only two *d.f.*'s, and a 4×4 has only six *d.f.*'s for the estimation of error. For this reason, it is generally recommended to use the Latin square design when the number of treatments is not less than four and no more than eight. Another limitation of the design is when there is reason to believe that there is interaction among the treatment, row, and column. In such circumstances, it is not appropriate to use this design.

In using the Latin square design, the experimenter must keep in mind that if one of the two factors under consideration does not have a substantial impact on the variate under study, elimination of its variance may not influence the experimental error. Thus, using the Latin square may not be an improvement over the RCB design. In agricultural field experiments, for example, in which there is a strong indication of a two-directional fertility gradient, adoption of a Latin square is desirable. Similarly, in an insecticide field trial, in which there is a predictable direction to insect migration that is perpendicular to the fertility gradient of the field, use of the Latin square design is appropriate. Horticulturists and agronomists who perform greenhouse experiments may find using this design helpful in situations in which experimental pots are arranged perpendicular to a source of light or shade such that the difference among rows of pots and the distance from the light or shade account for two sources of variability. In animal experiments the experimenter may use the animals themselves and time as the two sources of blocking, or age and weight may be considered as the two sources of variation that may have a major effect on the variate under study.

Randomization: The process of randomization in Latin square designs requires that a square be selected at random from a number of Latin squares of a given order. Appendix C presents examples of selected Latin squares. The procedure for the randomization of rows and columns is outlined in the following text.

Step 1: Let a $k \times k$ Latin square be first written by denoting treatments by Latin letters A, B, C, D, etc.; alternatively, randomly select a ready-made square given in Appendix C. Suppose we randomly select a 4×4 square from Appendix C, e.g.:

A	B	C	D
B	C	D	A
C	D	A	B
D	A	B	C

Step 2: Randomly reorder all rows (R_i) and columns (C_j), except the first, by using a table of random numbers such as the one given in Appendix D, or Table 4.2.1.1. Select four three-digit random numbers. For the present example, we randomly start at the intersection of the first row and the seventh column of Table 4.2.1.1. The first number selected is 315. Reading downward from 315, the following numbers are selected:

Random Number	Sequence
315	1
145	2
189	3
207	4

Step 3: Rank the numbers from smallest to the largest, i.e.:

Random Number	Sequence	Rank
315	1	4
145	2	1
189	3	2
207	4	3

Step 4: Use the rank to represent the existing row numbers of a selected square and the sequence of occurrence to represent the row numbers after reordering of the rows of the selected square. For the present example, the fourth row of the selected square becomes the first row of the reordered plan. After reordering all rows, the new plan will be:

D	A	B	C
A	B	C	D
B	C	D	A
C	D	A	B

Step 5: Reorder the columns obtained in Step 4, using the same procedures applied in the previous step. Use the rank to represent the column number of the plan obtained in Step 4 and the sequence of occurrence of numbers

to represent the column number of the final plan. The four selected numbers and the sequence of occurrence are the following:

Random Number	Sequence	Rank
168	1	2
094	2	1
897	3	4
243	4	3

Step 6: The final randomized layout for the experiment, based on the new row and column arrangement, is the following:

Row Number	Column Number			
	I	II	III	IV
1	A	D	B	C
2	B	A	C	D
3	C	B	D	A
4	D	C	A	B

ANOVA: As explained earlier, the intent of the design is to further partition the sources of variation in an experiment. This means that blocking by rows (R_i) and columns (C_j) allows for the removal of variation due to such sources from the experimental error. Hence, the design makes it possible to account for two more sources of variation than a CRD and one more source of variation than a randomized RCB design.

The analysis is conducted by following a similar procedure as described in Subsection 4.2.2. The following example illustrates the computational steps in performing the ANOVA.

Example 4.2.3.1

An animal scientist designed an experiment to determine the effect of four diets (A, B, C, and D) on liver cholesterol in sheep. Recognized sources of variation were body weight and age. The experimenter randomly selected a 4 × 4 Latin square and randomly arranged the results of the experiment from the different treatments as presented in the following table:

Weight Groups	Age Groups			
	I	II	III	IV
1	1.15A	1.75D	1.78B	2.10C
2	1.65B	1.20A	1.56C	2.05D
3	1.30C	1.79B	1.93D	1.25A
4	1.10D	1.50C	1.20A	2.01B

TABLE 4.2.3.1

Row, Column, and Treatment Totals and Means for Cholesterol Levels in Sheep Experiment

Row	Column I	II	III	IV	Row Total	Treatment Total
1	1.15A	1.75D	1.78B	2.10C	6.78	4.80A
2	1.65B	1.20A	1.56C	2.05D	6.46	7.23B
3	1.30C	1.79B	1.93D	1.25A	6.27	6.46C
4	1.10D	1.50C	1.20A	2.01B	5.81	6.83D
Column Total	5.20	6.24	6.47	7.41		
Grand Total (G)						25.32
Grand Mean						1.58

SOLUTION

Step 1: Calculate treatment totals (T), row totals (R_i), column totals (C_j), and the grand total (G) as shown in Table 4.2.3.1.

Step 2: Determine the *d.f.* for each source of variation, given that *t* represents treatments, as shown:

$$\text{Total } d.f. = t^2 - 1 = (4^2) - 1 = 15$$

$$\text{Treatment } d.f. = \text{Column } d.f. = \text{Row } d.f. = 4 - 1 = 3$$

$$\text{Error } d.f. = (t-1)(t-2) = (4-1)(4-2) = 6$$

Step 3: Calculate the correction factor (C) and the various sums of squares using the following equations:

$$C = \frac{G^2}{t^2} = \frac{(25.32)^2}{16} = 40.07 \tag{4.19}$$

$$\text{Total } SS = \sum X_{ijk}^2 - C \tag{4.20}$$

$$\text{Total } SS = [(1.15)^2 + (1.75)^2 + (1.78)^2 + \dots + (2.01)^2] - 40.07$$

$$= 41.88 - 40.07$$

$$= 1.81$$

$$\text{Treatment } SS = \frac{\sum T^2}{k} - C \qquad (4.21)$$

$$= 1/4 \, [(4.80)^2 + (7.23)^2 + (6.46)^2 + (6.83)^2] - 40.07$$

$$= 1/4 \, [163.69] - 40.07$$

$$= 40.92 - 40.07$$

$$= 0.85$$

$$\text{Row } SS = \frac{\sum R_i^2}{r} - C \qquad (4.22)$$

$$= 1/4 \, [(6.78)^2 + (6.46)^2 + (6.27)^2 + (5.81)^2] - 40.07$$

$$= 1/4 \, [160.77] - 40.07$$

$$= 40.19 - 40.07$$

$$= 0.12$$

$$\text{Column } SS = \frac{\sum C_j^2}{c} - C \qquad (4.23)$$

$$= 1/4 \, [(5.20)^2 + (6.24)^2 + (6.47)^2 + (7.41)^2] - 40.07$$

$$= 1/4 \, [162.75] - 40.07$$

$$= 40.69 - 40.07$$

$$= 0.62$$

$$\text{Error } SS = \text{Total } SS - \text{Treatment } SS - \text{Row } SS - \text{Col. } SS \qquad (4.24)$$

$$= 1.81 - 0.85 - 0.12 - 0.62$$

$$= 0.22$$

Step 4: Calculate the mean square for each source of variation by dividing the treatment, row, column, and error sum of squares by their respective *d.f.*'s.

$$\text{Treatment } MS = \frac{\text{Treatment } SS}{t-1} \tag{4.5}$$

$$= \frac{0.85}{3}$$

$$= 0.28$$

$$\text{Row } MS = \frac{\text{Row } SS}{t-1} \tag{4.25}$$

$$= \frac{0.12}{3}$$

$$= 0.04$$

$$\text{Column } MS = \frac{\text{Column } SS}{t-1} \tag{4.26}$$

$$= \frac{0.62}{3}$$

$$= 0.21$$

$$\text{Error } MS = \frac{\text{Error } SS}{(t-1)(t-2)} \tag{4.27}$$

$$= \frac{0.22}{(3)(2)}$$

$$= 0.04$$

Step 5: Calculate the F value, using the treatment and error mean squares as:

$$F = \frac{\text{Treatment } SS}{\text{Error } MS} \tag{4.7}$$

$$= \frac{0.28}{0.04}$$

$$= 7.0$$

Step 6: Compare the computed F value with the tabular F value given in Appendix E. For this example, the tabular F (for three *d.f.*'s for treatment and six *d.f.*'s for the error source of variation) is 4.76 for the 5% level of significance and 9.78 for the 1% level of significance. The computed F value is greater than the tabular F at the 5% level of significance; it is therefore concluded that there is a significant difference among the four treatments.

To determine the degree of precision with which the treatments are compared, we can compute the cv as follows:

$$cv = \frac{\sqrt{\text{Error } MS}}{\text{Grand Mean}} \times 100 \qquad (4.9)$$

$$= \frac{\sqrt{0.04}}{1.58} \times 100$$

$$= 12.65\%$$

The ANOVA table for the present example is shown in Table 4.2.3.2.

Efficiency of Latin square designs: To determine whether blocking by rows and columns is efficient in reducing experimental error, we will compute the F value for the row and column as follows:

$$F \text{ (row)} = \frac{\text{Row } MS}{\text{Error } MS} \qquad (4.28)$$

$$= \frac{0.04}{0.04}$$

$$= 1$$

$$F \text{ (column)} = \frac{\text{Column } MS}{\text{Error } MS} \qquad (4.29)$$

$$= \frac{0.21}{0.04}$$

$$= 5.25$$

TABLE 4.2.3.2

ANOVA Table of the Effect of Four Different Diets on Cholesterol Level of Sheep

Source of Variation	Degree of Freedom	Sum of Squares	Mean Square	F
Treatment	3	0.85	0.28	7.0*
Row	3	0.12	0.04	
Column	3	0.62	0.21	
Error	6	0.22	0.04	
Total	15	1.81		

Note: cv = 12.65 %; * = significant at 5% level.

Because the computed F value for both row and column are equal to or greater than 1, we will compare them with the tabular F from Appendix E. The tabular F at 5% and 1% levels of significance are 4.76 and 9.78, respectively. In comparing these results with the tabular F values, it appears that only column-blocking provides a significant difference. To determine the magnitude of the reduction in experimental error due to row- and column-blocking, the *R.E.* parameter of a Latin square design has to be computed. When computing the *R.E.* parameter for the Latin square, we are making a comparison with (1) a CRD; (2) an RCB design, in which the blocks are rows (first source of variation) of the Latin square; (3) a randomized block design, in which the blocks are columns (second source of variation) of the Latin square. The *R.E.* of the Latin square design as compared with each of the preceding three scenarios is computed as shown:

1. *R.E.* of a Latin square as compared to a CRD is:

$$R.E. = \frac{s_r^2 + s_c^2 + (t-1)s_e^2}{(t+1)s_e^2} \tag{4.30}$$

where

s_r^2 = row mean square

s_c^2 = column mean square

s_e^2 = error mean square

t = number of treatments

2. The efficiency of a Latin square design, relative to RCB design with rows as blocks, may be called *column efficiency* and is given as:

$$R.E. \text{ (Col.)} = \frac{s_c^2 + (t-1)s_e^2}{(t)s_e^2} \tag{4.31}$$

3. The efficiency of a Latin square design, relative to an RCB design in which columns are treated as blocks, is called *row efficiency*. Here the row variation is merged with the error variation and is computed as:

$$R.E. \text{ (Row)} = \frac{s_r^2 + (t-1)s_e^2}{(t)s_e^2} \tag{4.32}$$

Thus, for the present example, we have:

$$R.E. = \frac{0.04 + 0.21 + (4-1)0.04}{(4+1)0.04}$$

$$= 1.85$$

The result shows that by using a Latin square design instead of a CRD, the precision has increased by 85%.

Similarly, we compute the *R.E.* of a Latin square design compared to the RCB design as:

$$R.E. \text{ (RCB, Col.)} = \frac{0.21 + (4-1)0.04}{(4)0.04}$$

$$= 2.06$$

$$R.E. \text{ (RCB, Row)} = \frac{0.04 + (4-1)0.04}{(4)0.04}$$

$$= 1.00$$

It should be noted that the error *d.f.*'s is less than 20, and therefore the *R.E.* values should be multiplied by the adjustment factor k as shown:

$$k = \frac{\left[(t-1)(t-2)+1\right]\left[(t-1)^2+3\right]}{\left[(t-1)(t-2)+3\right]\left[(t-1)^2+1\right]} \tag{4.33}$$

$$k = \frac{\left[(4-1)(4-2)+1\right]\left[(4-1)^2+3\right]}{\left[(4-1)(4-2)+3\right]\left[(4-1)^2+1\right]}$$

$$= 0.93$$

Therefore, the adjusted *R.E.* values are:

$$R.E. \text{ (RCB, Col.)} = (2.06)(0.93)$$

$$= 1.92$$

$$R.E. \text{ (RCB, Row)} = (1.00)(0.93)$$

$$= 0.93$$

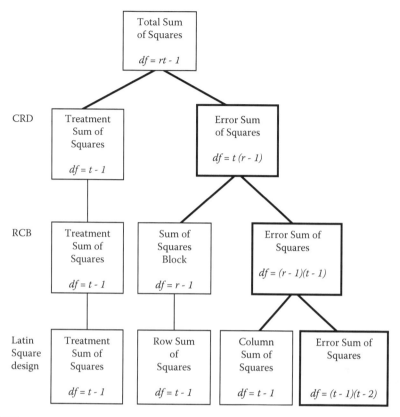

FIGURE 4.2.3.1

Schematic partitioning of the total sum of squares for three designs. Squares with bold outline identify the experimental errors for testing treatment effects.

As the results show, the column-blocking of the Latin square design did increase the experimental precision by 92%. However, the additional row-blocking of the Latin square did not increase the precision over the RCB design in which the columns were treated as blocks.

Schematic partitioning of the total sum of squares for the CRD, the RCB design, and the Latin square is shown in Figure 4.2.3.1.

4.3 Incomplete Block Designs

Many agricultural experiments involve varietal trials or other types of experiments in which the objective is to compare pairs of treatments for the purpose of selecting improved varieties of crops or breeds of animals. An agronomist, for example, may be interested in selecting a variety of a crop based on some genetic or economic characteristic. To determine which variety may be considered superior, an experiment could be to grow several varieties in a field, using an appropriate design. As mentioned in the previous sections of this chapter, when the number of varieties to be tested is small (less than ten), use of an ordinary randomized block design or a Latin square design may be appropriate. However, when the number of varieties tested is large, as is often the case with varietal trials, use of randomized block designs may not be appropriate because of the resultant increase in error variance due to the larger block sizes. In factorial experiments, use is made of the confounding device to reduce the block size and, hence, ensure more precise estimation of the lower-order interactions that are confounded with blocks. In experiments other than varietal trials, in which it is desired to make all comparisons among pairs of treatments with equal precision, use of complete block designs are also inefficient. That is, as the number of treatments increase, the block size increases, which contributes to the variability within the blocks. This loss of homogeneity contributes to increased experimental error and less precision. On the other hand, we have seen that the larger the number of replications of a treatment effect, the more the precision. Therefore, we need a set of designs that permits small groupings of subjects (fewer subjects per block than the number of treatments) for experiments that contain many treatments.

A set of designs for single-factor experiments were introduced by Yates (1936a) to accommodate large numbers of treatments with smaller blocks. These designs are called the *incomplete block designs*. In incomplete block designs, the blocks need not be of the same size and each treatment need not appear the same number of times. These characteristics of the designs are quite valuable, especially when there is a large number of treatment factors and when resources are limited and do not permit large replications, or when there are natural or practical restrictions on block size.

Generally, incomplete block designs have the advantage of providing improved precision through reduced block size, without loss of information on any treatment effect. This is accomplished by partial confounding of the effects of all treatments with the differences among incomplete blocks. That is, the variance between plots in the same block is smaller than the variance between plots in different blocks. This improved precision, although important, has drawbacks that should be carefully weighed before using these designs. For example, partial confounding leads to a more complex plan and analysis. Additionally, there is inflexibility in the number of treatments or replications, or both. To ensure equal, or nearly equal, precision of comparisons of different pairs of treatments, the treatments are allocated to different blocks in such a way that each pair of treatments has the same, or nearly the same, number of replications, and each treatment has an equal number of replications (r).

An incomplete block design should be used whenever block size is large enough to not allow the experimental units within the same block to be relatively homogeneous. So, in cases in which the experimental material is highly variable, but in which it is possible to form small groups that are homogeneous, the design may be advantageous even with a small number of treatments. It should also be kept in mind that because of the complex nature of these designs, the data analysis requires access to computing facilities and services. In research experiments in which missing or incomplete data (e.g., experimental units destroyed or injured during the course of the experiment) is a problem, the use of an incomplete design is not recommended. Such incomplete data present laborious computation of the block (or row and column) adjustments.

There are several types of incomplete block designs. The most commonly used in agricultural research is the *lattice design*. In the next two subsections of this chapter, we will discuss two lattice designs: namely, the *balanced lattice* and the *partially balanced lattice* designs.

4.3.1 Balanced Lattice Designs

In 1937, Yates introduced quasifactorial designs called the lattice designs. The main characteristics of these designs are that the number of treatments must be perfect squares ($t = k^2$) and that the block size k is the square root of the number of treatments ($k = \sqrt{t}$). And lastly, the number of replications is one more than the block size; that is, $r = (k + 1)$. The last criterion differentiates balanced lattice designs from the partially balanced designs that will be discussed later. To illustrate how to satisfy the first requirement of the design, the number of treatments may be 9, 16, 25, 36, 49, 81, etc. If the number of treatments that an experimenter wishes to work with is not a perfect square, it must be made a perfect square either by elimination or addition of treatments. Once the number of treatments is selected, the second and third requirement are easily met. For instance, if there are 49 treatments in the experiment, the block size is $k = \sqrt{t}$ or $k = 7$. The number of replications would be $r = (k + 1)$ or $r = 8$. The randomization and layout of the experimental plan, as well as the analysis, are discussed in the following text.

Randomization: The randomization process is illustrated in a field experiment using a simple case with nine treatments. The procedural steps for the layout of the design are the following:

Step 1: Determine the number of replications needed by dividing the experimental area into $r = (k + 1)$, such that each replication would have $t = k^2$ experimental plots. In the present case, $r = 4$, each containing 9 experimental plots, as shown in Figure 4.3.1.1.

Step 2: Divide each replication into k incomplete blocks. For this example, this would be three incomplete blocks each containing three experimental plots as shown in Figure 4.3.1.1.

Step 3: Select from Appendix F a basic plan that corresponds with the number of treatments tested in the experiment. For the present case, the 3×3 balanced lattice design selected is shown in Table 4.3.1.1.

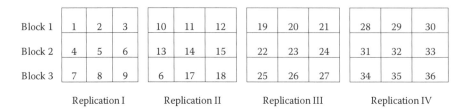

Block 1	1	2	3		10	11	12		19	20	21		28	29	30
Block 2	4	5	6		13	14	15		22	23	24		31	32	33
Block 3	7	8	9		6	17	18		25	26	27		34	35	36
	Replication I				Replication II				Replication III				Replication IV		

FIGURE 4.3.1.1

Division of the experimental area in a balanced design for nine treatments in blocks of three units.

TABLE 4.3.1.1

Basic Plan for a 3×3 Balanced Lattice Design for Nine Treatments in Blocks of Three Units

	Treatment Number											
Block Number	Rep. I			Rep. II			Rep. III			Rep. IV		
(1)	1	2	3	1	4	7	6	1	8	5	1	9
(2)	4	5	6	2	5	8	2	9	4	7	6	2
(3)	7	8	9	3	6	9	7	5	3	3	8	4

TABLE 4.3.1.2

A Selected Basic Plan with Rearranged Replications, Blocks, and Treatments in a Balanced Lattice Design

Block	Rep. I			Block	Rep. II			Block	Rep. III			Block	Rep. IV		
(1)	8	6	1	(4)	8	7	9	(7)	8	5	2	(10)	3	4	8
(2)	3	5	7	(5)	3	2	1	(8)	4	7	1	(11)	6	2	7
(3)	4	2	3	(6)	6	5	4	(9)	3	6	9	(12)	9	5	1

Step 4: Use an appropriate randomization technique, such as the one discussed in Subsection 4.2.1, to randomize the replication arrangement, the incomplete blocks within each replication, and the treatment arrangement within each block of the selected plan for field layout. To avoid repetition, we have not illustrated the randomization technique for the present example. The reader is directed to Subsection 4.2.1 for the technique of randomly rearranging the replication, treatments, and incomplete blocks. Following the randomization technique of Subsection 4.2.1, the replications, blocks, and treatments are rearranged as shown in Table 4.3.1.2.

Step 5: Use the rearranged plan to lay out the experiment in the field. This is shown in Figure 4.3.1.2. Note that every treatment appears in each of the three rows and every pair of treatments appears once in the same block. This feature allows each pair of treatments in a balanced lattice design to be compared with equal precision.

Block 1	T8	T6	T1		T8	T7	T9		T8	T5	T2		T3	T4	T8
Block 2	T3	T5	T7		T3	T2	T1		T4	T7	T1		T6	T2	T7
Block 3	T4	T2	T9		T6	T5	T4		T3	T6	T9		T9	T5	T1

Replication I   Replication II   Replication III   Replication IV

FIGURE 4.3.1.2
Allocation of nine treatments in a field layout with three blocks.

ANOVA: The difference in the analysis between a balanced lattice design and an RCB design is the added source of variation, which is due to the incomplete blocks of the same replication. Computation and statistical analysis of the design is not difficult. However, the use of such designs requires additional care and resources. In comparison to the RCB design, the potential source of the gain in accuracy may be the experimental material and the large number of treatments. The number of treatments in most incomplete block designs ranges from 6 to 200; this can increase up to 1000, as in the case of the cubic lattice designs.

Example 4.3.1.1
An animal scientist designed a 3×3 lattice design experiment to study the effect of nine different diets on nonprotein nitrogen in the colostrum of rats. The plan compares nine treatments in four replications. The results (mg/100 ml) are shown below.

Treatment No.	Rep. I	Rep. II	Rep. III	Rep. IV
1	69	89	90	80
2	92	72	132	72
3	74	81	127	98
4	68	74	98	56
5	80	83	92	63
6	83	72	87	76
7	86	55	102	84
8	66	76	85	81
9	78	71	96	95

Perform the ANOVA on the data.

SOLUTION
The steps involved in the ANOVA are as follows:

Step 1: Select a basic plan from Appendix F (basic plans for balanced and partially balanced lattice designs) and rearrange the replication, incomplete block, and treatment as shown:

Block	Rep. I			Block	Rep. II			Block	Rep. III			Block	Rep. IV		
(1)	1	2	3	(4)	1	4	7	(7)	1	5	9	(10)	1	8	6
(2)	4	5	6	(5)	2	5	8	(8)	7	2	6	(11)	4	2	9
(3)	7	8	9	(6)	3	6	9	(9)	4	8	3	(12)	7	5	3

Step 2: Using the randomization technique of Subsection 4.2.1, rearrange the replications, incomplete blocks, and treatments. The final layout for the plan is as shown:

Block	Rep. I			Block	Rep. II			Block	Rep. III			Block	Rep. IV		
(1)	8	6	1	(4)	8	7	9	(7)	8	5	2	(10)	3	4	8
(2)	3	5	7	(5)	3	2	1	(8)	4	7	1	(11)	6	2	7
(3)	4	2	9	(6)	6	5	4	(9)	3	6	9	(12)	9	5	1

Step 3: Allocate the treatments to the rearranged plan as shown:

Block	Rep. I			Block	Rep. II			Block	Rep. III			Block	Rep. IV		
(1)	66	83	69	(4)	76	55	71	(7)	85	92	132	(10)	98	56	81
(2)	74	80	86	(5)	81	72	89	(8)	98	102	90	(11)	76	72	84
(3)	68	92	78	(6)	72	83	74	(9)	127	87	96	(12)	95	63	80

Step 4: Calculate the replication totals (R), the block totals (B), the treatment totals (T), and the grand total (G) as shown in Table 4.3.1.3.

Step 5: Calculate the sum of block totals over all blocks (B_T) in which a particular treatment appears. For example, the (B_T) value for treatment 1 that appears in blocks 1, 5, 8, and 12 is $218 + 242 + 290 + 238 = 988$. Compute the ($B_T$) value for all nine treatments as shown in Table 4.3.1.3.

Step 6: Calculate the quantities (W) for each treatment whose sum must equal zero, using the following equation:

$$W = kT - (k + 1) B_T + G \qquad (4.34)$$

where

k = block size

T, B_T, and G as defined earlier

TABLE 4.3.1.3

Nonprotein Nitrogen (in mg/100 ml) in the Colostrum of Rats Fed Nine Different Diets

Block No.				Block Total (B)	Treatment Number	Treatment Total (T)	Block Total (B_T)	W
		Rep. I						
1	66(8)	83(6)	69(1)	218	1	328	988	15
2	74(3)	80(5)	86(7)	240	2	368	1,021	3
3	68(4)	92(2)	78(9)	238	3	380	1,027	15
	Rep. Total			696	4	296	992	97
					5	318	1,016	127
		Rep. II			6	318	989	19
4	76(8)	55(7)	71(9)	202	7	327	964	108
5	81(3)	72(2)	89(1)	242	8	308	964	51
6	72(6)	83(5)	74(4)	229	9	340	988	51
	Rep. Total			673	Total (G)	2,983	8,949	0
		Rep. III						
7	85(8)	92(5)	132(2)	309				
8	98(4)	102(7)	90(1)	290				
9	127(3)	87(6)	96(9)	310				
	Rep. Total			909				
		Rep. IV						
10	98(3)	56(4)	81(8)	235				
11	76(6)	72(2)	84(7)	232				
12	95(9)	63(5)	80(1)	238				
	Rep. Total			705				

Note: Treatment numbers are shown in parentheses.

The W values for each of the nine treatments are given in Table 4.3.1.3. Note that the sum of the W values is equal to zero.

Step 7: Determine the *d.f.*'s associated with each source of variation. The source of variation and the total *d.f.*'s depend on whether the treatments are adjusted or not. If the treatments are adjusted, the following sources of variation and the *d.f.*'s are identified:

Replication $d.f. = k = 3$

Treatment (unadj.) $d.f. = k^2 - 1 = 8$

Block (adj.) $d.f. = k^2 - 1 = 8$

Intrablock error $d.f. = (k - 1)(k^2 - 1) = 16$

Treatment (adj.) $.f. = k^2 - 1 = 8$

Effective error $d.f. = (k - 1)(k^2 - 1) = 16$

Total $d.f. = 59$

If the treatments are not adjusted (the reasons for which will be explained in the following text), then the sources of variation and the corresponding *d.f.*'s are:

Replication $d.f. = k = 3$

Treatment (unadj.) $d.f. = k^2 - 1 = 8$

Block (adj.) $d.f. = k^2 - 1 = 8$

Intrablock error $d.f. = (k - 1)(k^2 - 1) = 16$

Total $d.f. = 35$

Step 8: Calculate the various sums of squares by first computing the correction factor (C) as:

$$C = \frac{G^2}{\left(k\right)^2 \left(k + 1\right)} \tag{4.35}$$

$$= \frac{\left(2,983\right)^2}{\left(9\right)\left(4\right)}$$

$$= 247{,}174.7$$

$$\text{Total } SS = \sum X_{ijk}^2 - C \tag{4.36}$$

$$= [(66)^2 + (83)^2 + (69)^2 + \ldots + (80)^2] - 247{,}174.7$$

$$= 255{,}597 - 247{,}174.7$$

$$= 8{,}422.3$$

$$\text{Replication } SS = \frac{\sum R^2}{k^2} - C \tag{4.37}$$

$$= \frac{\left(696\right)^2 + \left(673\right)^2 + \left(909\right)^2 + \left(705\right)^2}{9} - 247{,}174.4$$

$$= 251{,}183.4 - 247{,}174.7$$

$$= 4{,}008.7$$

$$\text{Treatment (unadj.) } SS = \frac{\sum T^2}{(k+1)} - C \tag{4.38}$$

$$= \frac{(328)^2 + (368)^2 + \ldots + (340)^2}{4} - 247,174.4$$

$$= 248,666.3 - 247,174.7$$

$$= 1,491.6$$

The sum of squares for blocks within replications adjusted for treatment is computed as:

$$\text{Block (adj.) } SS = \frac{\sum W^2}{(k^3)(k+1)} \tag{4.39}$$

$$= \frac{(15)^2 + (3)^2 + (15)^2 + \ldots + (51)^2}{(27)(4)}$$

$$= 400.2$$

$$\text{Intrablock error } SS = \text{Total } SS \text{ Replication } SS \text{ Treatment (unadj.) } SS$$

$$- \text{Block (adj.) } SS \tag{4.40}$$

$$= 8,422.3 - 4,008.7 - 1,491.6 - 400.2$$

$$= 2,521.8$$

Step 9: Compute the mean square for treatment, block (adj.), and the intrablock error as:

$$\text{Treatment (unadj.) } MS = \frac{\text{Treatment (unadj.) } SS}{(k^2 - 1)} \tag{4.41}$$

$$= \frac{1,491.6}{8}$$

$$= 186.5$$

$$\text{Block (adj.) } MS = \frac{\text{Block (adj.) } SS}{(k^2 - 1)} \qquad (4.42)$$

$$= \frac{400.2}{8}$$

$$= 50.0$$

$$\text{Intrablock error } MS = \frac{\text{Intrablock error } SS}{(k - 1)(k^2 - 1)} \qquad (4.43)$$

$$= \frac{2,521.8}{16}$$

$$= 157.6$$

Step 10: Having obtained the values for the mean squares, we are now in a position to compute the adjusted treatment total $(\dot{T})$ as:

$$\dot{T} = T + \mu W \qquad (4.44)$$

where

$$\mu = \frac{\text{Block (adj.) } MS - \text{Intrablock error } MS}{k^2 \left[\text{Block (adj.) } MS \right]} \qquad (4.45)$$

This computation is necessary only if the intrablock error mean square is less than the block (adj.) mean square. In such conditions, the adjusted treatment totals $(\dot{T})$ for all treatments and the effective error mean square should be computed. They will in turn be used in performing the *F*-test of significance. The equation for computing the adjusted mean square and the effective error mean square is as shown:

$$\text{Treatment (adj.) } MS = \left[\frac{1}{(k+1)(k^2-1)} \right] \left[\sum \dot{T}^2 - \frac{G^2}{k^2} \right] \qquad (4.46)$$

$$\text{Effective error } MS = (\text{Intrablock error } MS)(1 + k\mu) \qquad (4.47)$$

$$F = \frac{\text{Treatment (adj.) } MS}{\text{Effective Error } MS} \qquad (4.48)$$

TABLE 4.3.1.4

ANOVA of Nonprotein Nitrogen in the Colostrum of Rats Fed Nine Different Diets. Experiment using a 3×3 Balanced Lattice Design

Source of Variation	Degrees of Freedom	Sum of Squares	Mean Square	F
Replication	3	4,008.7		
Block (adj.)	8	400.2	50.0	
Treatment (unadj.)	8	1,491.6	186.5	1.18[ns]
Intrablock error	16	2,521.8	157.6	
Total	35			

Note: $cv = 15.1\%$; ns = nonsignificant.

If the intrablock error mean square, on the other hand, is greater than the block (adj.) mean square, the value of μ is taken to be 0 and, therefore, there are no further adjustments necessary to the treatments. The F-test of significance is computed as the ratio of the treatment (unadj.) mean square to the intrablock error mean square. In the present example, the intrablock error mean square is greater than the block (adj.) mean square; therefore, μ is taken to be 0 and no adjustments are necessary.

Step 11: Compute the F value as:

$$F = \frac{\text{Treatment (unadj.) } MS}{\text{Effective Error } MS} \quad (4.49)$$

$$= \frac{186.5}{157.6}$$

$$= 1.18$$

Step 12: Compare the computed F with the tabular F value given in Appendix E. For this example, the tabular F (for eight treatment *d.f.*'s and 16 intrablock error *d.f.*'s) is 2.59 and 3.89 for 5% and 1% levels of significance, respectively, and the computed F as shown in Table 4.3.1.4 is 1.18. The result suggests that there is no significant difference among the diets.

As in the previous sections, we can determine the degree of precision with which the treatments are compared by computing the *cv* as:

$$cv = \frac{\sqrt{\text{Intrablock error } MS}}{\text{Grand mean}} \times 100 \quad (4.50)$$

$$= \frac{\sqrt{157.6}}{83} \times 100$$

$$= 15.1\%$$

Step 13: Compute the *R.E.* to estimate the precision relative to RCB designs as:

$$R.E. = \frac{\text{Block (adj.) } SS + \text{Intrablock error } SS}{\left(k - 1\right)\left(k^2 - 1\right)\left(\text{Intrablock } MS\right)} \times 100 \qquad (4.51)$$

$$= \frac{400.2 + 2{,}521.8}{\left(2\right)\left(8\right)\left(157.6\right)} \times 100$$

$$= 115.8\%$$

The result indicates that the experimental precision has increased by 15.8% over an RCB design.

4.3.2 Partially Balanced Lattice Designs

These designs were developed by Bose and Nair (1939) to overcome the problems associated with the restrictive assumptions of the balanced lattice designs, which require that the number of replications be determined by the number of treatments and the number of units per block. Partially balanced designs, although requiring treatments to be a perfect square, allow flexibility regarding the number of replications. Furthermore, all treatments need not be paired in the same block as often, which implies that sometimes, some treatments may never be paired in the same block. As a result, some treatments tested in the same incomplete block are compared with different levels of precision. This leads to a relatively more complicated data analysis. Such complications do not deter their use, especially when the number of replications required for a complete balance is too large for practical and economic reasons.

Partially balanced lattice designs are constructed similar to the balanced designs, with the exception that the former contain fewer replications. A partially balanced lattice design with two replications is called a *simple lattice*, that with three replications a *triple lattice*, and that with four replications a *quadruple lattice*.

Randomization and layout: With the exception of the modification in the number of replications, randomization and layout is similar to that of the balanced lattice designs. Modifications to a basic plan selected from Appendix F for designs with two or three replications follow the same procedures outlined in Subsection 4.3.1. However, when the number of replications (*r*) exceeds three and is an even number, the basic plan to use for randomization and layout is as follows:

- The first r replications of the basic plan (selected from Appendix F — balanced lattice design) that has the same number of treatments, or

- The first r/p replications of the basic plan (selected from Appendix F — balanced lattice design) that has the same number of treatments, repeated p times.

To apply the preceding criteria to a 5×5 quadruple balanced lattice design, for example, the basic plan to work with will be the first four replications of the selected plan, or the 5×5 simple lattice design plan repeated twice. Generally, it is preferable to work with the basic plans that do not involve repetition, as is the case with the first criterion. The advantage of using basic plans without repetition is the symmetry achieved by the balanced lattice designs in the arrangement of treatments over blocks. Examples are presented to illustrate the statistical analyses of designs with and without repetition.

ANOVA for a case without repetition: We will use a 5×5 triple lattice experiment on grain yield of corn hybrids double-cropped with winter wheat to illustrate the steps in performing the ANOVA. The treatments are 25 varieties of corn hybrids. Table 4.3.2.1 shows the yield data (bu/ac) rearranged according to the basic plan of Appendix F. Treatment numbers are shown in parentheses next to the yield data.

SOLUTION

Step 1: Compute the replication totals, the block totals (B), and the grand total as shown in Table 4.3.2.1.

Step 2: Compute the treatment totals (T) for all treatments. For example, the total for treatment 1 is

$$T_1 = 39 + 33 + 13 = 85 \tag{4.52}$$

Step 3: Calculate the C_b values for each block

where:

$$C_b = \text{total (over all replications) of all treatments in the block} - rB \tag{4.53}$$

where:

r = number of replications

B = block total

For example, the C_b value for block 1 in replication 1 is:

TABLE 4.3.2.1

Yield Data of 5×5 Triple Lattice Experiment on Hybrid Corn Double-Cropped with Winter Wheat (bu/ac, minus 20 bu)

Block No.						Block Total (B)	Treatment No.	Treatment Total (T)	C_b
			Rep. I				1	85	−108
1	39(1)	33(2)	40(3)	18(4)	22(5)	152	2	54	−112
2	24(6)	40(7)	16(8)	35(9)	29(10)	144	3	97	55
3	26(11)	32(12)	18(13)	7(14)	15(15)	98	4	63	42
4	36(16)	12(17)	21(18)	24(19)	40(20)	133	5	49	42
5	15(21)	24(22)	39(23)	9(24)	18(25)	105	6	51	−81
Rep. total						632	7	74	
							8	58	
			Rep. II				9	61	
1	33(1)	13(6)	18(11)	28(16)	26(21)	118	10	76	26
2	9(2)	20(7)	36(12)	35(17)	21(22)	121	11	74	5
3	36(3)	12(8)	19(13)	27(18)	25(23)	119	12	107	7
4	16(4)	19(9)	21(14)	22(19)	10(24)	88	13	57	65
5	19(5)	22(10)	9(15)	29(20)	38(25)	117	14	61	23
							15	50	126
Rep. total						563	16	99	
							17	76	
							18	72	
			Rep. III				19	80	
1	13(1)	7(9)	30(11)	24(18)	12(22)	86	20	114	91
2	12(2)	25(10)	26(15)	45(20)	29(25)	137	21	71	−32
3	21(3)	14(6)	20(13)	35(16)	45(24)	135	22	57	−37
4	29(4)	30(8)	39(12)	34(19)	30(21)	162	23	80	−107
5	8(5)	14(7)	33(14)	29(17)	16(23)	100	24	64	40
							25	85	45
Rep. total						620			
Grand total (G)						1815			

$$C_b = 85 + 54 + 97 + 63 + 49 - (3)(152) = -108$$

The C_b values for all the 15 blocks are computed and shown in Table 4.3.2.1. The replicate totals (R_c) of the C_b values are −81, 126, and −45. Note that these replicate totals sum to zero.

Step 4: Determine the d.f.'s associated with each source of variation. The source of variation and the total d.f.'s depend on whether the treatments are adjusted or not. If the treatments are adjusted, the following sources of variation and the d.f.'s are identified.

Replication d.f. $= r - 1 = 2$

Treatment (unadj.) d.f. $= k^2 - 1 = 24$

Blocks within replications (adj.) d.f. $= r (k - 1) = 12$

Intrablock error $d.f. = (k-1)(rk-k-1) = 36$

Treatment (adj.) $d.f. = k^2 - 1 = (24)$

Total $d.f. = (rk^2 - 1) = 74$

where:

r = number of replications

k = block size

Step 5: Calculate the various sums of squares by first computing the correction factor (C) as:

$$C = \frac{G^2}{(r)(k)^2} \tag{4.54}$$

$$= \frac{(1,815)^2}{(3)(5)^2} = 43,923$$

$$\text{Total } SS = \sum X_{ij}^2 - C \tag{4.55}$$

$$\text{Total } SS = \left[(39)^2 + (33)^2 + (40)^2 + \quad + (16)^2 \right] - 43,923$$

$$= 51,288 - 43,923$$

$$= 7,365$$

$$\text{Replication } SS = \frac{\sum R^2}{k^2} - C \tag{4.37}$$

$$= \frac{(632)^2 + (563)^2 + (620)^2}{25} - 43,923$$

$$= 44,031.7 - 43,923$$

$$= 108.7$$

$$\text{Treatment (unadj.) } SS = \frac{\sum T^2}{r} - C \qquad (4.56)$$

$$= \frac{(85)^2 + (54)^2 + \quad + (85)^2}{3} - 43,923$$

$$= 46,473.6 - 43,923$$

$$= 2,550.7$$

$$\text{Block } SS \text{ (adj.)} = \frac{\sum C_b^2}{kr(r\text{-}1)} - \frac{\sum R_c^2}{k^2 r(r-1)} \qquad (4.57)$$

$$= \frac{(-108)^2 + (-112)^2 + \quad + (40)^2}{(5)(3)(3-1)} - \frac{(-81)^2 + (126)^2 + (-45)^2}{(5)^2(3)(3-1)}$$

$$= 1,999.6 - 163.1$$

$$= 1,836.5$$

Step 6: Compute the intrablock error SS as:

Intrablock error $SS = TSS - \text{Rep. } SS - \text{Treatment (unadj.) } SS - \text{Block (adj.) } SS \,(4.40)$

$$= 7,365 - 108.7 - 2,550.7 - 1,836.5$$

$$= 2,869.1$$

Step 7: Calculate the mean square for the block (adj.) and the intrablock error as:

$$\text{Block } MS \text{ (adj.)} = \frac{\text{Block } SS \text{ (adj.)}}{r\,(k-1)} \qquad (4.58)$$

$$= \frac{1,836.5}{12}$$

$$= 153.0$$

$$\text{Intrablock Error } MS = \frac{\text{Intrablock Error } SS}{(k-1)(rk-k-1)} \tag{4.59}$$

$$= \frac{2,869.1}{(5-1)[(3)(5)-5-1)]}$$

$$= 79.70$$

Step 8: Calculate μ, the weighting factor, to adjust the treatment totals. μ is computed as shown:

$$\mu = \frac{\text{Block (adj.) } MS - \text{Intrablock Error } MS}{k(r-1)\left[\text{Block (adj.) } MS\right]} \tag{4.60}$$

If the block (adj.) mean square is less than the intrablock error mean square, μ is assumed to be 0 and there is no need for adjustments for the block effects. Thus, the data is analyzed as if in randomized blocks. However, for the present example the block (adj.) mean square is greater than the intrablock error mean square, so the adjustment factor is calculated as:

$$\mu = \frac{153 - 79.70}{5(3-1)(153)}$$

$$= \frac{73.3}{1,530}$$

$$= 0.05$$

Step 9: Calculate the adjusted treatment *SS* as:

$$\text{Treatment (adj.) } SS = \text{Treatment (unadj.) } SS - k(r-1)\mu\left\{\left[\frac{r}{(r-1)(1+k\mu)}\right]B_u - B_a\right\} \tag{4.61}$$

where

B_u = unadjusted sum of squares for blocks within replications

B_a = adjusted sum of squares for blocks within replications

To compute equation [4. 61], we first have to calculate B_u, which is defined as:

$$B_u = \frac{\sum B^2}{k} - \frac{\sum R^2}{k^2} \tag{4.62}$$

where

 B = block total

 R = replication total

 k = number of blocks

For the present example, we have:

$$B_u = \frac{(152)^2 + (144)^2 + \ldots + (100)^2}{5} - \frac{(632)^2 + (563)^2 + (620)^2}{(5)^2}$$

$$= 1{,}358.48$$

Substituting the value of B_u and B_a into Equation 4.61, we will have:

$$2{,}550.7 - 5(2)(0.05)\left\{\left[\frac{3}{(2)\,(1+(5)(.05)}\right](1{,}358.48) - (1{,}836.5)\right\} = 2{,}653.9$$

Step 10: Calculate the treatment-adjusted mean square as:

$$\text{Treatment } MS\,(\text{adj.}) = \frac{\text{Treatment } SS\,(\text{adj.})}{(k^2 - 1)} \tag{4.63}$$

$$= \frac{2{,}653.9}{24}$$

$$= 110.60$$

Step 11: Compute the F value as:

$$F = \frac{\text{Treatment } MS\,(\text{adj.})}{\text{Intrablock Error } MS} = \frac{110.60}{79.70} \tag{4.64}$$

$$= 1.39$$

Step 12: Compare the computed F shown in Table 4.3.2.2 with the tabular F value given in Appendix E. Because the computed F is less than the tabular F (for 24 treatment *d.f.*'s and 36 intrablock error *d.f.*'s), the F-test suggests that the experiment failed to show any significant difference among the treatments.

TABLE 4.3.2.2

ANOVA of an Experiment on Hybrid Corn Double-Cropped with Winter Wheat Using a 5×5 Triple Lattice Design

Source of Variation	Degrees of Freedom	Sum of Squares	Mean Square	F
Replication	2	108.7		
Block (adj.)	12	1,836.5	153.00	
Treatment (unadj.)	(24)	2,550.7		
Treatment (adj.)	24	2,654.3	110.60	
Intrablock error	36	2,869.1	79.70	1.39ns
Total	74	7,365.0		

Note: $cv = 24.79\%$; ns = nonsignificant.

As in the previous sections, we can determine the degree of precision with which the treatments are compared by computing the *cv* as:

$$cv = \frac{\sqrt{\text{Intrablock Error } MS}}{\text{Grand Mean}} \times 100 \tag{4.50}$$

$$= \frac{\sqrt{79.70}}{36} \times 100 = 24.79\%$$

Step 13: Estimate the gain in accuracy over randomized blocks. This is accomplished by comparing the error mean square with the effective error variance. The error mean square in the randomized block is the pooled mean square for blocks and intrablock error as shown in the following text:

$$\text{Randomized Block } MS = \frac{\text{Block } SS \text{ (adj.)} + \text{Intrablock Error } SS}{(k^2 - 1)(r - 1)} \tag{4.65}$$

The effective error variance is computed as:

$$\text{Intrablock Error } MS = \left[1 + \frac{rk\mu}{(k+1)} \right] \tag{4.66}$$

For our example, the respective values are:

$$\text{Randomized Block } MS = \frac{1,836.5 + 2,869.1}{48}$$

$$= 98.0$$

$$\text{Effective error variance} = 79.70\left[1+\frac{(3)(5)(0.05)}{(5+1)}\right]$$

$$= 89.7$$

Thus, the relative accuracy is:

$$\frac{98.0}{89.7} = 109.2\%$$

ANOVA for a case with repetition: In this design, n replicates contained in the basic plan are repeated p times, so the total number of replications $r = np$. To illustrate the case with repetition, we will apply the data from the triple lattice design used in the preceding section and change it to a quadruple lattice for the present case. This means that for a case with repetition we take the first two replications of the 5×5 plan selected from Appendix F and repeat each replication once to obtain the four replications needed for the experiment. Another way of stating this is that in a 5×5 quadruple lattice design, the basic plan is obtained by repeating a simple lattice twice (where $n = 2$ for a simple lattice). That is, replications III and IV are a repeat of replications I and II. The data are then rearranged according to the basic plan such that treatments (numbers in parentheses) follow the same order within the corresponding replications as shown in Table 4.3.2.3. The steps in performing the ANOVA are as follows:

Step 1: Compute the replication totals, the block totals (B), and the grand total (G) as shown in Table 4.3.2.3.

Step 2: Compute the treatment total (T) for all treatments. For this example, the total for treatment 1 is

$$T_1 = 39 + 33 + 19 + 28 = 119$$

Step 3: Determine the *d.f.*'s associated with each source of variation. The source of variation and the total *d.f.*'s depend on whether the treatments are adjusted or not. If the treatments are adjusted, the following sources of variation and *d.f.*'s are identified:

Replication *d.f.* $= r - 1 = 3$

Treatment (unadj.) *d.f.* $= k^2 - 1 = 24$

Blocks within replications (adj.) *d.f.* $= r (k - 1) = 16$

Component (a) *d.f.* $= n (p - 1) (k - 1) = 8$

Component (b) *d.f.* $= n (k - 1) = 8$

Intrablock error *d.f.* $= (k - 1) (rk - k - 1) = 56$

Total *d.f.* $= (rk^2 - 1) = 99$

TABLE 4.3.2.3

Yields of 5 × 5 Simple Lattice Experiment on Hybrid Corn Double-Cropped with Winter Wheat (bu/ac, minus 20 bu)

Block No.						Block Total B	Treatment No.	Treatment Total (T)
			Rep. I					
1	39(1)	33(2)	40(3)	18(4)	22(5)	152	1	119
2	24(6)	40(7)	16(8)	35(9)	29(10)	144	2	72
3	26(11)	32(12)	18(13)	7(14)	15(15)	98	3	121
4	36(16)	12(17)	21(18)	24(19)	40(20)	133	4	66
5	15(21)	24(22)	39(23)	9(24)	18(25)	105	5	82
							6	71
Rep. total						632	7	105
							8	70
			Rep. II				9	87
1	33(1)	13(6)	18(11)	28(16)	26(21)	118	10	92
2	9(2)	20(7)	36(12)	35(17)	21(22)	121	11	83
3	36(3)	12(8)	19(13)	27(18)	25(23)	119	12	112
4	16(4)	19(9)	21(14)	22(19)	10(24)	88	13	84
5	19(5)	22(10)	9(15)	29(20)	38(25)	117	14	60
							15	46
Rep. total						563	16	112
							17	83
							18	84
			Rep. III				19	90
1	19(1)	20(2)	15(3)	18(4)	25(5)	97	20	121
2	16(6)	30(7)	20(8)	15(9)	21(10)	102	21	80
3	24(11)	18(12)	28(13)	10(14)	12(15)	92	22	87
4	38(16)	20(17)	12(18)	14(19)	28(20)	112	23	104
5	25(21)	20(22)	19(23)	10(24)	30(25)	104	24	49
							25	106
Rep. total						507		
			Rep. IV					
1	28(1)	18(6)	15(11)	10(16)	14(21)	85		
2	10(2)	15(7)	26(12)	16(17)	22(22)	89		
3	30(3)	22(8)	19(13)	24(18)	21(23)	116		
4	14(4)	18(9)	22(14)	30(19)	20(24)	104		
5	16(5)	20(10)	10(15)	24(20)	20(25)	90		
Rep. total						484		
Grand total						2186		

where

r = total number of replications

k = block size

n = number of replication in the base design

p = number of repetition

Step 4: Calculate the various sums of squares by first computing the correction factor (C) as:

$$C = \frac{G^2}{(n)(p)(k^2)} \tag{4.67}$$

$$= \frac{(2,186)^2}{(2)(2)(25)}$$

$$= 47,785.96 \text{ or } 47,786$$

$$\text{Total } SS = \sum X^2 - C \tag{4.68}$$

$$= \left[(39)^2 + (33)^2 + (40)^2 + .. + (20)^2 \right] - 47,786$$

$$= 54,698 - 47,786$$

$$= 6,912$$

$$\text{Replication } SS = \frac{\sum R^2}{k^2} - C \tag{4.37}$$

$$= \frac{(632)^2 + (563)^2 + (507)^2 + (484)^2}{25} - 47,786$$

$$= 48,307.9 - 47,786$$

$$= 521.9$$

$$\text{Treatment } SS \text{ (unadj.)} = \frac{\sum T^2}{(n)(p)} - C \tag{4.69}$$

$$= 50,480.5 - 47,786$$

$$= 2,694.5$$

Step 5: For each block in each repetition, calculate the B_T value, which is the sum of block totals over all replications in that repetition. For example, the B_T value for block 1 from replications I and III is:

$$B_T = 152 + 97 = 249 \tag{4.70}$$

TABLE 4.3.2.4

B_T and C_b Values for Pairs of Blocks Containing the Same Set of Treatments

Block Number	Block Totals		Sum (B_T)	C_b
	Repetition 1			
	Rep. I	*Rep. III*		
1	152	97	249	−38
2	144	102	246	−67
3	98	92	190	5
4	133	112	245	0
5	105	104	209	8
Total	632	507	1,139	−92
	Repetition 2			
	Rep. II	*Rep. IV*		
1	118	85	203	59
2	121	89	210	39
3	119	116	235	−7
4	88	104	192	−32
5	117	90	207	33
Total	563	484	1,047	+92

The B_T value for the remainder of the blocks in each repetition is computed and shown in Table 4.3.2.4.

Step 6: Calculate for each block, the C_b values as:

$$C_b = \sum T - nB_T \qquad (4.71)$$

where

T = treatment total

n = number of replications in the base design

B_T = sum of block totals over all replications

For example, the C_b value for block 1 in replication I, which contains treatments 1, 2, 3, 4, and 5 is

$$C_b = 119 + 72 + 121 + 66 + 82 - (2)(249) = -38$$

The C_b value for all the blocks is computed and is shown in Table 4.3.2.4.

Step 7: Compute the block *SS*, which contains two components, (*a*) and (*b*). Component (*a*), which is the difference between the totals of blocks that contain the same set of treatments, arises only when the design is repeated. Component (*b*), on the other hand, is present when there are no repetitions involved. To compute each component, we will use the information in Table 4.3.2.4, which has k rows and p columns. Thus, the sum of squares for component (*a*) is the sum of the rows × columns interactions over all repetitions. Each component is computed as follows:

$$\text{Component } (a) \text{ } SS = \text{Total} - \text{Rows} - \text{Columns} \tag{4.72}$$

where

$$\text{Total} = \frac{\sum B^2}{k} - \frac{\sum R^2}{pk^2} \tag{4.73}$$

$$\text{Rows} = \frac{\sum B_T^2}{pk} - \frac{\sum R^2}{pk^2} \tag{4.74}$$

$$\text{Column} = \frac{\sum N^2}{k^2} - \frac{\sum R^2}{pk^2} \tag{4.75}$$

The term B is the block total, R is the sum of B_T values for each repetition, N is the sum of block totals for each replications, and k and p are as defined earlier. For our example, the values are:

$$\text{Total} = \frac{(152)^2 + (144)^2 + \cdots + (104)^2 + (90)^2}{5} - \frac{(1,139)^2 + (1,047)^2}{50}$$

$$= 49{,}094.4 - 47{,}870.6$$

$$= 1{,}223.8$$

$$\text{Row} = \frac{(249)^2 + (246)^2 + \cdots + (192)^2 + (207)^2}{10} - \frac{(1,139)^2 + (1,047)^2}{50}$$

$$= 48{,}257.0 - 47{,}870.6$$

$$= 386.4$$

$$\text{Column} = \frac{(632)^2 + (507)^2 + \cdots + (563)^2 + (484)^2}{25} - \frac{(1,139)^2 + (1,047)^2}{50}$$

$$= 48,307.9 - 47,870.6$$

$$= 437.3$$

$$\text{Component } (a)\ SS = 1,223.8 - 386.4 - 437.3$$

$$= 400.1$$

$$\text{Component } (b)\ SS = \frac{\sum C_b^2}{kr(n-1)} - \frac{\sum R_c^2}{k^2 r(n-1)} \qquad (4.76)$$

where

R_c = the total of C_b values over all blocks in a repetition

For the present example, we have two R_c values which are 92 and +92. These replicate totals should add to zero.

$$\text{Component } (b)\ SS = \frac{(-38)^2 + (-67)^2 + \cdots + (-32)^2 + (33)^2}{20} - \frac{(-92)^2 + (92)^2}{100}$$

$$= 659.3 - 169.3$$

$$= 490$$

Step 8: Calculate the adjusted sum of squares for blocks within replications as:

$$\text{Block } SS\ (\text{adj.}) = \text{Component } (a)\ SS + \text{Component } (b)\ SS \qquad (4.77)$$

$$= 400.1 + 490$$

$$= 890.1$$

Step 9: Calculate the intrablock error sum of squares as:

Intrablock error SS = Total SS – Rep. SS – Treatment SS (unadj.) – Block SS (adj.)

$$= 6,912 - 521.9 - 2,694.5 - 890.1 \qquad (4.78)$$

$$= 2,805.5$$

Step 10: Calculate the mean square for block (adj.) and the intrablock error as:

$$\text{Block } MS \text{ (adj.)} = \frac{\text{Block } SS \text{ (adj.)}}{r(k-1)} \qquad (4.79)$$

$$= \frac{890.1}{16}$$

$$= 55.6$$

$$\text{Intrablock error } MS \text{ (adj.)} = \frac{\text{Intrablock error } SS}{(k-1)(rk-k-1)} \qquad (4.80)$$

$$= \frac{2,805.5}{(5-1)\left[(3)(5)-5-1\right]}$$

$$= 77.9$$

Since the block (adj) mean square is less than the intrablock error mean square we do not need to make any adjustments to the treatments for the block effects. If this was not the case we compute μ, as shown below, the weighting factor and make the necessary correction.

$$u = \frac{P\left[\text{Block(adj.)}MS - \text{Intrablock error } MS\right]}{k\left[(r-p)\,\text{Block (adj.) } MS + (p-1)\,\text{Intrablock error } MS\right]} \qquad (4.81)$$

As before, *p* is the number of repetitions of the basic design, and is 2 for the present example. If block (adj.) mean square is less than the intrablock error mean square, μ is assumed to be 0 and there is no need for adjustments for the block effects.

Step 11: Compute the *F* value as:

$$F = \frac{\text{Treatment (unadj.) } MS}{\text{Effective error } MS} \qquad (4.48)$$

$$= \frac{112.3}{50.1}$$

$$= 2.24$$

TABLE 4.3.2.5

Analysis of Variance of Data for Hybrid Corn Using a 5 × 5 Quadruple Lattice Design

Source of Variation	Degrees of Freedom	Sum of Squares	Mean Square	F
Replication	3	521.9		
Block (adj.)	16	890.1	55.6	
Component (*a*)	(8)	400.1		
Component (*b*)	(8)	490.0		
Treatment (unadj.)	(24)	2,694.5	112.3	2.24**
Intrablock error	56	2,805.5	50.1	
Total	99	6,912.0		

Note: cv = 32.3%; ** = significant at 1% level.

The corresponding *cv* value is computed as:

$$cv = \frac{\sqrt{\text{Intrablock error } MS}}{\text{Grand mean}} \times 100 \qquad (4.50)$$

$$= \frac{\sqrt{50.1}}{21.86} \times 100$$

$$= 32.3\%$$

Step 12: Compare the computed *F* with the tabular *F* value given in Appendix E. For this example, the tabular *F* (for 24 treatment *d.f.*'s and 56 intrablock error *d.f.*'s) is less than the computed *F* value at the 1% level of significance. Thus, it is concluded that there is a highly significant difference among the treatments. Table 4.3.2.5 shows the summary of the results.

Step 13: Compute the error variance or the effective error *MS* for the difference between two treatment means in equation 4.8.2. To compute the error MS, we first need to compute μ using equation 4.81.

$$\mu = \frac{p[\text{Block(adj)}MS - \text{Intrablock error } MS]}{k[(r-p)\,\text{Block(adj.)}MS + (p1)\,\text{Intrablock error } MS]} \qquad (4.81)$$

$$\mu = \frac{2\,(55.6 - 77.9)}{5\,[(4-2)\,55.6 + (2-1)\,77.9]} = -0.05$$

1. The error mean square for two treatments in the same block:

$$\text{Error } MS = MS \text{ error } [1 + (n-1)\,\mu]$$

$$= 50.1[\,1 + (2-1)\,(-0.05)]$$

$$= 47.6$$

2. The error *MS* for two treatments not in the same block:

$$\text{Error } MS = MS \text{ error } (1 + n\,\mu) \qquad (4.82)$$

$$= 50.1[1 + 2(-0.05)]$$

$$= 45.1$$

3. The average effective error *MS* is computed as:

$$\text{Average error } MS = MS \text{ error } \left[1 + \frac{nk\mu}{k+1}\right] \qquad (4.83)$$

$$= 50.1\left[1 + \frac{2(5)(-0.05)}{5+1}\right]$$

$$= 54.3$$

Step 14: Calculate the *R.E.* to estimate the precision relative to RCB designs as:

$$R.E. = \left[\frac{\text{Block (adj.) } SS + \text{Intrablock error } SS}{r(k-1)+(k-1)(rk-k-1)}\right]\left[\frac{100}{\text{Error } MS}\right] \qquad (4.84)$$

We substitute the effective error *MS* computed in Step 13 for each of the conditions to calculate the *R.E.* as shown in the following text:

1. $R.E. = \left[\dfrac{890.1+2,805.5}{4(4)+(4)(20-5-1)}\right]\left[\dfrac{100}{47.6}\right]$

$$= 107.8\%$$

2. $R.E. = \left[\dfrac{890.1+2,805.5}{4(4)+(4)(20-5-1)}\right]\left[\dfrac{100}{45.1}\right]$

$$= 113.7\%$$

3. $R.E. = \left[\dfrac{890.1+2,805.5}{4(4)+(4)(20-5-1)}\right]\left[\dfrac{100}{54.3}\right]$

$$= 94.5\%$$

In summary, this chapter has introduced single-factor experiments using the complete and incomplete block designs. Three different types of complete block designs — the CRD, the RCB design, and the Latin square design — were discussed. For the incomplete block designs, the two most commonly used designs — the lattice design and the partially balanced lattice designs — were detailed. Each design has its advantages and disadvantages. Using the *cv* as a measure of the degree of precision, the researcher is able to determine whether a certain design gives more precision than others.

References and Suggested Readings

Bose, R.C. and Nair, K.R. 1939. On construction of balanced incomplete designs. *Ann. Eugenics* 9: 353–399.

Fisher, R.A. 1974. *The Design of Experiments*. London: Collier Macmillan.

Gomez, K.A. and Gomez, A.A. 1984. *Statistical Procedures for Agricultural Research*. New York: John Wiley & Sons.

Turan, M. and Likar Angin. 2004. Organic chelate assisted phytoextraction of B, Cd, Mo, and Pb from contaminated soils using two agricultural crop species. *Acta Agriculture Scand. Section B, Soil and Plant Science*. 54: 221–31.

Wang, L.F., Cai, Z.C., and Yan, H. 2004. Nitrous oxide emission and reduction in a laboratory-incubated paddy soil response to pretreatment and water regime. *J. Environ. Sci.* 16: 353–357.

Yates, F. 1936(a). Incomplete randomized blocks. *Ann. Eugenics* 7: 121–140.

Yates, F. 1936(b). A new method of arranging variety trials involving a large number of varieties. *J. Agric. Sci.* 26: 424–455.

Yates, F. 1937. *The Design and Analysis of Factorial Experiments*. Technical Communication No. 35. Rothamsted: Commonwealth Bureau of Soil Science.

Exercises

1. Field measurements were made to study the response of field-grown cassava (*Manihot esculenta* Crantz) to changes in the application of a fertilizer. The researcher conducted a CRD to find out if there were any differences in the amount of the dry matter produced under five

different fertilizer regimes. The following data were collected from the experiment with four replications:

Dry Matter Production (t/ha) of Cassava as a Result of Five Different Fertilizer Applications

Treatment	Rep. I	Rep. II	Rep. III	Rep. IV
Control	2.20	2.10	2.25	2.01
50 kg/ha	2.40	2.56	2.66	2.52
100 kg/ha	2.60	2.68	2.79	2.66
150 kg/ha	3.00	3.56	4.00	4.66
200 kg/ha	3.50	4.98	5.00	4.20

(a) Perform the ANOVA.

(b) Are there differences between the treatments?

(c) Compute the *cv*. What does this value mean?

2. Assume that the data collected in Exercise 1 have some observations missing, as shown in the following table.

Dry Matter Production (t/ha) of Cassava as a Result of Five Different Fertilizer Applications

Treatment	Rep. I	Rep. II	Rep. III	Rep. IV
Control	2.20	2.10	2.25	2.01
50 kg/ha	2.40	—	2.66	2.52
100 kg/ha	2.60	2.68	—	2.66
150 kg/ha	—	3.56	4.00	4.66
200 kg/ha	3.50	4.98	5.00	4.20

(a) Perform an ANOVA on the data.

(b) What can you say about the results of this exercise as compared to Exercise 1?

(c) Compute the *cv*.

(d) Is there any difference between the *cv* computed here and in Exercise 1? What can you say about the differences?

3. Seed yield of early-maturing high-protein soybean lines adapted to the Mid-Atlantic area of the U.S. were evaluated in a field test in an RCB with three replications. Each plot contained four 20-ft rows, 30

in. apart. Each plot was evaluated for seed yield, and the following data were recorded:

Seed Yield (bu/ac) of Early-Maturing Soybean Lines

Line	Rep. I	Rep. II	Rep. III
CX797-115	32.2	33.5	33.3
CX797-21	35.8	36.2	36.8
CX804-3	34.5	35.4	36.6
K1085	37.2	36.4	38.3
K1091	39.8	36.2	40.2
Williams	37.8	38.2	41.1
Douglas	36.4	38.6	40.2

(a) Perform an ANOVA.

(b) Compute the *cv*.

(c) Determine the *R.E.* of this experiment.

4. An ornamental horticulturist conducted a fertilizer experiment in a greenhouse in which five fertilizer treatments (A, B, C, D, and E) were tested by arranging plants in a Latin square design. Thus, the rows and columns in the table represent the rows and columns in the greenhouse. The data in the table shows the yield from the experiment.

A22	B23	C19	D12	E14
B20	C13	D16	E19	A18
C14	D10	E12	A26	B23
D19	E18	A20	B18	C14
E15	A24	B20	C17	D10

(a) Using a 1% level of significance, determine if the mean yields are not equal for the five fertilizers.

(b) Compute the *cv* for the data.

(c) What can be said about the *R.E.* of the Latin square design?

5. Plant breeders are interested in determining the spikelet initiation differences among nine winter wheat cultivars. The number of spikelets per plant from a field experiment which followed a 3×3 balanced lattice design with four replications are given in the following table. Each cultivar is given a treatment number and they are: Turkey (1), Pawnee (2), Scout (3), Larned (4), Newton (5), Hawk (6), Vona

(7), HW 1010 (8), and Bounty 100 (9). The data collected from each replication are presented in the table.

Incomplete Block Number	Spikelet Number			Incomplete Block Number	Spikelet Number		
		Rep. I				*Rep. II*	
1	18.1(8)	18.4(6)	17.6(1)	4	18.2(8)	20.2(7)	16.5(9)
2	16.5(3)	18.7(5)	17.9(7)	5	15.2(3)	19.9(2)	17.8(1)
3	16.0(4)	18.0(2)	16.0(9)	6	17.8(6)	18.1(5)	16.4(4)
		Rep. III				*Rep. IV*	
7	17.1(8)	18.4(5)	18.6(2)	10	16.2(3)	15.9(4)	18.5(8)
8	16.2(4)	17.7(7)	16.9(1)	11	17.2(6)	18.9(2)	17.6(7)
9	16.5(3)	18.9(6)	16.2(9)	12	15.4(9)	18.9(5)	17.4(1)

(a) Perform the ANOVA.

(b) From the analysis in (a), is it necessary to compute the adjusted treatment totals for all treatments?

(c) Compute the *cv*.

(d) Compute the *R.E.* coefficient for this design. What do the results indicate?

6. An animal scientist is interested in evaluating the role of progestrone in stimulating sexual receptivity in estrogen-treated gilts. She has used a 4×4 triple lattice design to conduct an experiment in which 16 ovariectomized gilts were treated with estradiol benzoate (EB). After EB treatment the gilts were moved to an evaluation pen where boars were brought in. Gilts remained in the evaluation pen for 5 min, during which time the number of mounts attempted by the boar were recorded. The following data were collected from the

experiment with three replications. The treatment numbers appear in parentheses.

Incomplete Block Number	Mounts (Number/5 min)			
	Replication I			
1	7(01)	5(02)	4(03)	2(04)
2	5(05)	2(06)	1(07)	3(08)
3	4(09)	3(10)	2(11)	2(12)
4	1(13)	3(14)	1(15)	5(16)
	Replication II			
1	6(01)	6(05)	6(09)	2(13)
2	4(02)	2(06)	3(10)	3(14)
3	3(03)	3(07)	1(11)	2(15)
4	1(04)	1(08)	3(12)	5(16)
	Replication III			
1	7(01)	5(06)	4(11)	6(16)
2	4(05)	4(02)	1(15)	3(12)
3	5(09)	2(14)	4(03)	2(08)
4	1(13)	3(10)	2(07)	2(04)

(a) Perform the ANOVA.

(b) Estimate the gain in accuracy over randomized blocks.

7. Suppose the animal scientist in the previous exercise chose to conduct her experiment as a case with repetition. As you will recall from your reading in this chapter, when the replicates contained in the basic plan are repeated p times so that the total number of replication $r = np$, then we have a case of data analysis with repetition. For the present case, the basic plan of the first two replicates has been

repeated. This changes a triple lattice to a quadruple lattice. The data collected from this experiment are shown in the following table.

Incomplete Block Number	Mounts (Number/5 min)				Incomplete Block Number	Mounts (Number/5 min)			
	Rep. I					*Rep. II*			
1	7(01)	5(02)	4(03)	2(04)	5	6(01)	6(05)	5(09)	2(13)
2	5(05)	2(06)	1(07)	3(08)	6	4(02)	2(06)	3(10)	3(14)
3	4(09)	3(10)	2(11)	2(12)	7	3(03)	3(07)	1(11)	2(15)
4	1(13)	3(14)	1(15)	5(16)	8	1(04)	1(08)	3(12)	5(16)
	Rep. III					*Rep. IV*			
1	6(01)	4(02)	3(03)	1(04)	5	4(01)	5(05)	5(09)	1(13)
2	4(05)	1(06)	2(07)	4(08)	6	3(02)	3(06)	4(10)	4(14)
3	5(09)	2(10)	1(11)	2(12)	7	2(03)	4(07)	2(11)	3(15)
4	3(13)	5(14)	2(15)	4(16)	8	1(04)	3(08)	2(12)	4(16)

(a) Perform the ANOVA.

(b) Estimate the gain in accuracy of this design over randomized blocks.

8. An agronomist is interested in the grain yield of rice after the application of postemergence herbicides. He has designed a CRD with unequal number of replications. The data gathered are as follows:

Herbicides	Time of Application[a]	Rate (kg a.i./ha[b])	Treatment Grain Yield (kg/ha)			
			Rep. I	Rep. II	Rep. III	Rep. IV
Propanil/2, 4-D-B	21	3.0	3,200	2,976	2,657	3,125
Propanil/Ioxynil	28	2.0	2,850	2,980	2,865	
Propanil/Bromoxynil	24	2.0	2,900	2,985	2,545	2,255
Propanil/CHCH	14	3.0	2,178	2,543	2,654	
Phenyedipham	14	1.5	2,875	2,455	2,495	2,214
Control			1,150	1,125	1,089	1,055

[a] Time of application is measured in days after seeding.
[b] a.i. = active ingredient.

(a) Perform the ANOVA to determine if there are differences between the treatments.

(b) Are there differences between the treatments?

(c) Compute the *cv* and interpret its meaning.

9. A researcher at the experimental station of a university has performed an RCB design to study the effect of different seeding rates

with a high-yielding variety of rice. The scientist has used five replications to conduct this experiment. The results are as follows:

Treatment (kg of seed/ha)	Yield (kg/ha)				
	Rep. I	Rep. II	Rep. III	Rep. IV	Rep. V
25	5,295	5,150	5,450	4,957	5,050
50	5,355	5,798	5,850	4,670	5,230
75	5,400	5,650	5,239	4,860	5,321
100	5,240	5,456	5,109	4,700	4,896
125	5,375	4,950	4,841	4,300	4,750

(a) Perform the ANOVA to determine if there are differences between the treatments.

(b) Are there differences between the treatments?

(c) Compute the *cv* and interpret its meaning.

10. Soil contamination, naturally and from human activities, has been a concern of the environmental scientists. A recent study by Turan and Angin (2004) indicates that adding different rates (0, 2.5, 5.0, and 10 mmol kg^{-1}) of various organic complexifying agents (OCAs) such as ethylene diamine tetra acetate (EDTA), diethylene triamine penta acetate (DTPA), citric acid (CA), and humic acid (HA) will have an impact on the uptake of heavy metals such as B, Cd, Mo, and Pb by corn plant (*Zea Mays* L). The following data were gathered by the scientists. Perform an ANOVA and determine if the chelate-assisted phytoextraction increased heavy-metal availability and uptake by corn.

Effects of Organic Complexifying Agents on Dry Matter Weight of Corn (dw g pot^1)

Species	OCA (mmol kg^1)	Organic Complexifying Agents			
		HA	CA	DTPA	EDTA
Corn	0.0 (Control)	30.7	30.7	31.2	30.6
	2.5	20.6	18.5	21.2	18.4
	5.0	17.5	11.3	14.3	11.1
	10.0	13.6	6.4	11.0	5.3

Source: Adapted from Turan, M. and Llker Angin, 2004. Organic chelate assisted phytoextraction of B, Cd, Mo, and Pb from contaminated soils using two agricultural crop species, *Acta Agriculture Scandinavia*. Section B, soil and plant science. 54:221–31.

(a) Perform an ANOVA on the data.

(b) Are there differences between the treatments?

(c) Compute the *cv.* What does this value mean?

11. Scientists have determined that nitrous oxide (N_2O) is detrimental to the ozone layer. Because agricultural soils are a major source of N_2O, a study was conducted to investigate N_2O emission and reduction in a paddy soil response to the pretreatment of water regime. The following data were gathered by the scientists:

KCl Extractable NO_3-N and NH_4-N Content at the End of Incubation of the Paddy Soil

N Species	Treatments	Moisture (% Water Holding Capacity)				
		20	40	60	80	100
NO_3-N, mg/kg	F	27.2	30.6	31.6	3.88	1.74
	D	25.8	31.4	31.2	1.71	1.73
	F + ACE	20.5	22.2	21.6	2.89	1.86
	D + ACE	20.6	20.7	21.9	2.88	1.86
NH_4-N, mg/kg	F	85.4	88.6	78.5	86.1	89.2
	D	80.9	81.3	77.7	81.3	87.7
	F + ACE	72.9	80.9	81.0	80.3	88.9
	D + ACE	71.4	80.2	82.3	80.9	86.6

Source: Adapted from Wang, L.F., Cai, Z.C., and Yan, H. 2004. Nitrous oxide emission and reduction in a laboratory-incubated paddy soil response to pretreatment and water regime. *J. Environ. Sci.* 16: 353–357.

(a) Perform an ANOVA on the data.

(b) Are there differences between the treatments?

5

Two-Factor Experimental Designs

5.1 Factorial Experiments

In agricultural experiments in which a number of independent variables or factors often have an impact on one another, it is not appropriate to use single-factor experiments (discussed in the previous chapter.) This chapter deals with the experimental designs that permit the experimenter to study a number of independent variables within the same experiment. Such experiments that consist of two or more combinations of different factors are referred to as *factorial experiments*. Another way of looking at factorial experiments is that they are ones in which all, or nearly all, factor combinations are of interest to the researcher.

Factorial experiments are different from varietal trials. In factorial experiments, comparisons are based on the main and interaction effects. Varietal trials, on the other hand, involve comparisons of different levels of only one factor, which is the treatment variable.

When a factor-by-factor experiment is conducted, we are unable to investigate the interaction effects. In such experiments the levels of one factor are changed one at a time while holding other factors constant. In agricultural experiments in which factors are likely to interact with each other, factorial experiments are appropriate investigative tools. It should be kept in mind that even when interactions do not occur, the results of factorial experiments are more widely applicable, as the main treatment effects have been shown to hold over a wide range of conditions.

There are advantages and disadvantages in using factorial experiments. Advantages of performing factorial experiments lie in the purpose of the experiment. If the purpose of the experiment is to investigate the effect of each factor, rather than the combination of levels of the factors that produce a maximum response, then the researcher could either conduct separate experiments in which each deals with a single factor, or include all factors simultaneously, performing a factorial experiment.

The main feature of factorial experiments is that they economize on experimental resources. For example, if factors are independent, all the simple effects are equal to the experiment's main effect. This implies that under such circumstances, the main effects are the only quantities needed to measure the consequence of variations in the other factor. Furthermore, factorial experiments allow us to estimate each main effect with the same precision as if we had performed a whole experiment on each factor alone. This means that if we have an experiment with n factors, all at two levels and all independent, the single-factor approach would require n times as much experimental resources as would a factorial experiment keeping the same level of precision. Thus, factorial experiments offer savings in time and material resources. An additional advantage of the factorial experiment is its ability to extend the range of validity of conclusions in a convenient way. For instance, in a fertilizer trial in which the presence or absence of nitrogen (N), phosphorous (P), and potassium (K) are of primary interest to us, we add three varieties of corn in the experiment. Here, the object is not a comparison of the varieties of corn as much as the effects of fertilizers on varieties of corn. Should the experiment point to the fact that the fertilizer effects are essentially the same for the three varieties, we can apply the conclusions to a new variety, with greater confidence than if the experiment had been confined to one variety.

It is also important to keep in mind that indiscriminate use of factorial experiments can lead to an increase in complexity, size, and cost of an experiment. Therefore, if the research is still in the exploratory stages and especially if the number of potentially relevant factors is large, the experimenter may wish to use only two levels of each factor (Gill, 1978). For practical reasons, others (Gomez and Gomez, 1984) have suggested that an experimenter may wish to use an initial single-factor experiment in order to avoid using large experimental areas and large expenditures.

Another disadvantage of factorial experiments lies in the less precise estimates that result from problems associated with heterogeneity of factors as the number of treatments increase in size. The larger the number of treatments, the more difficult it is to measure the influence of the primary trait of interest (even when good statistical design such as blocking is used to overcome the problems arising from natural variation). Further, it should be pointed out that in comparison to a single-factor experiment, a factorial experiment of comparable size would have a larger standard error. As the number of treatment combinations increases in an experiment, the standard error per unit also increases. This increase in the standard error can usually be kept small by forming blocks or by a device known as confounding.

The following examples illustrate the concepts related to factorial experiments in which simple cases are involved.

Example 5.1.1

Consider an experiment in which a horticulturist is interested in the application of a fertilizer trial with N, P, and K as the three factors of interest, each at two levels. In the simplest case the horticulturist will have eight

TABLE 5.1.1

The Eight Fertilizer Treatment Combinations

Treatment	N	P	K
1	No	No	No
2	No	No	Yes
3	No	Yes	No
4	No	Yes	Yes
5	Yes	No	No
6	Yes	No	Yes
7	Yes	Yes	No
8	Yes	Yes	Yes

treatment combinations in which either N, P, or K is present or absent in a treatment. This is shown in Table 5.1.1.

Because this is an experiment with three factors, each at two levels, the experiment is a $2 \times 2 \times 2$ or a (2^3) experiment. In this situation, if we use all eight combinations in which the treatments include the selected levels of variable factors, we can refer to the experiment as a complete factorial experiment. On the other hand, if we use a fraction of all the combinations, we refer to this as an incomplete factorial experiment. This is shown in Example 5.1.2.

Example 5.1.2

Consider a situation similar to Example 5.1.1, in which the three fertilizers (treatment factors), each being applied at two levels, is considered by an agronomist in a fertilizer trial on corn. Suppose the agronomist considers the treatment factors to be substitutes for one another. In such a situation, the experiment is conducted with four treatments: a control and the above three factors. In this case, the experiment cannot be called a factorial as not all combinations of factors are included in the experiment.

In a factorial experiment, the total number of treatments is the product of the levels in each factor. By way of illustration, in Example 5.1.1 we had three factors, each at two levels. If we are to add an additional factor such as the application of a mineral, the number of treatments increases to $2 \times 2 \times 2 \times 2 = 16$ or (2^4) factorial. You will note that the number of treatments increases rapidly with an increase in either the number of factors or the level of each factor. For example, if we have a factorial experiment with four cultivars, four fertilizer application rates, three irrigation methods, and three weed control practices, the total number of treatments would be $4 \times 4 \times 3 \times 3 = 144$.

Example 5.1.3

In a comparison study to determine the main and interaction effects, an animal scientist has planned a dietary experiment in which two factors, (1) fiber at three levels, denoted f_0, f_1, and f_2, and (2) protein at two levels, p_0 and p_1, are included. In this experiment the animal scientist forms the following six combinations taking one level from each factor p_0f_0, p_0f_1, p_0f_2, p_1f_0, p_1f_1, and p_1f_2. These combinations form treatments in factorial experiments. The researcher's interest lies in determining the main and interaction effects of this comparison. To determine the main effect of protein, for example, the researcher wants to observe the difference of the totals between the first and the last three of the above six combinations. Hence, the total of the first three treatment combinations represents the effect of protein at its p_0 level, whereas the sum of the last three represents its effect at the p_1 level, given the same levels of fiber. Thus, the difference between the first and the last three treatment combinations provides a comparison between the responses from the two levels of protein. To determine the main effect of fiber, two independent comparisons among the totals are made, that is, $p_0f_0 + p_1f_0$, $p_0f_1 + p_1f_1$, and $p_0f_2 + p_1f_2$.

To determine the interaction effects, the animal scientist needs to make a comparison that would indicate whether factors act independently or the results are influenced by their interaction. To make such a comparison, the effect of protein when all levels of fiber are present are first determined. Thus, the following three contrasts are written:

$$p_0f_0 - p_1f_0$$

$$p_0f_1 - p_1f_1$$

$$p_0f_2 - p_1f_2$$

The interaction effects are determined by taking the difference between the preceding contrasts in the following manner:

$$(p_0f_0 - p_1f_0) - (p_0f_1 - p_1f_1)$$

In the present case, there are a total of two interaction contrasts.

The preceding examples have provided a brief view of the concepts of factorial experiments. Additional examples in Section 5.2 show the detailed computation of the main and interaction effects.

The advantages of the factorial designs are in their efficiency (saving time and money) and, more importantly, their ability to enable study of the joint effects of variables. Two-factor experiments allow the effect on the response due to increasing the level of each factor to be estimated at each level of the other factor. This permits conclusions that are valid over a wider range of experimental conditions. Such is not the case with the single-factor experiments. Simultaneous investigation of two factors is necessary when factors interact with each other. Such interactions imply that a factor effect depends on the level of the other factor.

The major characteristic of the single-factor designs is that treatments consist solely of different levels of a single-variable factor. This means that only one factor varies while all other factors are held constant. If, in conducting an experiment, the experimenter recognizes that the variable of interest is affected by the different levels of other factors, then it is important to use a design that allows for two or more variable factors.

In conducting an experiment in which we compare v treatments using $N = vb$ experimental units, the design is a two-way or two-factor classification if the N experimental units are divided into b homogeneous groups (e.g., parcels of land), and the v treatments are allocated to the experimental units, at random in each group and independently from group to group. In the two-factor designs any observation is classified by the treatment it receives and the group to which it belongs. As mentioned in the previous chapter, groups are called *blocks* in agricultural field experiments, and the experimental plan is referred to as the *block design*. In such designs, treatment differences are what the experimenter is interested in, and block contributions, while accounted for, are eliminated from experimental error. Thus, the experiment deals with one factor (treatments), and blocking is a restriction on the randomization.

When dealing with two or more factors in an experiment, one needs to take account of interactions between factors and measure the main and simple effects of each of the factors. A *replicated factorial design* is a two-way arrangement in which the main effects and interactions of two factors are of equal interest. Let us consider a factorial experiment involving two factors A and B, with respective a and b levels to explain the different components of such designs. Table 5.1.2 shows an example of a hypothetical two-factor experiment in which rows of the table correspond to the level of factor A and the columns to the levels of factor B, and X_{ijk} represents the observation taken under the ith level of factor A and the jth level of factor B in the kth replicate. The ab experimental groups of n observations may be considered as ab random samples, one from each of ab treatment populations. The treatment populations are assumed to have been drawn from the same infinitely large parent population; thus, any differences among ab distributions are attributable solely to differences among the treatment effects.

TABLE 5.1.2

Data Matrix for a Two-Factor Design

		Factor B				
	Levels	1	2	3	...	b
Factor A	1	X_{111}	X_{112}	X_{113}	..	X_{11b}
	2	X_{211}	X_{212}	X_{213}	..	X_{21b}
	3	X_{311}	X_{312}	X_{313}	..	X_{31b}
	.	.	.	.	..	
	.	.	.	.	..	
	a	X_{a11}	X_{a12}	X_{a13}		X_{a1b}

The underlying model of the two-way classification is of the form

$$X_{ijk} = \mu + \alpha_i + \beta_j + (\alpha\beta)_{ij} + \varepsilon_{ijk}$$

$$\text{For } i = 1, \ldots, a; j = 1, \ldots, b; k = 1, \ldots, n$$

(5.1)

where

X_{ijk} = kth observation at the ith level of A and jth level of B

μ = overall mean response; that is, the average of the mean responses for the ab treatments

α_i = effect of the ith level of the first factor (A), averaged over the b levels of the second factor (B). The ith level of the first factor adds α_i to the overall mean μ

β_j = effect of the jth level of the second factor

$(\alpha\beta)_{ij}$ = interaction between the ith level of the first factor and the jth level of the second factor; the population mean for the ijth treatment minus $\mu + \alpha_i + \beta_j$

ε_{ijk} = the random error component

The effects α_i and β_j are usually referred to as *main effects* in contrast with the *interaction effect* $(\alpha\beta)_{ij}$. Implicit in the statement of the model is the assumption that the random error component ε_{ijk} is independently and normally distributed with a mean of 0 and a variance σ_e^2 within each treatment population defined by a combination of levels of A and B. This is the only assumption we make for the design. The interaction effect and its relation to the main effect will be discussed further in Section 5.2.

5.2   Main Effects and Interactions in a Two-Factor Experiment

As we discuss factorial experiments, it is important to elaborate on the concepts of simple effects, main effects, interactions, and how each is computed and represented in the analysis of variance of the results. To explain each of these concepts, we will use a 2×2 factorial experiment with factors A and B each at two levels. This factorial is called 2^2 as there are four treatment combinations in the experiment. The treatment combinations can be written as a_0b_0, a_0b_1, a_1b_0, and a_1b_1. Suppose factor A is two cultivars of soybean (Hobbit and Mead) and factor B is two different levels of fertilizer application (no fertilizer application, and application of fertilizer at 100 lb/ ac). Assume that there is no uncontrolled variation, and we obtain the observations shown in Table 5.2.1.

TABLE 5.2.1

Yields of Soybean from Two Cultivars and Their Means (bu/ac)

Cultivar	Fertilizer Levels		Row Average	Response Due to b_1
	(b_0) 0 lb/ac	(b_1)100 lb/ac		
Hobbit (a_0)	30	40	35	+10
Mead (a_1)	40	60	50	+20
Column Average	35	50	42.5	
Mead–Hobbit	10	20		

Looking at these results, we are able to explain the effect of the application of fertilizer on yield as well as the superiority of one cultivar over another. It appears that the yield of Hobbit and Mead cultivars increased by 10 and 20 bu/ac, respectively, when fertilizer was applied. These increases are called the *simple effects* of fertilizer application. The simple effects of the cultivars could also be reported similarly. That is, Mead cultivar is superior to Hobbit: 10 bu/ac, when no fertilizer is applied, and 20 bu/ac increase in yield when fertilizer is applied.

When factors are *independent*, the *main effects* are the effect of the *i*th level of the first factor averaged over the *b* levels of the second factor. Hence, the main effect for fertilizer application in the present example is the average of the two simple effects (10 + 20)/2, or 15 bu/ac. The main effect can alternatively be derived as the difference between two column means 35 and 50, which is 15. Factors are said to be independent when, for example, the response to fertilizer is the same whether Hobbit or Mead cultivar is used, and when the difference between Hobbit and Mead cultivar is the same whether fertilizer is applied or not. To determine whether factors are independent or not, experimenters may use their knowledge of the processes by which factors produce their effects, or simply test the assumption of independence using the information provided by the factorial experiment. For example, if using different cultivars affects the response to fertilizer, the difference between 20 and 10 bu/ac, or 10, is an estimate of this effect. We can use a *t*-test for this purpose. If the difference proves significant, then we can say that the assumption of independence is rejected by the data. This would mean that factors are not independent in their effects.

Interaction between factors is defined as a condition in which observations obtained under the various levels of one treatment behave differently under the various levels of the other treatment. For the present example, we can measure interaction by taking the difference between 20 (superiority in yield when fertilizer is added) and 10 bu/ac (superiority in yield of Mead over Hobbit when no fertilizer is added) and averaging it over the 2 factors. The difference of 10 bu/ac is averaged over the 2 factors and the result of 5 bu/ac is the interaction between fertilizer application and the cultivars used. The interaction effect for each factor is computed as:

$$A \times B = (1/2)(\text{simple effect of } A \text{ at } b_1 - \text{simple effect of } A \text{ at } b_0) \quad (5.2)$$

$$= (1/2)\,[(a_1 b_1 - a_0 b_1) - (a_1 b_0 - a_0 b_0)]$$

or

$$A \times B = (1/2)\,(\text{simple effect of } B \text{ and } a_1 - \text{simple effect of } B \text{ at } a_0) \quad (5.3)$$

$$= (1/2)\,[(a_1 b_1 - a_1 b_0) - (a_0 b_1 - a_0 b_0)]$$

Thus, the interaction for our example using the preceding equation is:

$$A \times B = (1/2)[(60 - 40) - (40 - 30)$$

$$= (1/2)(20 - 10)$$

$$= 5\,\text{bu/ac}$$

or

$$A \times B = (1/2)\,[(60 - 40) - (40 - 30)]$$

$$= (1/2)\,(20 - 10)$$

$$= 5\,\text{bu/ac}$$

If there is no interaction between factors as is shown by the hypothetical data in Table 5.2.2 and computed in the following text, the results of an experiment are simply described in terms of the main effects and not the effects of all treatment combinations. The interaction effect between cultivar and fertilizer application for a data set with no interaction is:

TABLE 5.2.2

Yields of Soybean from Two Cultivars and Their Means (bu/ac)

| Cultivar | Fertilizer Levels | | Row Average | Response Due to b_1 |
	(b_0) 0 lb/ac	(b_1) 100 lb/ac		
Hobbit (a_0)	30	50	40	+20
Mead (a_1)	40	60	50	+20
Column Average	35	55	45	
Mead–Hobbit	10	10		

$$A \times B = (1/2)[(60 - 50) - (40 - 30)]$$

$$= (1/2)(10 - 10)$$

$$= 0 \text{ bu/ac}$$

or

$$A \times B = (1/2)[(60 - 40) - (50 - 30)]$$

$$= (1/2)(20 - 20)$$

$$= 0 \text{ bu/ac}$$

When factors are independent in their effects, as in the case of data in Table 5.2.2, all the simple effects of a factor equal its main effect. This means that main effects are the only quantities we need to describe fully the consequences of all variations in the factor. This gain in describing the results is especially advantageous in experiments with more than two factors. The main effect of cultivar or factor A in the present example is computed as the average of the simple effects of cultivar over all levels of fertilizer or factor B as shown in the following text.

Main effect of cultivar $= (1/2)$ (simple effect of A at b_0 + simple effect of A at b_1)

$$= (1/2)[(a_1b_0 - a_0b_0) + (a_1b_1 - a_0b_1)] \tag{5.4}$$

$$= (1/2)[(40 - 30) + (60 - 50)]$$

$$= (1/2)(20)$$

$$= 10 \text{ bu/ac}$$

Similarly, the main effect of fertilizer (factor B) is computed as:

Main effect of fertilizer $= (1/2)$ (simple effect of B and a_0 + simple effect of B at a_1)

$$= (1/2)[(a_0b_1 - a_0b_0) + (a_1b_1 - a_1b_0)] \tag{5.5}$$

$$= (1/2)[(50 - 30) + (60 - 40)]$$

$$= (1/2)(40)$$

$$= 20 \text{ bu/ac}$$

The data in the last column and row of Table 5.2.1 show the same results obtained here.

5.3　Interpretation of Interactions

Many of the agricultural experiments will show some type of interaction. To determine the presence of interaction among factors, we need to apply the appropriate significance test (Cochran and Cox, 1957; Goulden, 1952). Once the presence of interaction is established, we compute the interaction effect and account for it in the analysis of variance. The advantage of no interaction, as mentioned before, is the considerable economy in describing the results of an experiment. Furthermore, absence of interaction may shed light on the experimental condition that may have special physical significance.

The interpretation of a two-factor interaction may best be clarified by graphical illustration. Figure 5.3.1 shows factor B to be represented along the x axis, and we have plotted each set of points joined by a light line to correspond to different levels of factor A.

When factors are not independent, as is the case with Figure 5.3.1(a), the simple effect of a factor changes as the level of the other factor changes. In this situation we have interaction between factors. When there is no interaction between factors, as is the case in Figure 5.3.1(b), the simple effect of a factor is the same for all levels of the other factors and equals the main effect. In this case the curves are parallel, indicating zero interaction among the factors.

It should be pointed out that although the separation of the treatment comparisons into main effects and interaction is a powerful tool in analyzing cases in which interactions are small relative to main effects, much care in summarizing the results and detailed examination of the nature of interaction is required when interactions are large (Cochran and Cox, 1957).

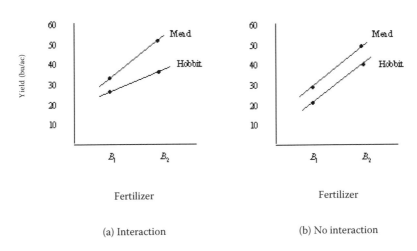

(a) Interaction　　　　　　　　(b) No interaction

FIGURE 5.3.1
Graphical representation of data in a two-factor experiment (a) with interaction, (b) without interaction.

In summary, interaction between factors can only be measured if two factors are tested together in the same experiment. Interaction between factors is absent when, for instance, the *difference* between the observations at two levels of A is the same for all levels of B. This implies that the simple effect of a factor equals the main effect. Under such circumstances the results from separate single-factor experiments are equivalent to factorial experiments with all factors tested together. Finally, when interaction is present we should examine the simple effects, and not the main effects, in reporting the results.

5.4 Factorials in Complete Block Designs

Factorial experiments may be laid out in any convenient design, such as complete block designs. In particular, if the total number of treatment combinations is not large, the designs discussed in Chapter 4 are frequently used. In Chapter 4 we discussed the procedures for the randomization and layout of such designs. The same procedures apply here with some modifications in factor composition. In factorial experiments we consider all factorial treatments to be unrelated to each other. In computing the treatment sum of squares for the analysis of variance, we take account of the factorial components corresponding to the main effects of individual factors and their interactions. Because the procedure for such partitioning is the same for all complete block designs, we will use an example for randomized complete block design to illustrate the step-by-step computation of the analysis of variance.

Example 5.4.1

Livestock breeders have continually changed animal types to meet perceived market demands and to adjust to changing economic pressures. To obtain data on measures of economic importance, such as average daily weight gain on three breeds of swine fed three different diets, an animal scientist used the following factorial treatment combination with three replications in conducting a performance test.

	Factorial Treatment Combinations		
Diet	Duroc (B_1)	Yorkshire (B_2)	Landrace (B_3)
D_1	B_1D_1	B_2D_1	B_3D_1
D_2	B_1D_2	B_2D_2	B_3D_2
D_3	B_1D_3	B_2D_3	B_3D_3

The randomized layout and the data recorded from the experiment are shown in Figure 5.4.1 and Table 5.4.1.

Rep. I		
D_1B_1	D_2B_1	D_3B_2
D_2B_3	D_3B_3	D_1B_2
D_3B_1	D_2B_2	D_1B_3

Rep. II		
D_3B_2	D_2B_2	D_3B_3
D_3B_1	D_2B_3	D_1B_1
D_2B_1	D_1B_3	D_1B_2

Rep. III		
D_2B_2	D_2B_3	D_1B_1
D_1B_3	D_3B_1	D_1B_2
D_2B_1	D_3B_3	D_3B_2

FIGURE 5.4.1

A sample layout of a 3×3 factorial experiment with three replications.

TABLE 5.4.1

The Average Daily Weight Gain of Three Breeds of Swine

Diet	Average Daily Weight Gain (lb)		
	Rep. I	Rep. II	Rep. III
Duroc (B_1)			
D_1	1.80	2.00	2.20
D_2	1.50	1.74	1.85
D_3	2.05	2.08	2.25
Yorkshire (B_2)			
D_1	2.00	1.80	1.95
D_2	1.85	1.65	1.58
D_3	1.90	2.22	2.15
Landrace (B_3)			
D_1	2.15	2.12	2.18
D_2	1.75	1.86	1.95
D_3	2.17	2.05	2.28
Rep. Total (R)	17.17	17.52	18.39
Grand Total (G)			53.08

Before proceeding to the computation of the various components, let us denote the number of replications as (r), the level of factor A (breeds of swine) as a, and the level of factor B (diets) as b.

SOLUTION

Step 1: Calculate the replication totals (R) and the grand total (G) as shown in Table 5.4.1, and the treatment totals (T) as shown in Table 5.4.2.

Step 2: Determine the degrees of freedom associated with each source of variation. The treatment component accounts for the variation due to breeds

TABLE 5.4.2

Treatment and Diet Totals (lb) in a 3×3 Daily Weight Gain Experiment

Diet	Breed			Total
	B_1	B_2	B_3	
D_1	6.00	5.75	6.45	18.20
D_2	5.09	5.08	5.56	15.73
D_3	6.38	6.27	6.50	19.15
Total	17.47	17.10	18.51	53.08

(factor A), diets (factor B) and the interaction between these two factors. The following sources of variation and the degrees of freedom are identified.

Replication $d.f. = r - 1 = 2$
Treatment $d.f. = ab - 1 = 8$
Breed (A) $d.f. = a - 1 = 2$
Diet (B) $d.f. = b - 1 = 2$
$A \times B$ $d.f. = (a - 1)(b - 1) = 4$
Error $d.f. = (r - 1)(ab - 1) = 16$
Total $d.f. = (rab - 1) = 26$

Step 3: Calculate the various sums of squares by first computing the correction factor (C) as:

$$C = \frac{G^2}{rab} \tag{5.6}$$

$$= \frac{(53.08)^2}{3 \times 3 \times 3}$$

$$= \frac{2,817.49}{27}$$

$$= 104.35$$

$$\text{Total } SS = \sum X^2 - C \tag{5.7}$$

$$= \left[(1.80)^2 + (1.50)^2 + ... + (2.28)^2 \right] - 104.35$$

$$= 105.52 - 104.35$$

$$= 1.17$$

$$\text{Replication } SS = \frac{\sum R^2}{ab} - C \tag{5.8}$$

$$= \frac{\left(17.17\right)^2 + \left(17.52\right)^2 + \left(18.39\right)^2}{9} - 104.35$$

$$= \frac{294.81 + 306.95 + 338.19}{9} - 104.35$$

$$= 104.44 - 104.35$$

$$= 0.09$$

$$\text{Treatment } SS = \frac{\sum T^2}{r} - C \tag{5.9}$$

$$= \frac{\left(6.00\right)^2 + \left(5.09\right)^2 + \dots + \left(6.50\right)^2}{3} - 104.35$$

$$= 105.19 - 104.35$$

$$= 0.84$$

$$\text{Error } SS = \text{Total } SS - \text{Rep. } SS - \text{Treatment } SS \tag{5.10}$$

$$= 1.17 - 0.09 - 0.84$$

$$= 0.24$$

A preliminary analysis of variance is shown in Table 5.4.3.

Comparing the results with the tabular F values given in Appendix E shows that there exists a difference among the treatment means. We continue with our analysis to determine which component of the treatment is contributing to the difference.

TABLE 5.4.3

Preliminary Analysis of Variance of the Three Diets on Three Breeds of Swine

Source of Variation	Degree of Freedom	Sum of Squares	Mean Square	F
Replication	2	0.09	0.045	3.0^{ns}
Treatment	8	0.84	0.105	7.0^{**}
Error	16	0.24	0.015	
Total	26	1.17		

Note: ^{ns} = nonsignificant; ** = significant at 1% level.

Step 4: Calculate the three factorial components of the treatment sum of squares. In our example, factor A is the breeds of swine. Thus, to calculate the sum of squares we take the treatment totals shown in Table 5.4.2 and use them as follows:

$$SS_A = \frac{\sum A^2}{rb} - C \tag{5.11}$$

$$= \frac{(17.47)^2 + (17.10)^2 + ... + (18.51)^2}{3 \times 3} - 104.35$$

$$= \frac{305.20 + 292.41 + 342.62}{9} - 104.35$$

$$= 104.47 - 104.35$$

$$= 0.12$$

To calculate the sum of squares for factor B, which represents the three diets, we use the treatment totals shown in Table 5.4.2 in the following manner:

$$SS_B = \frac{\sum B^2}{ra} - C \tag{5.12}$$

$$= \frac{(18.20)^2 + (15.73)^2 + (19.15)^2}{9} - 104.35$$

$$= \frac{331.24 + 247.43 + 366.72}{9} - 104.35$$

$$= 105.04 - 104.35$$

$$= 0.69$$

$$SS_{AB} = \text{Treatment } SS - SS_A - SS_B \tag{5.13}$$

$$= 0.84 - 0.12 - 0.69$$

$$= 0.03$$

Step 5: Calculate the mean square for each component of the treatment sum of squares and the error mean square.

$$MS_A = \frac{SS_A}{a-1} = \frac{0.12}{3-1} = 0.060 \tag{5.14}$$

$$MS_B = \frac{SS_B}{b-1} = \frac{0.69}{2} = 0.345 \tag{5.15}$$

$$MS_{AB} = \frac{SS_{AB}}{(a-1)(b-1)} = \frac{0.03}{4} = 0.008 \tag{5.16}$$

$$\text{Error } MS = \frac{\text{Error } SS}{(r-a)(ab-1)} = \frac{0.24}{16} = 0.015 \tag{5.17}$$

Step 6: Compute the F value for each source of variation as shown in Table 5.4.4.

Step 7: Compare the computed F with the tabular F value given in Appendix E. It appears that both the breed main effect and the diet main effect are significant. Furthermore, the result suggests that there is no significant interaction between breeds of swine and the diets. This indicates that the yield difference among the breeds was not significantly affected by the diets.

As in the previous sections, we can determine the degree of precision with which the treatments are compared by computing the coefficient of variation *cv* as:

$$cv = \frac{\sqrt{\text{Error } MS}}{\text{Grand mean}} \times 100 \tag{5.18}$$

$$= \frac{\sqrt{0.015}}{2.056} \times 100 = 5.96\%$$

TABLE 5.4.4

Analysis of Variance of Three Diets on Three Breeds of Swine

Source of Variation	Degree of Freedom	Sum of Squares	Mean Square	F
Replication	2	0.09	0.045	3.00[ns]
Treatment	8	0.84	0.105	7.00**
Breed (A)	(2)	0.12	0.060	4.00*
Diet (B)	(2)	0.69	0.345	23.00**
Breed x Diet	(4)	0.03	0.008	0.53[ns]
Error	16	0.24	0.015	
Total	26	1.17		

Note: Coefficient of variation (*cv*) = 5.96%.; [ns] = nonsignificant; * = significant at 5% level; ** = significant at 1% level.

As mentioned at the beginning of this section, most types of experimental plans are suitable for factorial experiments. However, as the number of factors or the levels of a factor increase, so do the number of treatment combinations. This creates problems of maintaining homogeneous replications in randomized blocks which, in turn, leads to an increase in the experimental error. To overcome this problem, a number of designs have been developed. Most of these designs tend to reduce the block size, thus controlling increases in error. However, such reductions in block size are accompanied by a sacrifice in the accuracy of certain treatment comparisons. The split-plot experiments discussed in Section 5.5 belong to this group of designs, in which a factor or a group of factors and their interactions may be sacrificed.

5.5 Split-Plot or Nested Designs

In two-factor experiments in which the number of treatments is significantly large and cannot easily be accommodated by the complete block design, the split-plot design is used. In this design the *main plot* is being split into smaller *subplots* so as to accommodate a more precise measurement of the subplot factor and its interaction with the main plot factor. This means that the precision achieved by the subplot factor is at the expense of precision for the measurement of the main factor. These designs are also called *nested* designs because of the repeated sampling and subsampling that occur in the design.

To better understand how subplot factor is measured more precisely than the main factor, let us denote replications by r and the subplot treatments by a. In performing a test, each main plot is tested r times, whereas each subplot treatment is tested $(r \times a)$ times. For example, if we have 4 replications and 3 subplots within each replication, then the main plot gets tested 4 times while the subplot is tested 12 times. This increase in the number of times a subplot is tested contributes to the increased precision of the subplots. For this reason, it is extremely important to keep in mind which factor should be assigned to the main plot and which to the subplot. For example, if the experimenter is mainly interested in more precision in factor A than B, then factor A should be assigned to the subplot, and factor B to the main plot.

As a feature of the split-plot design, the subplot treatments are not randomized over the whole large block, but only over the main plots as shown in Figure 5.5.3. This means that randomization of the subtreatments is newly done in each main plot, and the main treatments are randomized in the large blocks. As a consequence of this difference in randomization, the experimental error for the comparison between the levels of one treatment is not the same as that for the comparison between the levels of the other treatment. Thus, the experimental error of the subtreatment is smaller than that for the main treatments.

Randomization and layout: Given the nature of the design, in which main plots and subplots are the essential features, the randomization process requires that the main plot treatments be randomly assigned to the main plots first, followed by the assignment of the subplot treatments to each main plot. We may use any of the randomization schemes discussed in Chapter 4.

To provide an example of how such randomization and layout is carried out, let us denote the main plot treatments as *a*, the subplot treatments as *b*, and the number of replications as *r*. Assume that we are investigating the properties of three varieties of cotton bred for resistance to wilt as the subplot treatment and four dates of sowing as the main plot treatments, and the experiment is carried out in four replications. The steps in randomization and layout are given in the following text.

Step 1: Divide the experimental area into four blocks ($r = 4$). Further divide each block into four main plots as shown in Figure 5.5.1.

Step 2: Randomly assign the main plot treatments (four dates of sowing: D_1, D_2, D_3, and D_4) to the main plots among the four blocks or replications as shown in Figure 5.5.2.

Step 3: Divide each of the main plots into three subplots and randomly assign the cotton varieties (Factor *B*) to each subplot as shown in Figure 5.5.3.

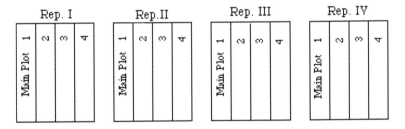

FIGURE 5.5.1
Layout of an experimental area into four blocks and four main plots.

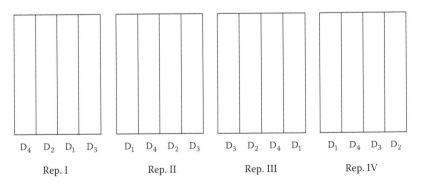

FIGURE 5.5.2
Random assignment of the four main plot treatments to each of the four replications.

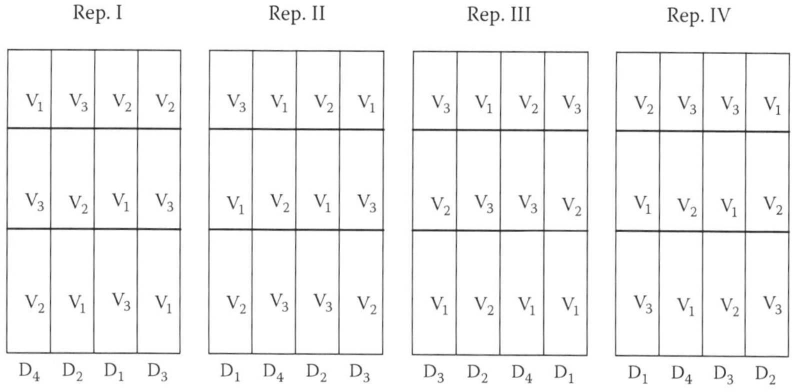

| Rep. I | Rep. II | Rep. III | Rep. IV |

FIGURE 5.5.3
A possible layout of a split-plot experiment with four main plot treatments, three subplot treatments, and four replications.

Analysis of variance: In performing the analysis of variance on an experiment in which the split-plot design is used, the experimenter must consider a separate analysis for the main plot (factor A) and the subplot factor (factor B), respectively. As an example of a split-plot experiment, consider the following study.

Example 5.5.1

In a study carried out by agronomists to determine if major differences in yield response to N fertilization exist among widely grown hybrids in the northern "Corn Belt," the subplot treatments were 5 hybrids of 105- to 110-day relative maturity (Pioneer 3747, 3732, Mo 17 x A634, A632 x LH 38, and LH 74 x LH 51), and the main plot treatments were N rates of 0-, 70-, 140-day and 210-lb/ac broadcast applied before planting. The study was replicated twice, and the findings are as shown in Table 5.5.1.

SOLUTION

Step 1: Calculate the replication totals (R), and the grand total (G) by first constructing a table for the nitrogen × replication totals shown in Table 5.5.2, and then a second table for the nitrogen × hybrid totals as shown in Table 5.5.3.

Step 2: Determine the degrees of freedom associated with each source of variation. The following sources of variation and the degrees of freedom are identified:

Replication *d.f.* $= r - 1 = 1$

Main plot factor (A) *d.f.* $= a - 1 = 3$

Error (a) *d.f.* $= (r - 1)(a - 1) = 3$

TABLE 5.5.1

Grain Yield Data of Five Corn Hybrids Grown with Four Levels of Nitrogen in a Split-Plot Experiment with Two Replications

Replication	Hybrid	N Rate, lb/ac			
		0	70	140	210
		Yield, bu/ac			
I.	P3747	130	150	170	165
	P3732	125	150	160	165
	Mo17 × A634	110	140	155	150
	A632 × LH38	115	140	160	140
	LH74 × LH51	115	170	160	170
II.	P3747	135	170	190	185
	P3732	150	160	180	200
	Mo17 × A634	135	155	165	175
	A632 × LH38	130	150	175	170
	LH74 × LH51	145	180	195	200

TABLE 5.5.2

The Replication x Nitrogen (Factor *A*) Totals Computed from Data in Table 5.5.1

Replication	N Rate, lb/ac				Rep. Total (*R*)
	0	70	140	210	
I.	595	750	805	790	2,940
II.	695	815	905	930	3,345
Nitrogen Total (*A*)	1,290	1,565	1,710	1,720	
Grand Total (*G*)					6,285

TABLE 5.5.3

The Hybrid x Nitrogen (Factor *B*) Totals Computed from Data in Table 5.5.1

Hybrid	Yield Total (*AB*)				Hybrid Total (*B*)
	N Rate, lb/ac				
	0	70	140	210	
P3747	265	320	360	350	1,295
P3732	275	310	340	365	1,290
Mo17 × A634	245	295	320	325	1,185
A632 × LH38	245	290	335	310	1,180
LH74 × LH51	260	350	355	370	1,335

Subplot factor (*B*) $d.f. = b - 1 = 4$

$A \times B$ $d.f. = (a - 1)(b - 1) = 12$

Error (*b*) $d.f. = a(r - 1)(b - 1) = 16$

Total $d.f. = rab - 1 = 39$

Step 3: Compute the various sums of squares for the main-plot analysis by first computing the correction factor, using Equation 5.6 as follows:

$$C = \frac{G^2}{rab} \tag{5.6}$$

$$= \frac{(6,285)^2}{2 \times 4 \times 5}$$

$$= 987{,}530.62$$

$$\text{Total } SS = \sum X^2 - C \tag{5.7}$$

$$= \left[(130)^2 + (125)^2 + \ldots + (200)^2 \right] - 987{,}530.62$$

$$= 1{,}007{,}775 - 987{,}530.62$$

$$= 20{,}244.38$$

$$\text{Replication } SS = \frac{\sum R^2}{ab} - C \tag{5.8}$$

$$= \frac{(2{,}940)^2 + (3{,}345)^2}{20} - 987{,}530.62$$

$$= \frac{8{,}643{,}600 + 11{,}189{,}025}{20} - 987{,}530.62$$

$$= 991{,}631.25 - 987{,}530.62$$

$$= 4{,}100.63$$

$$SS_A \text{(Nitrogen)} = \frac{\sum A^2}{rb} - C \tag{5.19}$$

$$= \frac{(1{,}290)^2 + (1{,}565)^2 + (1{,}710)^2 + (1{,}720)^2}{2 \times 5} - 987{,}530.62$$

$$= 999{,}582.50 - 987{,}530.62$$

$$= 12{,}051.88$$

$$\text{Error } SS(a) = \frac{\sum (RA)^2}{b} - \left(C + \text{Rep. } SS + SS_A\right) \qquad (5.20)$$

$$= \frac{(595)^2 + (750)^2 + \ldots + (930)^2}{5} - \left(987,530.62 + 4,100.63 + 12,051.88\right)$$

$$= 1,003,965 - 1,003,683.13$$

$$= 281.87$$

Step 4: Compute the various sums of squares for the subplot (hybrid) analysis as shown below:

$$SS_B(\text{Hybrid}) = \frac{\sum B^2}{ra} - C \qquad (5.21)$$

$$= \frac{(1,295)^2 + (1,290)^2 + (1,185)^2 + (1,180)^2 + (1,335)^2}{2 \times 4} - 987,530.62$$

$$= \frac{7,919,975}{8} - 987,530.62$$

$$= 989,996.87 - 987,530.62$$

$$= 2,466.25$$

$$SS_{AB}(\text{Nitrogen} \times \text{Hybrid}) = \frac{\sum (AB)^2}{r} - \left(C + SS_A + SS_B\right) \qquad (5.22)$$

$$= \frac{(265)^2 + (320)^2 + \ldots + (370)^2}{2} - \left(987,530.62 + 12,051.88 + 2,466.25\right)$$

$$= \frac{2,005,825}{2} - 1,002,048.75$$

$$= 1,002,912.50 - 1,002,048.75$$

$$= 863.75$$

Error SS (b) = TSS All other sum of squares $\qquad\qquad$ (5.23)

$$= 20{,}244.38 - (4{,}100.63 + 12{,}051.88 + 281.87 + 2{,}466.25 + 863.75)$$

$$= 480$$

Step 5: Calculate the mean square for each variation as:

$$\text{Replication } MS = \frac{\text{Rep. } SS}{r-1} = \frac{4{,}100.63}{1} = 4{,}100.63 \qquad (5.24)$$

$$MS_A = \frac{SS_A}{a-1} = \frac{12{,}051.88}{4-1} = 4{,}017.29 \qquad (5.25)$$

$$\text{Error }(a)MS = \frac{\text{Error } SS\ (a)}{(r-1)(a-1)} = \frac{281.87}{(1)(3)} = 93.96 \qquad (5.26)$$

$$MS_B = \frac{SS_B}{b-1} = \frac{2{,}466.25}{5-1} = 616.56 \qquad (5.27)$$

$$MS_{AB} = \frac{SS_{AB}}{(a-1)(b-1)} = \frac{863.75}{(3)(4)} = 71.98 \qquad (5.28)$$

$$\text{Error }(b)MS = \frac{\text{Error } SS\ (b)}{a(r-1)(b-1)} = \frac{480}{4(1)(4)} = 30 \qquad (5.29)$$

Step 6: Compute the F value for each source of variation as shown in Table 5.5.4.

TABLE 5.5.4

Analysis of Variance of Hybrid $\times$ N Response Experiment in a Split-Plot Design

Source of Variation	Degree of Freedom	Sum of Squares	Mean Square	F
Replication	1	4,100.63	4,100.63	
Nitrogen (A)	3	12,051.88	4,017.29	42.76**
Error (a)	3	281.87	93.96	
Hybrid (B)	4	2,466.25	616.56	20.55**
Nitrogen $\times$ Hybrid (A $\times$ B)	12	863.75	71.98	2.40ns
Error (b)	16	480	30.00	
Total	39	20,244.38		

Note: $cv(a) = 6.17\%$, $cv(b) = 3.49\%$; ns = nonsignificant; ** = significant at 1% level.

Step 7: Compare the computed F with the tabular F value given in Appendix E. For example, the tabular F for the effect of nitrogen is 9.28 at the 5% level of significance and 29.46 at the 1% level of significance. This indicates that grain yield was significantly affected by nitrogen. Similarly, the results show that in both replications grain yield was significantly affected by the hybrid. However, the interaction between N rate and hybrid were not significant.

Finally, we will compute the coefficient of variation (*cv*) for the main plot and the subplot using Equation 5.17 as shown below:

$$cv = \frac{\sqrt{\text{Error } (a)MS}}{\text{Grand Mean}} \times 100$$

$$= \frac{\sqrt{93.96}}{157.13} \times 100 = 6.17\%$$

$$cv = \frac{\sqrt{\text{Error } (b)MS}}{\text{Grand Mean}} \times 100 \qquad (5.18)$$

$$= \frac{\sqrt{30}}{157.13} \times 100 = 3.49\%$$

As before, the value of the coefficient of variation (*cv*) indicates the degree of precision with which the factors are compared and is a good measure of the reliability of the experiment. Once again, the coefficient of variation expresses the experimental error as a percentage of the mean. Thus, the smaller the value of *cv*, the greater the reliability of the experiment. As is apparent from our computations, the *cv*(*b*) is smaller than the *cv*(*a*), as would be expected. We had mentioned earlier that a feature of the subplot design is the sacrifice of precision on the main plot in order to achieve greater precision in the measurement of the factor assigned to the subplot.

Missing data: If for any reason (loss of animal during the experiment, crop destroyed by accident or a natural disaster, or errors made in recording the data), data on an observation are missing when an experiment is conducted, use is made of the following formula developed by Anderson (1946):

$$\hat{M} = \frac{rU + b\left(A_i B_j\right) - \left(A_i\right)}{(r-1)(b-1)} \qquad (5.30)$$

where

$\hat{M}$ = the estimated value of the missing observation

r = number of replications

U = total for unit containing the missing observation

b = the level of subplot factors

$A_i B_j$ = total of observed values of the treatment combination that contain the missing observation

A_i = total of all observations that receive the *i*th level of *A*

To see how the above formula is applied, assume that in Example 5.5.1 the observation for hybrid P3732 receiving 140 lb of nitrogen fertilizer in replication I (which is 160 bu) is missing. To find the estimated value for the missing observation, we have to first recompute the totals by subtracting the value of the missing observation as shown below:

$$U = 805 - 160 = 645$$
$$A_i = 1{,}710 - 160 = 1{,}550$$
$$A_i B_j = 340 - 160 = 180$$
$$r = 2 \text{ replications}$$
$$b = 5, \text{ the level of the subplot factor}$$

$$\hat{M} = \frac{2(645) + 5(180) - (1{,}550)}{(2-1)(5-1)}$$

$$= \frac{1{,}290 + 900 - 1{,}550}{4}$$

$$= 160$$

Interestingly, the estimated value is the same as the actual value. If several observations such as *a, b, c, d, ...,* etc., are missing, we first guess the values of all the missing observations except *a*. We then use the preceding equation to find an approximation for *a*. With the estimated value for *a*, we use the estimating equation again to find the value for *b*, and so on. This iterative procedure is continued until we have found an estimated value for all the missing observations. Keep in mind that in performing the analysis of variance, one degree of freedom is subtracted from the total and error sum of squares for each missing observation. For a more elaborate explanation on how to deal with more than one missing observation, the reader is referred to Tocher (1952) and Bennett and Franklin (1954).

5.6   Strip-Plot Design

It is sometimes desirable to get more precise information on the interaction between factors than on their main effect. The strip-plot is a two-factor design that allows for greater precision in the measurement of the interaction effect while sacrificing the degree of precision on the main effects. In measuring the interaction effect between two factors, the experimental area is divided into three plots, namely, the *vertical-strip plot*, the *horizontal-strip plot*, and the *intersection plot*. We allocate factors A and B, respectively, to the vertical- and horizontal-strip plots, and allow the intersection plot to accommodate the interaction between these two factors. As in the split-plot design, the vertical and the horizontal plots are perpendicular to each other. However, in the strip-plot design the relationship between the vertical and the horizontal plot sizes is not as distinct as the main and subplots were in the split-plot design. The intersection plot, which is one of the characteristics of the design, is the smallest in size. In the text following, the randomization and layout of the strip-plot design is explained.

Randomization and layout: Given the nature of the design, in which we have specifically divided experimental area into three separate plots, the randomization procedure calls for the allocation of each factor to its respective plots separately. This means that the vertical and the horizontal factor each would have to be randomly assigned to their plots. To illustrate the steps in the randomization and layout, we have denoted factor A as the horizontal factor and B as the vertical factor. The levels of each factor and the numbers of replications are denoted by a, b, and r, respectively.

Assume we have a two-factor experiment testing four cultivars of wheat (horizontal factor) and four nitrogen rates (vertical factor) in a strip-plot design involving four replications. The steps in randomization are:

Step 1: Divide the experimental area into four blocks or replications ($r = 4$), in which each block is further divided into four horizontal strips. Randomly assign the four treatments ($a = 4$) to each of the strips, following any of the randomization schemes outlined in Chapter 4. Figure 5.6.1 shows the layout for this step.

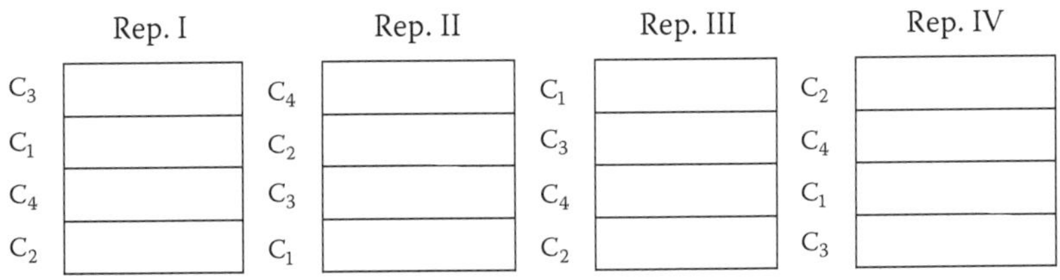

FIGURE 5.6.1
Random allocation of the four cultivars (C_1, C_2, C_3, and C_4) to the horizontal strip with four replications.

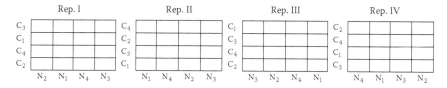

FIGURE 5.6.2

A strip-plot layout with random allocation of the four cultivars (C_1, C_2, C_3, and C_4) to the horizontal strip, and four nitrogen treatments (N_1, N_2, N_3, and N_4) to the vertical strip with four replications.

Step 2: Divide each block or replication into four vertical strips, and randomly assign the four nitrogen treatments ($b = 4$) following a randomization scheme discussed in Chapter 5. This is shown below in Figure 5.6.2.

Analysis of variance: To perform the analysis of variance on a strip-plot experiment, we must analyze the data for the horizontal and vertical factors, and take account of the interaction. Thus, the analysis would involve the horizontal-factor analysis, vertical-factor analysis, and the interaction analysis. Each of these analyses will be performed in Example 5.6.1.

Example 5.6.1

In a study to determine the effect of N rate on soft winter wheat yields, agronomists have used a strip-plot design with four soft red winter wheat cultivars — Arthur 71, Auburn, Caldwell, and Compton — with four replications. The study was conducted on Parr silt loam soil, and the yield in bushels per acre is given in Table 5.6.1.

TABLE 5.6.1

Grain Yields of Four Soft Red Winter Wheat Cultivars Grown at Four Nitrogen Fertilizer Rates

Cultivar	Nitrogen Rate (lb/ac)	Grain Yield (bu/ac)			
		Rep. I	Rep. II	Rep. III	Rep. IV
	40	72	74	76	70
Arthur 71	80	76	75	74	78
	120	72	74	73	75
	160	74	76	82	86
	40	60	62	64	65
Auburn	80	61	63	69	68
	120	70	72	69	70
	160	72	70	82	86
	40	75	73	72	80
Caldwell	80	77	78	77	82
	120	80	82	86	88
	160	84	82	84	89
	40	65	68	63	72
Compton	80	68	72	74	76
	120	69	68	70	72
	160	73	75	74	76

TABLE 5.6.2

The Replication × Cultivar (Factor A) Table of Yield Totals Computed from Data in Table 5.6.1

Cultivar	Rep. I	Rep. II	Rep. III	Rep. IV	Cultivar Total (A)
		Yield Total (RA)			
Arthur 71	294	299	305	309	1,207
Auburn	263	267	284	289	1,103
Caldwell	316	315	319	339	1,289
Compton	275	283	281	296	1,135
Rep. Total (R)	1,148	1,164	1,189	1,233	
Grand Total (G)					4,734

TABLE 5.6.3

The Replication × Nitrogen (Factor B) Totals Computed from Data in Table 5.6.1

Nitrogen	Rep. I	Rep. II	Rep. III	Rep. IV	Nitrogen Total (B)
		Yield Total (RB)			
40	272	277	275	287	1,111
80	282	288	294	304	1,168
120	291	296	298	305	1,190
160	303	303	322	337	1,265

TABLE 5.6.4

The Cultivar × Nitrogen Table of Totals Computed from Data in Table 5.6.1

	Yield Total (AB)			
	N Rate, lb/ac			
Cultivar	40	80	120	160
Arthur 71	292	303	294	318
Auburn	251	261	281	310
Caldwell	300	314	336	339
Compton	268	290	279	298

SOLUTION

Step 1: Calculate the replication totals (R), the grand total (G), and the cultivar total (A) by first constructing a table for the replication × horizontal factor (cultivar) totals (RA) as shown in Table 5.6.2, and then a second table for the replication × vertical factor (nitrogen) totals (RB) as shown in Table 5.6.3. Finally, construct a horizontal × vertical factor table of totals (AB) as shown in Table 5.6.4.

Step 2: Determine the degrees of freedom associated with each source of variation. The following sources of variation and the degrees of freedom are identified.

Replication *d.f.* = $r - 1 = 3$
Horizontal factor (A) *d.f.* = $a - 1 = 3$
Error (a) *d.f.* = $(r - 1)(a - 1) = 9$
Vertical factor (B) *d.f.* = $b - 1 = 3$
Error (b) *d.f.* = $(r - 1)(b - 1) = 9$
A x B *d.f.* = $(a - 1)(b - 1) = 9$
Error (c) = $(r - 1)(a - 1)(b - 1) = 27$
Total *d.f.* = $rab - 1 = 63$

Step 3: Compute the correction factor and the total sum of squares as follows:

$$C = \frac{G^2}{rab} \tag{5.6}$$

$$= \frac{(4,734)^2}{4 \times 4 \times 4}$$

$$= 350,168.1$$

$$\text{Total } SS = \sum X^2 - C \tag{5.7}$$

$$= \left[(72)^2 + (74)^2 + \dots + (78)^2 \right] - 350,168.1$$

$$= 353,034 - 350,168.1$$

$$= 2,865.90$$

Step 4: Calculate the replication, cultivar, and error sum of squares for the horizontal factor analysis, using Equation 5.8 as given below:

$$\text{Replication } SS = \frac{\sum R^2}{ab} - C \tag{5.8}$$

$$= \frac{(1,148)^2 + (1,164)^2 + (1,189)^2 + (1,233)^2}{16} - 350.168.1$$

$$= \frac{5,606,810}{16} - 350,168.1$$

$$= 350,425.62 - 350,168.1$$

$$= 257.52$$

$$SS_A(\text{Cultivar}) = \frac{\sum A^2}{rb} - C \tag{5.31}$$

$$= \frac{(1,207)^2 + (1,103)^2 + (1,289)^2 + (1,135)^2}{16} - 350.168.1$$

$$= 351{,}450.25 - 350{,}168.1$$

$$= 1{,}282.15$$

$$\text{Error } SS\,(a) = \frac{\sum (RA)^2}{b} - (C + \text{Rep. } SS + SS_A) \tag{5.20}$$

$$= \frac{(294)^2 + (263)^2 + \ldots + (296)^2}{4} - (350{,}168.1 + 257.52 + 1{,}282.15)$$

$$= 351{,}758 - 351{,}707.77$$

$$= 50.23$$

Step 5: Compute the various sums of squares for the vertical factor analysis as shown below:

$$SS_B(\text{Nitrogen}) = \frac{\sum B^2}{ra} - C \tag{5.32}$$

$$= \frac{(1,111)^2 + (1,168)^2 + (1,190)^2 + (1,265)^2}{16} - 350.168.1$$

$$= 350{,}929.37 - 350{,}168.1$$

$$= 761.27$$

$$\text{Error } SS\,(b) = \frac{\sum (RB)^2}{a} - (C + \text{Rep. } SS + SS_B) \tag{5.33}$$

$$= \frac{(272)^2 + (282)^2 + \ldots + (337)^2}{4} - (350{,}168.1 + 257.52 + 761.27)$$

$$= 351{,}256 - 351{,}186.89$$

$$= 69.11$$

Step 6: Compute the sums of squares for the interaction analysis as shown in the following text.

$$SS_{AB}(\text{Cultivar} \times \text{Nitrogen}) = \frac{\sum AB^2}{r} - \left(C + SS_A + SS_B\right) \tag{5.34}$$

$$= \frac{(292)^2 + (251)^2 + \ldots + (298)^2}{4} - \left(350,168.1 + 1,282.15 + 761.27\right)$$

$$= \frac{1,409,798}{4} - 352,211.52$$

$$= 352,449.50 - 352,211.52$$

$$= 237.98$$

Error SS (c) = TSS – All other sum of squares(5.35)

$$= 2,865.90 - (257.52 + 1,282.15 + 50.23 + 761.27 + 69.11 + 237.98)$$

$$= 2,865.90 - 2,658.26$$

$$= 207.64$$

Step 7: Calculate the mean square for each source of variation as:

$$\text{Replication } MS = \frac{\text{Rep. } SS}{r-1} = \frac{257.52}{3} = 85.84$$

$$MS_A = \frac{SS_A}{a-1} = \frac{1,282.15}{4-1} = 427.38$$

$$\text{Error } (a)MS = \frac{\text{Error } SS\ (a)}{(r-1)(a-1)} = \frac{50.23}{9} = 5.58$$

$$MS_B = \frac{SS_B}{b-1} = \frac{761.27}{3} = 253.76$$

$$\text{Error } (b)MS = \frac{\text{Error } SS\ (b)}{(r-1)(b-1)} = \frac{69.11}{9} = 7.68$$

TABLE 5.6.5

Analysis of Variance of Wheat Cultivar × N Response Experiment in a Strip-Plot Design

Source of Variation	Degree of Freedom	Sum of Squares	Mean Square	F
Replication	3	257.52	85.84	
Cultivar (A)	3	1,282.15	427.38	76.59**
Error (*a*)	9	50.23	5.58	
Nitrogen (B)	3	761.27	253.76	33.04**
Error (*b*)	9	69.11	7.68	
Cultivar × Nitrogen $(A \times B)$	9	237.98	26.44	3.43**
Error (*c*)	27	207.64	7.69	
Total	63	2,865.90		

Note: *cv*(*a*) = 3.19%, *cv*(*b*) = 3.75%, *cv*(*c*) = 3.75%; ** = significant at 1% level.

$$MS_{AB} = \frac{SS_{AB}}{(a-1)(b-1)} = \frac{237.98}{9} = 26.44$$

$$\text{Error } (c)MS = \frac{\text{Error } SS\ (c)}{(r-1)(a-1)(b-1)} = \frac{207.64}{27} = 7.69$$

Step 8: Compute the *F* value for each source of variation as shown in Table 5.6.5.

Step 9: Compare the computed *F* with the tabular *F* value given in Appendix E. For example, the tabular *F* for the effect of cultivar is 3.86 at 5% level of significance and 6.99 at 1% level of significance. This indicates that grain yield was significantly affected by the cultivar. The results also show that grain yield was significantly affected by nitrogen fertilizer. Similarly, the tabular *F* for the interaction between cultivar and nitrogen is 2.25 at the 5% level of significance and 3.14 at the 1% level of significance. Thus, the results show a significant interaction between N rate and cultivar, indicating that the N recommendation developed for one cultivar cannot be applied to the other cultivars tested in this experiment.

Finally, we will compute the coefficient of variation (*cv*) for the three error mean squares using Equation 5.18 as shown in the following text.

$$cv = \frac{\sqrt{\text{Error } (a)MS}}{\text{Grand mean}} \times 100 \qquad (5.18)$$

$$= \frac{\sqrt{5.58}}{73.96} \times 100 = 3.19\%$$

$$cv\left(b\right) = \frac{\sqrt{\text{Error }(b)MS}}{\text{Grand Mean}} \times 100$$

$$= \frac{\sqrt{7.68}}{73.96} \times 100 = 3.75\%$$

$$cv\left(c\right) = \frac{\sqrt{\text{Error }(c)MS}}{\text{Grand Mean}} \times 100$$

$$= \frac{\sqrt{7.69}}{73.96} \times 100 = 3.75\%$$

As before, the value of the coefficient of variation (cv) indicates the degree of precision with which the factors are compared and is a good measure of the reliability of the experiment. $cv(a)$ is a measure of the precision associated with the horizontal factor, and $cv(b)$ with the vertical factor. The degree of precision in the measurement of the interaction between horizontal and vertical factor is shown by $cv(c)$.

Missing data: To illustrate how the value of a missing observation (a_ib_j) is estimated in a strip-plot experiment, assume that the value is missing on the Auburn cultivar in Example 5.6.1, receiving 120 lb of nitrogen per acre in replication II, which is 72. To find the missing value, the following equation is used.

$$\hat{M} = \frac{a\{b[A_iB_j] - A_i\} + r(aH + bV - B) - bv + G}{(a-1)(b-1)(r-1)} \tag{5.36}$$

where

$\hat{M}$ = estimated value of the missing observation

a = level of horizontal factor

b = level of vertical factor

r = number of replications

A_iB_j = total of observed values of the treatment combination that contain the missing observation

A_i = total of all observations that receive the ith level of horizontal factor A

H = total of observed values of the horizontal strip that contain the missing observation

B = total of observed values of the replication that contain the missing observation

V = total of observed values of the vertical strip that contain the missing observation

v = total of observed value of the specific level of the vertical factor that contains the missing observation

G = total of all observed values

To estimate this missing value, the parameters to work with are:

$$a = 4$$
$$b = 4$$
$$r = 4$$
$$A_iB_j = 281 - 72 = 209$$
$$A_i = 1{,}103 - 72 = 1031$$
$$H = 267 - 72 = 195$$
$$B = 1{,}164 - 72 = 1{,}092$$
$$V = 296 - 72 = 224$$
$$v = 1{,}190 - 72 = 1{,}118$$
$$G = 4{,}734 - 72 = 4{,}662$$

Thus, we have:

$$\hat{M} = \frac{4\{[4(209) - 1{,}031]\} + 4[4(195) + 4(224) - 1{,}092] - (4)(1{,}118) + 4{,}662}{(4-1)(4-1)(4-1)}$$

$$= \frac{6998 - 5251}{27} = 64.66$$

Once estimated, the missing value is placed in the table of observed values, and the analysis of variance is performed in the usual way. Once again, remember to subtract one degree of freedom from the total and error before computing the F value.

References and Suggested Readings

Anderson, R.L. 1946. Missing-plot techniques. *Biomed. Bull.* 2: 41–47.

Bennett, C.A. and Franklin, N.L. 1954. *Statistical Analysis in Chemistry and the Chemical Industry.* New York: John Wiley & Sons.

Bundy, L.G. and Carter, P.R. 1988. Corn hybrid response to nitrogen fertilization in the northern corn belt. *J. Prod. Agric.* 1(2): 99–104.

Cochran, W.G. and Cox, G.M. 1957. *Experimental Designs.* 2nd ed. New York: John Wiley & Sons. pp. 168–170. chap. 5.

Goulden, C.H. 1952. *Methods of Statistical Analysis.* 2nd ed. New York: John Wiley & Sons. 94 pp.

Tocher, K.D. 1952. The design and analysis of block experiments. *J.R. Stat. Soc. Ser. B* 14: 45–100.

Exercises

1. An animal scientist wishes to conduct a split-plot design experiment in which two factors (two breeds of swine and four different diets) are to be arranged in four blocks. The objective of the experiment is to see if the four diets are different from one another. Show the layout of this experiment, keeping the objective in mind.

2. The omnivorous looper, larva of the moth *Sabulodes aegrotata* (Guenee), is a sporadic pest of California's avocado. Researchers are testing four chemicals to control the omnivorous looper in avocado orchards. The treatments were assigned to the subplots, whereas the concentration rates of active ingredients were assigned to the main plots. The results of the experiment are shown in the following table.

Insecticide Used for Omnivorous Looper Control in San Diego County

Treatment	Number of Larvae per Tree Sample at 14-d Posttreatment Interval		
	Rep. I	Rep. II	Rep. III
Active Ingredient 1 lb/ac			
Dylox 80SP	8	10	11
Kryocide 8F	12	9	10
Lannate L	2	4	3
Orthene 75SP	1	2	4
Control	14	18	20
Active Ingredient 2 lb/ac			
Dylox 80SP	5	9	8
Kryocide 8F	6	4	5
Lannate L	1	3	4
Orthene 75SP	1	2	4
Control	17	19	24
Active Ingredient 4 lb/ac			
Dylox 80SP	3	3	5
Kryocide 8F	5	4	7
Lannate L	2	1	3
Orthene 75SP	1	2	1
Control	12	15	22

(a) Perform the analysis of variance on the data.

(b) What conclusions can be drawn from this analysis?

(c) Determine the degree of precision with which the treatments are compared.

3. In a study to determine the viability of triticale as a feedstuff for poultry, animal researchers conducted a split-plot experiment. They gathered the following data on the weight gain of the birds fed triticale and control.

Cumulative Growth of Broilers Fed Control or Triticale Diets

Diet	Weight Gain over Days (lb/bird)			
	Rep. I	Rep. II	Rep. III	Rep. IV
7 d				
Control	0.25	0.31	0.22	0.28
Low-triticale	0.26	0.28	0.29	0.27
Medium-triticale	0.22	0.21	0.24	0.22
14 d				
Control	0.65	0.70	0.59	0.68
Low-triticale	0.73	0.75	0.79	0.69
Medium-triticale	0.70	0.69	0.65	0.68
21 d				
Control	1.25	1.31	1.24	1.27
Low-triticale	1.29	1.28	1.29	1.30
Medium-triticale	1.22	1.21	1.23	1.22
28 d				
Control	1.35	1.33	1.29	1.28
Low-triticale	1.46	1.38	1.40	1.39
Medium-triticale	1.42	1.31	1.34	1.36

(a) Perform the analysis of variance on the data.

(b) Compute the two coefficients of variation, one corresponding to the main plot and the other to the subplot analysis.

4. The western flower thrip, *Frankliniella occidentalis* (Pergande), is a major threat to the floricultural crops around the world. Researchers have conducted a split-plot experiment in which biological control has been used to control the thrips. In this experiment the objective is to determine if using predaceous mites such as *A. cucumeris* and *A. barkeri*, separately and in combination, would reduce the thrip population. Using chrysanthemum plants, the researchers have gathered the following data.

Impact of Predaceous Mites on the Number of Western Flower Thrips Infesting Chrysanthemum

Replication	Day	*Amblyseius cucumeris*	*Amblyseius barkeri*	*A. cucumeris* + *A. barkeri*	Control
I	Day 10	3	2	1	7
	Day 17	2	4	2	9
	Day 24	1	2	2	21
	Day 31	6	3	8	38
II	Day 10	2	1	2	9
	Day 17	3	4	1	10
	Day 24	5	2	3	25
	Day 31	3	5	7	34
III	Day 10	1	3	1	8
	Day 17	3	5	1	14
	Day 24	2	1	3	23
	Day 31	5	2	9	32
IV	Day 10	2	4	2	10
	Day 17	1	2	2	11
	Day 24	4	5	5	21
	Day 31	6	6	10	39

(a) Analyze the data.

(b) Compute the coefficient of variation for the two categories of data.

5. Suppose in the preceding experiment, the researchers lost some of the observations as shown in the following table. The missing data is shown as (—) in each column of the data set.

Impact of Predaceous Mites on the Number of Western Flower Thrips Infesting Chrysanthemum

Replication	Day	Amblyseius cucumeris	Amblyseius barkeri	A. cucumeris + A. barkeri	Control
I	Day 10	3	2	1	7
	Day 17	2	4	2	9
	Day 24	1	2	2	—
	Day 31	—	3	8	38
II	Day 10	2	1	2	9
	Day 17	3	4	1	10
	Day 24	5	2	—	25
	Day 31	3	5	7	34
III	Day 10	1	3	1	8
	Day 17	3	—	1	14
	Day 24	2	1	3	23
	Day 31	5	2	9	32
IV	Day 10	2	4	2	10
	Day 17	1	2	—	11
	Day 24	4	5	5	21
	Day 31	6	6	10	39

(a) How would you estimate the missing values, and perform the analysis of variance?

(b) Is there a significant difference between the results obtained in Problem 3 and the present case?

6. In a strip-plot design experiment, researchers were interested more in the interaction between the amount of fertilizer and rainfall than in either of these factors alone. They have hypothesized that fertilizers produce more range forage in drought than normal years. The data collected from the experiment is shown in the following table.

Range Forage Yields per Acre from Fertilizer Application in Two Normal Precipitation Years and Two Drought Years

Rainfall (inches)	Fertilizer Application (lb/ac)	Forage Yield (lb/ac)				
		Rep. I	Rep. II	Rep. III	Rep. IV	Rep. V
	Ammonium Sulfate					
17.5	(300 lb)	4500	4355	4100	4600	4250
12.8		4100	4235	4005	4095	4050
6.0		1400	1325	1200	1375	1390
6.5		1000	1025	995	1020	1000
	Ammonium Nitrate					
17.5	(180 lb)	4300	4235	4025	4165	4120
12.8		4050	4110	4033	4195	4250
6.0		1630	1624	1595	1675	1595
6.5		1060	1028	1000	1029	1016
	16-20-0					
17.5	(375 lb)	4250	4151	4170	4300	4290
12.8		3700	3935	3205	3495	3850
6.0		1400	1302	1296	1315	1390
6.5		1100	1025	991	1022	1019
	Control					
17.5	(None)	3950	3801	3905	3390	3890
12.8		4101	4035	4205	4007	4100
6.0		901	897	906	942	899
6.5		698	674	688	700	645

(a) Perform the analysis of variance.

(b) Compute the coefficient of variation for the problem.

(c) Assume that the observation for the plot receiving ammonium nitrate when rainfall is 12.8 in. in replication II (which is 4110) is missing. How do you estimate this missing value before the analysis is performed?

6

Three (or More)-Factor Experimental Designs

6.1 Introduction

In the previous chapter, we discussed how two factors could be studied in the same environment simultaneously. The concepts developed in Chapter 4 and Chapter 5 could be extended to include three or more factors. Studying several factors together offers the advantage of simultaneously examining the interaction between factors. Such interactions, if statistically significant, take precedence over the main effects when reporting the results of an experiment. It should be kept in mind, however, that the number of treatments to study increases rapidly as more factors are added in the experiment. This point was made in Chapter 5 when discussing factorial experiments. Additionally, as the number of factors increases in an experiment, so does the interaction between factors. This complicates the analysis and great care is required in interpreting results.

In this chapter, we shall consider the simplest arrangement of a factorial, or 2^3 experiment, in which the experiment includes three factors (A, B, and C) each at two levels (a_1, a_2, b_1, b_2, and c_1, c_2). In a three-factor experiment, we have to estimate and test three main effects A, B, and C; three first-order (two-factor) interactions, AB, BC, and AC; and one second-order (three-factor) interaction, ABC. Additionally, we shall consider blocking in cases in which the blocks are not large enough to contain all the treatments as was the case with split-plot design, in which each whole plot contained only one level of each of the main treatments.

Inclusion of an additional factor to a 2^2 factorial experiment does not change the principle of the design or the randomization and layout of the experiment. The data matrix for a three-factor design is presented in Table 6.1.1.

The underlying model of the three-factor experiment is of the form:

$$X_{ijkl} = \mu + \alpha_i + \beta_j + \gamma_k + \left(\alpha\gamma\right)_{ij} + \left(\beta\gamma\right)_{jk} + \left(\alpha\beta\gamma\right)_{ijk} + \varepsilon_{ijkl} \tag{6.1}$$

TABLE 6.1.1

Data Matrix for a Three-Factor Design

		Factor C				
Factor A	Factor B	1	2	3	...	c
1	1	X_{1111}	X_{1112}	X_{1113}	..	X_{111c}
	2	X_{1121}	X_{1122}	X_{1123}	..	X_{112c}
	3	X_{1131}	X_{1132}	X_{1133}	..	X_{113c}
	.	.	.		..	.
	b	X_{11b1}	X_{11b2}	X_{11b3}	..	X_{11bc}
2	1	X_{1211}	X_{1212}	X_{1213}	..	X_{121c}
	2	X_{1221}	X_{1222}	X_{1223}	..	X_{122c}
	3	X_{1231}	X_{1232}	X_{1233}	..	X_{123c}
	.	.	.	.	..	.
	..		..			..
	b	X_{12b1}	X_{12b2}	X_{12b3}	..	X_{12bc}
a	1	X_{1a11}	X_{1a12}	X_{1a13}	..	X_{1a1c}
	2	X_{1a21}	X_{1a22}	X_{1a23}	..	X_{1a2c}
	3	X_{1a31}	X_{1a32}	X_{1a33}	..	X_{1a3c}
	.	.	.	.	..	.
	b	X_{1ab1}	X_{1ab2}	X_{1ab3}	..	X_{1abc}

For $i = 1, \ldots, a$; $j = 1, \ldots, b$; $k = 1, \ldots, c$, $l = 1, \ldots, n$

where

X_{ijkl} = the lth yield observation of the ijkth treatment combination

μ = overall mean, that is, the average of the population mean responses for abc treatments

α_i = effect of the ith level of the first factor (A), averaged over b levels of the second factor (B). The ith level of the first factor adds α_i to the overall mean μ

βj = average effect of the jth level of factor B

γ_k = average effect of the kth level of factor C

$(\alpha\beta)_{ij}$ = interaction between the ith level of factor A and the jth level of factor B, average of the population means for the c treatments that involve the ith level of factor A and jth level of factor B minus $\mu + \alpha_i + \beta_j$

$(\alpha\gamma)_{ik}$ = interaction of the ith level of factor A with the kth level of factor C, the average of the population means for the b treatments involving the ith level of factor A and the kth level of factor C minus $\mu + \alpha_i + \beta_j$

$(\beta\gamma)_{jk}$ = interaction of the jth level of factor B with the kth level of factor C, the average of the population means for the a treatments involving the jth level of factor B and the kth level of factor C minus $\mu + \alpha_i + \beta_j$

$(\alpha\beta\gamma)_{ijk}$ = interaction between the ith level of factor A, the jth level of factor B, and the kth level of factor C, population mean response for the ijkth treatment minus $\mu + \alpha_i + \beta_j + \gamma_k + (\alpha\beta)_{ij} + (\alpha\gamma)_{ik} + (\beta\gamma)_{jk}$

ε_{ijkl} = the random error component

The preceding equation specifies that there are n observations on each of the abc treatment combinations, a levels of factor A, b levels of factor B, and c levels of factor C. The error component, ε_{ijkl}, is assumed to be independently and normally distributed with zero mean and variance (σ_e^2) within each of the abc treatment combinations. The aforementioned quantities have all been encountered in Chapter 5 with the exception of $(\alpha\beta\gamma)_{ijk}$, which we designate as a second-order interaction (three independent variables) to distinguish it from first-order interactions that involve only two independent variables.

In the following sections of this chapter, we will illustrate different designs with three or more factors and how they are analyzed.

6.2 Split-Split-Plot Design

In Chapter 5 (Section 5.5) you were introduced to a two-factor design called the split-plot design. An extension of this design is called the *split-split-plot design*, in which the subplot is further divided to include a third factor in the experiment. The design allows for three levels of precision associated with the three factors. That is, the degree of precision associated with the main factor is lowest, whereas the degree of precision associated with the sub-subplot is highest. Thus, the design is suited for three-factor experiments in which different levels of precision are desired with different factors.

Randomization: In the split-split-plot design, the main plot is divided into subplots, and the subplots are further divided into sub-subplots. The randomization process requires that the levels of the main plot factor be applied at random to the blocks. Similarly, the levels of the subplot factor should be applied randomly within the main plots, using a separate randomization in each main plot. Finally, the levels of the last factor should be assigned to the sub-subplots at random, using a separate randomization in each subplot. To illustrate randomization and layout, we will use a $2 \times 2 \times 3$ factorial experiment replicated four times in a randomized complete block design arranged in a split-split-plot layout. Factor A (two hybrids, denoted as H_1 and H_2) is assigned to the main plot. Factor B (two row spacings, denoted as S_1 and S_2) is assigned to the subplot. Factor C (three plant densities, denoted as D_1, D_2, and D_3) is assigned to the sub-subplot. The steps followed are given below:

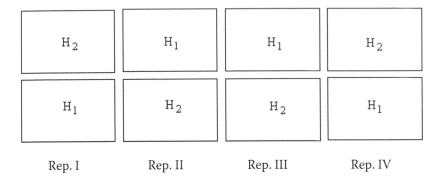

FIGURE 6.2.1
Random assignment of factor *A* (hybrids, H_1 and H_2) to the two main plots in each of the four replications.

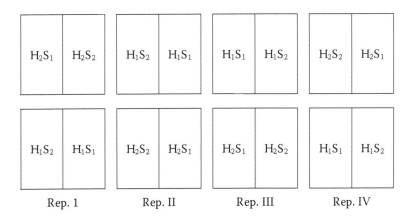

FIGURE 6.2.2
Random assignment of factor *B* (rows spacing, S_1 and S_2) to the two subplots in each of the two main plots in the four replications.

Step 1: Divide the experimental area into four replications and each replication into two main plots. Factor *A* (two hybrids: H_1 and H_2) is randomly assigned to the main plot as shown in Figure 6.2.1.

Step 2: Subdivide each main plot into two subplots, and assign randomly the two levels of factor *B* (row spacings: *S*1 and *S*2) to the subplots within the main plots using separate randomization in each main plot as shown in Figure 6.2.2.

Step 3: Divide each subplot into three sub-subplots, and assign at random the three levels of factor *C* (plant densities: D_1, D_2, and D_3) to the sub-subplots using a separate randomization in each subplot as shown in Figure 6.2.3.

$H_2S_1D_3$	$H_2S_2D_1$		
$H_2S_1D_1$	$H_2S_2D_2$		
$H_2S_1D_2$	$H_2S_1D_3$		

$H_1S_2D_1$	$H_1S_1D_2$
$H_1S_2D_3$	$H_1S_1D_1$
$H_1S_2D_2$	$H_1S_1D_3$

$H_1S_1D_2$	$H_1S_2D_2$
$H_1S_1D_1$	$H_1S_2D_1$
$H_1S_1D_3$	$H_1S_2D_3$

$H_2S_2D_1$	$H_2S_1D_1$
$H_2S_2D_3$	$H_2S_1D_2$
$H_2S_2D_2$	$H_2S_1D_3$

$H_1S_2D_1$	$H_1S_1D_2$
$H_1S_2D_3$	$H_1S_1D_1$
$H_1S_2D_2$	$H_1S_1D_3$

$H_2S_2D_3$	$H_2S_1D_1$
$H_2S_2D_2$	$H_2S_1D_3$
$H_2S_2D_1$	$H_2S_1D_2$

$H_2S_1D_3$	$H_2S_2D_1$
$H_2S_1D_2$	$H_2S_2D_3$
$H_2S_1D_1$	$H_2S_2D_2$

$H_1S_1D_1$	$H_1S_2D_2$
$H_1S_1D_3$	$H_1S_2D_1$
$H_1S_1D_2$	$H_1S_2D_3$

Rep. 1   Rep. II   Rep. III   Rep. IV

FIGURE 6.2.3

A 2 × 2 × 3 factorial experiment layout plan arranged in a split-split-plot design with two hybrids assigned to the main plot, two row spacing assigned to the subplots, and three plant densities assigned to the sub-subplots with four replications.

Analysis of variance: The analysis of variance performed on data from a split-split-plot design is similar to one carried out with the split-plot design. However, the number and type of comparisons have increased because of the added factor and the need for a second split to obtain the split-split-plot design. To illustrate the steps in computation, the following example is given.

Example 6.2.1

A study was conducted in Michigan to determine the influence of plant density and hybrids on corn (*Zea mays* L.) yield. The experiment was a 2 × 2 × 3 factorial replicated four times in a randomized complete block design arranged in a split-split-plot layout. In this experiment, factor *A* is the two corn hybrids (P3730 and B70 × LH55) assigned to the main plots, factor *B* is the two row spacings (12 and 25 in.) assigned to the subplots, and factor *C* is the 3 target plant densities (12,000; 16,000; and 20,000 plants/ac) assigned to the sub-subplots. The data gathered for the experiment are shown in Table 6.2.1.

SOLUTION

The statistical analyses are performed on the main plot, the subplot, and the sub-subplot. The steps are:

Step 1: Perform the main-plot analysis by computing the replication totals (*R*), the grand total (*G*), and the hybrid total (*A*) by first constructing a table for the replication × hybrid totals (*RA*) as shown in Table 6.2.2.

TABLE 6.2.1

Yield of Two Corn Hybrids (bu/ac) with Two Row Spacings and
Three Plant Densities

Hybrid	Row Spacing (in.)	Plant Density (plants/ac)	Grain Yield (bu/ac) Replications			
			I	II	III	IV
P3730	12	12,000	140	138	130	142
		16,000	145	146	150	147
		20,000	150	149	146	150
			435	433	426	439
	25	12,000	136	132	134	138
		16,000	140	134	136	140
		20,000	145	138	138	142
			421	404	408	420
B70 × LH55	12	12,000	142	132	128	140
		16,000	146	136	140	141
		20,000	148	140	142	140
			436	408	410	421
	25	12,000	132	130	136	134
		16,000	138	132	130	132
		20,000	140	134	130	136
			410	396	396	402

TABLE 6.2.2

The Replication × Hybrid (Factor A) Table of Yield Totals Computed
from Data in Table 6.2.1

Hybrid	Yield Total (RA)				Hybrid Total (A)
	Rep. I	Rep. II	Rep. III	Rep. IV	
P3730	856	837	834	859	3,386
B70 × LH55	846	804	806	823	3,279
Rep. Total (R)	1,702	1,641	1,640	1,682	
Grand Total (G)					6,665

Step 2: Determine the degrees of freedom associated with each source of
variation. The following sources of variation and degrees of freedom are
identified:

Replication $d.f. = r - 1 = 3$

Main-plot factor (A) $d.f. = a - 1 = 1$

Error (a) $d.f. = (r - 1)(a - 1) = 3$

Subplot factor (B) $d.f. = b - 1 = 1$

$A \times B$ $d.f. = (a - 1)(b - 1) = 1$

Error (b) $d.f. = a(r - 1)(b - 1) = 6$

Sub-subplot factor (C) $d.f. = c - 1 = 2$

$A \times C$ *d.f.* $= (a - 1) (c - 1) = 2$
$B \times C$ *d.f.* $= (b - 1) (c - 1) = 2$
$A \times B \times C$ *d.f.* $= (a - 1) (b - 1)(c - 1) = 2$
Error (c) *d.f.* $= ab (r - 1) (c - 1) = 24$
Total *d.f.* $= rabc - 1 = 47$

Step 3: Compute the correction factor and the total sum of squares for the main plot as:

$$C = \frac{G^2}{rabc} \tag{6.1}$$

$$= \frac{(6,665)^2}{4 \times 2 \times 2 \times 3}$$

$$= 44{,}422{,}225/48$$

$$= 925{,}463.02$$

$$\text{Total } SS = \sum X^2 - C \tag{6.2}$$

$$= \left[(140)^2 + (138)^2 + \ldots + (136)^2 \right] - 925{,}463.02$$

$$= 927{,}177 - 925{,}463.02$$

$$= 1{,}713.98$$

Step 4: Calculate the replication, hybrid, and error sum of squares for factor A, for the main-plot analysis as:

$$\text{Replication } SS = \frac{\sum R^2}{abc} - C \tag{6.3}$$

$$= \frac{(1{,}702)^2 + (1{,}641)^2 + (1{,}640)^2 + (1{,}682)^2}{12} - 925{,}463.02$$

$$= 11{,}108{,}409/12 - 925{,}463.02$$

$$= 925{,}700.75 - 925{,}463.02$$

$$= 237.73$$

$$SS_A\left(\text{Hybrid}\right) = \frac{\sum A^2}{rbc} - C \tag{6.4}$$

$$= \frac{\left(3,386\right)^2 + \left(3,279\right)^2}{\left(4\right)\left(2\right)\left(3\right)} - 925,463.02$$

$$= 925,701.54 - 925,463.02$$

$$= 238.52$$

$$\text{Error } SS(a) = \frac{\sum\left(RA\right)^2}{bc} - \left(C + \text{Rep.SS} + SS_A\right) \tag{6.5}$$

$$= \frac{\left(856\right)^2 + \left(846\right)^2 + \ldots + \left(823\right)^2}{\left(2\right)\left(3\right)} - \left(925,463.02 + 237.73 + 238.52\right)$$

$$= 5,555,839/6 - 925939.27$$

$$= 33.90$$

Step 5: Compute the various sums of squares for the subplot analysis by constructing a summary table of yield totals for factor $A \times$ factor B or (AB), and a three-way table of totals for replication $\times$ factor $A \times$ factor B or (RAB) as shown in Table 6.2.3 and Table 6.2.4.

TABLE 6.2.3

Summary Table of Yield Totals for the Hybrid $\times$ Row Spacing (Factor $A \times$ Factor B) Computed from Data in Table 6.2.1

	Yield Total (AB)	
	Row Spacing (in.)	
Hybrid	**12**	**25**
P3730	1,733	1,653
B70 × LH55	1,675	1,604
Row Spacing Total (B)	3,408	3,257

TABLE 6.2.4

Summary Table of Yield Totals for the Replication $\times$ Hybrid $\times$ Row Spacing or (RAB) Computed from Data in Table 6.2.1

Hybrid	Row Spacing (in.)	Yield Total (RAB)			
		Rep. I	**Rep. II**	**Rep. III**	**Rep. IV**
P3730	12	435	433	426	439
	25	421	404	408	420
B70 × LH55	12	436	408	410	421
	25	410	396	396	402

$$SS_B \left(\text{Row Spacing} \right) = \frac{\sum B^2}{rac} - C \tag{6.6}$$

$$= \frac{\left(3{,}408\right)^2 + \left(3{,}257\right)^2}{\left(4\right)\left(2\right)\left(3\right)} - 925{,}463.02$$

$$= 22{,}222{,}513/24 - 925{,}463.02$$

$$= 925{,}938.04 - 925{,}463.02$$

$$= 475.02$$

$$SS_{AB} \left(\text{Hybrid} \times \text{Row Spacing} \right) = \frac{\sum \left(AB \right)^2}{rc} - \left(C + SS_A + SS_B \right) \tag{6.7}$$

$$= \frac{\left(1{,}733\right)^2 + \left(1{,}675\right)^2 + \left(1{,}653\right)^2 + \left(1{,}604\right)^2}{\left(4\right)\left(3\right)} - \left(925{,}463.02 + 238.52 + 475.02\right)$$

$$= 926{,}178.25 - 926{,}176.56$$

$$= 1.69$$

$$\text{Error } (b)SS = \frac{\sum \left(RAB \right)^2}{c} - \left(C + \text{Rep.}SS + SS_A + \text{Error}\left(a\right)SS + SS_B + SS_{AB} \right) \tag{6.8}$$

$$= \frac{\left(435\right)^2 + \left(421\right)^2 + \ldots + \left(402\right)^2}{\left(3\right)} - \left(\begin{array}{l} 925{,}463.02 + 237.73 + 238.52 + 33.90 + \\ 475.02 + 1.69 \end{array} \right)$$

$$= 2{,}779{,}469/3 - 926{,}449.88$$

$$= 926{,}489.67 - 926{,}449.88$$

$$= 39.79$$

Step 6: Perform the sub-subplot analysis by constructing three summary tables of yield totals for: (1) factor $A \times$ factor C or (AC), (2) factor $B \times$ factor C or (BC), and (3) for factor $A \times B \times C$ or (ABC) as shown in Table 6.2.5, Table 6.2.6, and Table 6.2.7, respectively.

In computing the sum of squares for factor C, it should be kept in mind that the squared C value is different from the correction factor notation (C) shown in the following text.

TABLE 6.2.5

Summary Table of Yield Totals for Hybrid × Plant Density
(Factor A × Factor C) Computed from Data in Table 6.2.1

| | Yield Total (AC) | | |
| | Plant Density/ac | | |
Hybrid	12,000	16,000	20,000
P3730	1,090	1,138	1,158
B70 × LH55	1,074	1,095	1,110
Plant Density Total (C)	2,164	2,233	2,268

TABLE 6.2.6

Summary Table of Yield Totals for Row Spacing × Plant Density
(Factor B × Factor C) Computed from Data in Table 6.2.1

| | Yield Total (BC) | | |
| | Plant Density/ac | | |
Row Spacing	12,000	16,000	20,000
12	1,092	1,151	1,165
25	1,072	1,082	1,103

TABLE 6.2.7

Summary Table of Yield Totals for Hybrid × Row Spacing ×
Plant Density (Factor A × Factor B × Factor C) Computed from
Data in Table 6.2.1

| | | Yield Total (ABC) | | |
| | | Plant Density/ac | | |
Hybrid	Row Spacing	12,000	16,000	20,000
P3730	12	550	588	595
	25	540	550	563
B70 × LH55	12	542	563	570
	25	532	532	540

$$SS_C\left(\text{Plant Density}\right) = \frac{\sum C^2}{rab} - C \tag{6.9}$$

$$= \frac{\left(2,164\right)^2 + \left(2,233\right)^2 + \left(2,268\right)^2}{\left(4\right)\left(2\right)\left(2\right)} - 925,463.02$$

$$= 14,813,009/16 - 925,463.02$$

$$= 925,813.06 - 925,463.02$$

$$= 350.04$$

$$SS_{AC} = \frac{\sum (AC)^2}{rb} - (C + SS_A + SS_C) \tag{6.10}$$

$$= \frac{(1,090)^2 + (1,074)^2 + \dots + (1,110)^2}{(4)(2)} - (925,463.02 + 238.52 + 350.04)$$

$$= 7,408,709/8 - 926,051.58$$

$$= 926,088.62 - 926,051.58$$

$$= 37.04$$

$$SS_{BC} = \frac{\sum (BC)^2}{ra} - (C + SS_B + SS_C) \tag{6.11}$$

$$= \frac{(1,092)^2 + (1,072)^2 + \dots + (1,103)^2}{(4)(2)} - (925,463.02 + 475.02 + 350.04)$$

$$= 7,411,007/8 - 926,288.08$$

$$= 926,375.86 - 926,288.08$$

$$= 87.80$$

$$SS_{ABC} = \frac{\sum (ABC)^2}{r} - (C + SS_A + SS_B + SS_C + SS_{AB} + SS_{AC} + SS_{BC}) \tag{6.12}$$

$$= \frac{(550)^2 + (540)^2 + \dots + (540)^2}{4}$$

$$- (925,463.02 + 238.52 + 475.02 + 350.04 + 1.69 + 37.04 + 87.80)$$

$$= 3,706,619/4 - 926,653.13$$

$$= 926,654.75 - 926,653.13$$

$$= 1.62$$

Error (c) SS = Total SS – All other sum of squares   (6.13)
$$= 1{,}713.98 - (237.73 + 238.52 + 33.90 + 475.02 + 1.69 +$$
$$39.79 + 350.04 + 37.04 + 87.80 + 1.62)$$
$$= 1{,}713.98 - 1{,}503.15$$
$$= 210.83$$

Step 7: Calculate the mean square for each source of variation as:

$$\text{Replication } MS = \frac{\text{Rep. } SS}{r-1} = \frac{273.73}{3} = 91.24 \tag{6.14}$$

$$MS_A = \frac{SS_A}{a-1} = \frac{238.53}{2-1} = 238.53 \tag{6.15}$$

$$\text{Error (a) } MS = \frac{Error(a)SS}{(r-1)(a-1)} = \frac{33.90}{(3)(1)} = 11.30 \tag{6.16}$$

$$MS_B = \frac{SS_B}{b-1} = \frac{475.02}{2-1} = 475.02 \tag{6.17}$$

$$MS_{AB} = \frac{SS_{AB}}{(a-1)(b-1)} = \frac{1.69}{1} = 1.69 \tag{6.18}$$

$$\text{Error (b) } MS = \frac{Error(b)SS}{a(r-1)(b-1)} = \frac{39.79}{2(3)(1)} = 6.63 \tag{6.19}$$

$$MS_C = \frac{SS_C}{(c-1)} = \frac{350.40}{3-1} = 175.20 \tag{6.20}$$

$$MS_{AC} = \frac{SS_{AC}}{(a-1)(c-1)} = \frac{37.04}{(1)(2)} = 18.52 \tag{6.21}$$

$$MS_{BC} = \frac{SS_{BC}}{(b-1)(c-1)} = \frac{87.80}{(1)(2)} = 43.90 \tag{6.22}$$

TABLE 6.2.8

Analysis of Variance of a Three-Factor Corn Experiment in a Split-Split-Plot Design

Source of Variation	Degrees of Freedom	Sum of Squares	Mean Square	F
Replication	3	273.73	91.24	
Hybrid (*A*)	1	238.53	238.53	21.11**
Error (*a*)	3	33.90	11.30	
Row Spacing (*B*)	1	475.02	475.02	71.64**
AB	1	1.69	1.69	<1
Error (*b*)	6	39.79	6.63	
Plant Density (*C*)	2	350.40	175.20	19.95**
AC	2	37.04	18.52	2.11ns
BC	2	87.80	43.90	5.00*
ABC	2	1.62	0.81	<1
Error (*c*)	24	210.83	8.78	
Total	47	1,713.98		

Note: $cv(a) = 2.42\%$, $cv(b) = 1.85\%$, $cv(c) = 2.13\%$; ** = significant at 1% level; * = significant at 5%; ns = nonsignificant.

$$MS_{ABC} = \frac{SS_{ABC}}{(a-1)(b-1)(c-1)} = \frac{1.62}{(1)(1)(2)} = 0.81 \qquad (6.23)$$

$$\text{Error (c) } MS = \frac{Error(c)SS}{ab(r-1)(c-1)} = \frac{210.83}{(4)(3)(2)} = 8.78 \qquad (6.24)$$

Step 8: Compute the *F* value for each source of variation as shown in Table 6.2.8.

Step 9: Compare the computed *F* value in Table 6.2.8 with the tabular *F* value given in Appendix E. The results indicate that the interaction between row spacing and plant density is significant, whereas other interactions are all nonsignificant.

Finally, we will compute the coefficient of variation (*cv*) for the three error mean squares as shown below and reported at the bottom of Table 6.2.8.

$$cv(a) = \frac{\sqrt{\text{Error (a) } MS}}{\text{Grand Mean}} \times 100 \qquad (6.25)$$

$$= \frac{\sqrt{11.30}}{138.85} \times 100 = 2.42\%$$

$$cv(b) = \frac{\sqrt{\text{Error (b) } MS}}{\text{Grand Mean}} \times 100 \tag{6.26}$$

$$= \frac{\sqrt{6.63}}{138.85} \times 100 = 1.85\%$$

$$cv(c) = \frac{\sqrt{\text{Error (c) } MS}}{\text{Grand Mean}} \times 100 \tag{6.27}$$

$$= \frac{\sqrt{8.87}}{138.85} \times 100 = 2.14\%$$

As before, the value of the coefficient of variation (cv) indicates the degree of precision with which the factors are compared and is a good measure of the reliability of the experiment. $cv(a)$ is a measure of the degree of precision associated with the main effect of the main-plot factor, and $cv(b)$ shows the degree of precision of the subplot factor in relation to the main-plot factor and their interactions. The degree of precision in the measurement of the main effect of the sub-subplot factor and its interaction with all other factors is given by $cv(c)$. As indicated in previous sections of this chapter, the degree of precision in the main and subplot factors is sacrificed in order to have greater precision in the measurement of the sub-subplot factor. Thus, we would expect $cv(a)$ to have the largest value, and $cv(c)$, the smallest. However, in this particular example, $cv(a)$ is the largest, but the value for $cv(c)$ is not the smallest. Such unexpected results are not uncommon.

6.3 Strip-Split-Plot Design

This design, which is an extension of the strip-plot design discussed in Section 5.6, is a three-factor design that accommodates the inclusion of a third factor by dividing the intersection plot into subplots. Essentially, the design has four plot sizes — the horizontal strip, the vertical strip, the intersection plot, and the subplot, each measured with different levels of precision. As in the case of the strip-plot design, the precision associated with the subplot factor and its interactions with other factors is the highest.

Randomization and layout: To illustrate the randomization and layout, let us take an example of a $4 \times 2 \times 4$ factorial experiment replicated four times in a randomized complete block design arranged in a strip-split-plot layout. Factor A (first harvest time denoted as T_1, T_2, T_3, and T_4) is assigned to the horizontal plot. Factor B (herbicide application denoted as H_0 and H_1) is assigned to the vertical plot. Factor C (four alfalfa cultivars denoted as C_1,

Rep. I	Rep. II	Rep. III	Rep. IV
T_3	T_2	T_1	T_4
T_1	T_4	T_3	T_2
T_4	T_3	T_4	T_1
T_2	T_1	T_2	T_3

FIGURE 6.3.1
Random assignment of the horizontal factor (first time T_1, T_2, T_3, T_4) to the horizontal plots with four replications.

C_2, C_3, and C_4) is assigned to the subplot. As before, the treatment levels and number of replications are denoted as a, b, c, and r. The steps followed are as follows:

Step 1: Divide the experimental area into four blocks or replications ($r = 4$) in which each block is further divided into four horizontal strips. Randomly assign the four treatments ($a = 4$) to each of the strips, following any of the randomization schemes outlined in Chapter 5. Figure 6.3.1 shows the layout for this step.

Step 2: Divide each block or replication into two vertical strips, and randomly assign the two herbicide treatments ($b = 2$), following a randomization scheme discussed in Chapter 5. This is shown in Figure 6.3.2.

Step 3: Subdivide each of the intersection plots into four subplots and randomly assign the four alfalfa cultivars ($c = 4$) to them separately and independently. The final layout of the experiment is shown in Figure 6.3.3.

Analysis of variance: As in the split-plot design, we need to perform separate analyses for the vertical, horizontal, and intersection factors. These analyses are performed for each factor as illustrated in the following example.

Example 6.3.1

Four cultivars newly planted of alfalfa (*Medicago sativa* L.) were studied with or without the use of preemergence herbicide, to determine whether variable dates of first cutting, followed by harvests after it had reached the bloom stage, would influence yield. The forage dry matter production results are shown in Table 6.3.1. Perform the analysis of variance.

SOLUTION

Step 1: Calculate the treatment totals (*ABC*) as shown in Table 6.3.1.

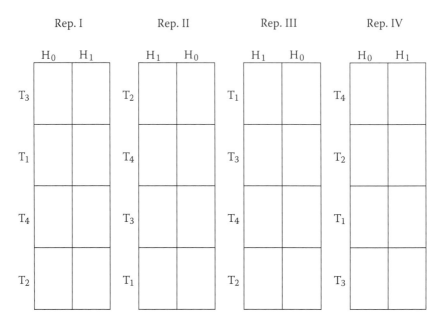

FIGURE 6.3.2
Random assignment of the horizontal factor (first harvest T_1, T_2, T_3, and T_4) to the horizontal plots and the vertical factor (herbicide, H_0 and H_1) to the vertical plots with four replications.

Step 2: Determine the *d.f.*'s associated with each source of variation. The following sources of variation and the *d.f.*'s are identified.

Replication *d.f.* = $r - 1 = 3$
Horizontal factor (*A*) *d.f.* = $a - 1 = 3$
Error (*a*) *d.f.* = $(r - 1)(a - 1) = 9$
Vertical factor (*B*) *d.f.* = $b - 1 = 1$
Error (*b*) *d.f.* = $(r - 1)(b - 1) = 3$
$A \times B$ *d.f.* = $(a - 1)(b - 1) = 3$
Error (*c*) *d.f.* = $(r - 1)(a - 1)(b - 1) = 9$
Subplot factor (*C*) *d.f.* = $c - 1 = 3$
$A \times C$ *d.f.* = $(a - 1)(c - 1) = 9$
$B \times C$ *d.f.* = $(c - 1)(b - 1) = 3$
$A \times B \times C$ *d.f.* = $(a - 1)(b - 1)(c - 1) = 9$
Error (*d*) *d.f.* = $ab(r - 1)(c - 1) = 72$
Total *d.f.* = $rabc - 1 = 127$

Step 3: Perform the vertical analysis by calculating the replication totals (*R*), the grand total (*G*), and the first harvest time total (*A*). To do this, we

Rep. I   Rep. II   Rep. III   Rep. IV

	H_0	H_1		H_1	H_0		H_1	H_0		H_0	H_1
	C_1	C_4		C_2	C_1		C_1	C_1		C_4	C_2
T_3	C_4	C_2	T_2	C_3	C_4	T_1	C_4	C_3	T_4	C_2	C_1
	C_2	C_3		C_4	C_3		C_2	C_4		C_3	C_4
	C_3	C_1		C_1	C_2		C_3	C_2		C_1	C_3
	C_2	C_4		C_1	C_4		C_4	C_3		C_3	C_1
T_1	C_1	C_1	T_4	C_4	C_3	T_3	C_1	C_4	T_2	C_2	C_3
	C_4	C_2		C_3	C_1		C_2	C_1		C_4	C_2
	C_3	C_3		C_2	C_2		C_3	C_2		C_1	C_4
	C_2	C_3		C_2	C_2		C_2	C_4		C_4	C_3
T_4	C_1	C_2	T_3	C_1	C_4	T_4	C_4	C_2	T_1	C_1	C_2
	C_3	C_1		C_3	C_1		C_3	C_3		C_3	C_4
	C_3	C_4		C_4	C_3		C_1	C_1		C_2	C_1
	C_1	C_4		C_3	C_2		C_4	C_3		C_2	C_3
T_2	C_2	C_3	T_1	C_4	C_1	T_2	C_1	C_4	T_3	C_3	C_4
	C_4	C_2		C_1	C_4		C_2	C_1		C_4	C_1
	C_3	C_1		C_2	C_3		C_3	C_2		C_1	C_2

FIGURE 6.3.3
Layout of a $4 \times 2 \times 4$ factorial experiment arranged in a strip-split-plot design with four replications.

need to construct a two-way table for the replication × first harvest time totals (RA) as shown in Table 6.3.2. To show how we have computed the yield total for all the replications in Table 6.3.2, we have used the early bud in the first replication (replication I) as an example, and the total is computed as:

$$1.9 + 5.5 + 1.8 + 5.1 + 2.0 + 5.1 + 1.6 + 7.4 = 30.4$$

Step 4: Compute the correction factor and the various sums of squares as:

TABLE 6.3.1

Yield of Dry Matter Forage from 4 Alfalfa Cultivars with Varied First-Time Harvest in the Absence and Presence of Herbicide

| First Harvest Time | Herbicide | Cultivar | Dry Matter Yield (t/ac) Replications | | | | Total (ABC) |
			I	II	III	IV	
Early bud	Absence	Kanza	1.9	1.8	1.6	1.8	7.1
Late bud			2.0	2.1	1.9	2.2	8.2
Early bloom			2.4	2.3	2.1	2.2	9.0
Late bloom			2.6	2.4	2.5	2.4	9.9
Early bud	Presence		5.5	5.2	5.7	5.3	21.7
Late bud			5.8	5.2	5.6	5.1	21.7
Early bloom			6.4	6.4	6.2	6.1	25.1
Late bloom			6.2	6.0	6.4	6.1	24.7
Early bud	Absence	Liberty	1.8	1.8	1.0	1.5	6.1
Late bud			2.4	2.5	2.8	2.4	10.1
Early bloom			2.6	2.3	2.5	2.4	9.8
Late bloom			2.9	2.7	2.6	2.4	10.6
Early bud	Presence		5.1	5.4	5.2	5.5	21.2
Late bud			5.0	5.1	5.1	4.9	20.1
Early bloom			5.3	5.6	5.2	5.3	21.4
Late bloom			5.4	5.7	5.5	5.3	21.9
Early bud	Absence	Arc	2.0	2.2	2.1	2.0	8.3
Late bud			2.4	2.1	2.3	2.2	9.0
Early bloom			2.4	2.2	2.6	2.4	9.6
Late bloom			2.6	2.8	2.4	2.2	10.0
Early bud	Presence		5.1	5.5	5.3	5.2	21.1
Late bud			5.7	5.9	5.1	5.4	22.1
Early bloom			5.3	5.5	5.1	5.3	21.2
Late bloom			5.7	5.6	5.7	5.5	22.5
Early bud	Absence	CA90	1.6	1.2	1.5	1.3	5.6
Late bud			2.5	2.2	2.1	2.6	9.4
Early bloom			2.3	2.2	2.6	2.1	9.2
Late bloom			2.8	2.9	2.4	2.3	10.4
Early bud	Presence		7.4	7.1	7.7	7.3	29.5
Late bud			7.8	7.3	7.9	7.2	30.2
Early bloom			7.5	7.1	7.6	7.3	29.5
Late bloom			7.7	7.1	7.4	7.9	30.1

$$C = \frac{G^2}{rabc} \tag{6.1}$$

$$C = \frac{(526.3)^2}{4 \times 4 \times 2 \times 4}$$

$$= 276991.69/128$$

$$= 2164.00$$

TABLE 6.3.2

Summary Table of Replication × First Harvest Time (Factor *A*) Yield Totals
Computed from Data in Table 6.3.1

| First Harvest | Yield Total (*RA*) | | | | First-Time Harvest |
Time	Rep. I	Rep. II	Rep. III	Rep. IV	Total (*A*)
Early bud	30.4	30.2	30.1	29.9	120.6
Late bud	33.6	32.4	32.8	32.0	130.8
Early bloom	34.2	33.6	33.9	33.1	134.8
Late bloom	35.9	35.2	34.9	34.1	140.1
Rep. Total (*R*)	134.1	131.4	131.7	129.1	
Grand total (*G*)					526.3

$$\text{Total } SS = \sum X^2 - C \tag{6.2}$$

$$= [(1.9)^2 + (1.8)^2 + \ldots + (7.9)^2] - 2{,}164.00$$

$$= 2{,}684.39 - 2{,}164.00$$

$$= 520.39$$

Step 5: Calculate the replication, factor *A* (first harvest time), and the error
sum of squares for factor *A* as:

$$\text{Replication } SS = \frac{\sum R^2}{abc} - C \tag{6.3}$$

$$= \frac{(134.1)^2 + (131.4)^2 + (131.7)^2 + (129.1)^2}{(3)} - 2{,}164$$

$$= 69{,}260.47/32 - 2{,}164$$

$$= 2{,}164.39 - 2{,}164$$

$$= 0.39$$

$$SS_A\left(\text{First Harvest Time}\right) = \frac{\sum A^2}{rbc} - C \tag{6.4}$$

$$= \frac{\left(120.6\right)^2 + \left(130.8\right)^2 + \left(134.8\right)^2 + \left(140.1\right)^2}{\left(4\right)\left(2\right)\left(4\right)} - 2,164$$

$$= 69,452.05/32 - 2,164$$

$$= 2,170.38 - 2,164$$

$$= 6.38$$

$$\text{Error } SS\left(a\right) = \frac{\sum \left(RA\right)^2}{bc} - \left(C + \text{Rep.}SS + SS_A\right) \tag{6.5}$$

$$= \frac{\left(30.4\right)^2 + \left(33.6\right)^2 + \ldots + \left(34.1\right)^2}{\left(2\right)\left(4\right)} - \left(2,164 + 0.39 + 6.38\right)$$

$$= 17,366.87/8 - 2,170.77$$

$$= 2,170.86 - 2,170.77$$

$$= 0.09$$

Step 6: Perform the horizontal analysis by constructing a two-way table for the replication × herbicide totals (*RB*) as shown in Table 6.3.3, and then compute the various sums of squares as:

$$SS_B\left(\text{Herbicide}\right) = \frac{\sum B^2}{rac} - C \tag{6.6}$$

$$= \frac{\left(142.3\right)^2 + \left(384.0\right)^2}{\left(4\right)\left(4\right)\left(4\right)} - 2,164$$

$$= 167,705.29/64 - 2,164$$

$$= 2,620.4 - 2,164$$

$$= 456.4$$

TABLE 6.3.3

Summary Table of Yield Totals for the Replication × Herbicide (*RB*) Computed from Data in Table 6.3.1

Herbicide	Yield Total (*RB*)				Herbicide Total (*B*)
	Rep. I	**Rep. II**	**Rep. III**	**Rep. IV**	
Absent	37.2	35.7	35.0	34.4	142.3
Present	96.9	95.7	96.7	94.7	384.0

TABLE 6.3.4

Summary Table of Yield Totals for First Harvest Time × Herbicide (Factor A × Factor B) Computed from Data in Table 6.3.1

Herbicide	Yield Total (*AB*)			
	Early Bud	**Late Bud**	**Early Bloom**	**Late Bloom**
Absent	27.1	36.7	37.6	40.9
Present	93.5	94.1	97.2	99.2

$$\text{Error } SS(b) = \frac{\sum (RB)^2}{ac} - \left(C + \text{Rep.}SS + SS_B \right) \tag{6.28}$$

$$= \frac{(37.2)^2 + (96.9)^2 + \ldots + (94.7)^2}{16} - \left(2,164 + 0.39 + 456.4 \right)$$

$$= 41{,}933.77/16 - 2{,}620.79$$

$$= 2{,}620.86 - 2{,}620.79$$

$$= 0.07$$

Step 7: Perform the interaction analysis by constructing a two-way table for the vertical × horizontal factor totals (*AB*), and a three-way table of totals for the replication × vertical × horizontal factors (*RAB*) as shown in Table 6.3.4 and Table 6.3.5. Using these summary tables, compute the various sums of squares as:

TABLE 6.3.5

Summary Table of Yield Totals for the Replication × First Harvest Time ×Herbicide or (RAB) Computed from Data in Table 6.3.1

First Harvest Time	Herbicide	Yield Total (RAB)			
		Rep. I	Rep. II	Rep. III	Rep. IV
Early bud	Absent	7.3	7.0	6.2	6.6
	Present	23.1	23.2	23.9	23.3
Late bud	Absent	9.3	8.9	9.1	9.4
	Present	24.3	23.5	23.7	22.6
Early bloom	Absent	9.7	9.0	9.8	9.1
	Present	24.5	24.6	24.1	24.0
Late bloom	Absent	10.9	10.8	9.9	9.3
	Present	25.0	24.4	25.0	24.8

$$SS_{AB} = \frac{\sum (AB)^2}{rc} - (C + SS_A + SS_B) \tag{6.7}$$

$$= \frac{(27.1)^2 + (93.5)^2 + \ldots + (99.2)^2}{(4)(4)} - (2,164 + 6.38 + 456.4)$$

$$= 42,053.41/16 - 2,626.78$$

$$= 2,628.34 - 2,626.78$$

$$= 1.56$$

$$\text{Error } SS\ (c) = \frac{\sum (RAB)^2}{c} - (C + \text{Rep.}SS + SS_A + \text{Error}SS(a) + SS_B \tag{6.29}$$

$$+ \text{Error}SS(b) + SS_{AB})$$

$$= 10,518.81/4 - (2,164 + 0.39 + 6.38 + 0.09 + 456.4 + 0.07 + 1.56)$$

$$= 2,629.70 - 2,628.89$$

$$= 0.81$$

Step 8: Perform the subplot analysis by constructing two summary tables of yield totals. The first table is for vertical factor *A* × subplot factor *C* and is presented as Table 6.3.6. The second table is for the horizontal factor *B* × subplot factor *C*, shown in Table 6.3.7. From these tables compute the various sums of squares as follows.

In computing the sum of squares for factor *C* it should be kept in mind that the squared *C* value represents factor *C* and is different from the correction factor notation (*C*) as shown:

TABLE 6.3.6

Summary Table of Yield Totals for Factor $A \times$ Factor C
Computed from Data in Table 6.3.1

First Harvest Time	Yield Total (AC)			
	Kanza	Liberty	Arc	CA90
Early bud	28.8	27.3	29.4	35.1
Late bud	29.9	30.2	31.1	39.6
Early bloom	34.1	31.2	30.8	38.7
Late bloom	34.6	32.5	32.5	40.5
First-time harvest total (C)	127.4	121.2	123.8	153.9

TABLE 6.3.7

Summary Table of Yield Totals for Factor $B \times$ Factor C
Computed from Data in Table 6.3.1

Herbicide	Yield Total (BC)			
	Kanza	Liberty	Arc	CA90
Absent	34.2	36.6	36.9	34.6
Present	93.2	84.6	86.9	119.3

$$SS_C\left(\text{Cultivar}\right) = \frac{\sum C^2}{rab} - C \tag{6.9}$$

$$= \frac{\left(127.4\right)^2 + \left(121.2\right)^2 + \left(123.8\right)^2 + \left(153.9\right)^2}{\left(4\right)\left(4\right)\left(2\right)} - 2,164$$

$$= 69,931.85/32 - 2,164$$

$$= 2,185.37 - 2,164$$

$$= 21.37$$

$$SS_{AC} = \frac{\sum \left(AC\right)^2}{rb} - \left(C + SS_A + SS_C\right) \tag{6.10}$$

$$= \frac{\left(28.8\right)^2 + \left(27.3\right)^2 + ... + \left(40.5\right)^2}{\left(4\right)\left(2\right)} - \left(2,164 + 6.38 + 21.37\right)$$

$$= 17,545.01/8 - 2,191.75$$

$$= 2,193.13 - 2,191.75$$

$$= 1.38$$

$$SS_{BC} = \frac{\sum (BC)^2}{ra} - (C + SS_B + SS_C) \tag{6.11}$$

$$= \frac{(34.2)^2 + (36.6)^2 + ... + (119.3)^2}{(4)(4)} - (2,164 + 456.4 + 21.37)$$

$$= 42,695.47/16 - 2,641.77$$

$$= 2,668.47 - 2,641.77$$

$$= 26.70$$

$$SS_{ABC} = \frac{\sum (ABC)^2}{r} - (C + SS_A + SS_B + SS_C + SS_{AB} + SS_{AC} + SS_{BC}) \tag{6.12}$$

$$= \frac{(7.1)^2 + (8.2)^2 + ... + (30.1)^2}{4} - (2,164 + 6.38 + 456.4 + 21.37 + 1.56 + 1.38 + 26.70)$$

$$= 10,719.61/4 - 2,677.79$$

$$= 2,679.90 - 2,677.79$$

$$= 2.11$$

Error SS (d) = Total SS – All other sum of squares

$$= 520.39 - (0.39 + 6.38 + 0.09 + 456.4 + 0.07 + 1.56 +$$

$$0.81 + 21.37 + 1.38 + 26.70 + 2.11)$$

$$= 520.39 - 517.26$$

$$= 3.13 \tag{6.30}$$

Step 9: Calculate the mean square for each source of variation as:

$$\text{Replication } MS = \frac{\text{Rep.SS}}{r-1} = \frac{0.39}{3} = 0.13 \tag{6.14}$$

$$MS_A = \frac{SS_A}{a-1} = \frac{6.38}{4-1} = 2.13 \tag{6.15}$$

$$\text{Error}(a)\,MS = \frac{\text{Error}(a)\,SS}{(r-1)(a-1)} = \frac{0.09}{(3)(3)} = 0.01 \qquad (6.16)$$

$$MS_B = \frac{SS_B}{b-1} = \frac{456.4}{2-1} = 456.4 \qquad (6.17)$$

$$\text{Error}(b)\,MS = \frac{\text{Error}(b)\,SS}{(r-1)(b-1)} = \frac{0.07}{(3)(1)} = 0.02 \qquad (6.31)$$

$$MS_{AB} = \frac{SS_{AB}}{(a-1)(b-1)} = \frac{1.56}{(3)(1)} = 0.52 \qquad (6.18)$$

$$\text{Error}(c)\,MS = \frac{\text{Error}(c)\,SS}{(r-1)(a-1)(b-1)} = \frac{0.81}{(3)(3)(1)} = 0.09 \qquad (6.32)$$

$$MS_C = \frac{SS_C}{c-1} = \frac{21.37}{4-1} = 7.12 \qquad (6.20)$$

$$MS_{AC} = \frac{SS_{AC}}{(a-1)(c-1)} = \frac{1.38}{(3)(3)} = 0.15 \qquad (6.21)$$

$$MS_{BC} = \frac{SS_{BC}}{(b-1)(c-1)} = \frac{26.70}{(1)(3)} = 8.9 \qquad (6.22)$$

$$MS_{ABC} = \frac{SS_{ABC}}{(a-1)(b-1)(c-1)} = \frac{2.11}{(3)(1)(3)} = 0.23 \qquad (6.23)$$

$$\text{Error}(d)\,MS = \frac{\text{Error}(d)\,SS}{ab(r-1)(c-1)} = \frac{3.13}{(8)(3)(3)} = 0.04 \qquad (6.33)$$

TABLE 6.3.8

Analysis of Variance of Alfalfa Dry Matter Yield in Strip-Split-Plot Design

Source of Variation	Degree of Freedom	Sum of Squares	Mean Square	F
Replication	3	0.39	0.13	
First harvest time (A)	3	6.38	2.13	213.00**
Error (a)	9	0.09	0.01	
Herbicide (B)	1	456.40	456.40	g
Error (b)	3	0.07	0.02	
AB	3	1.56	0.52	5.78*
Error (c)	9	0.81	0.09	
Hybrid (C)	3	21.37	7.12	178.00**
AC	9	1.38	0.15	3.75**
BC	3	26.70	8.90	222.50**
ABC	9	2.11	0.23	5.75**
Error (d)	72	3.13	0.04	
Total	127	520.39		

Note: $cv(a) = 2.43\%$, $cv (c) = 7.30\%$, $cv(d) = 4.87\%$; ** = significant at 1% level; * = significant at 5%; g = since there is only 1 degree of freedom associated with the use of herbicide, it is not adquate for a valid test of significance

Step 10: Compute the F value for each source of variation as shown in Table 6.3.8.

Step 11: Compare the computed F value in Table 6.3.8 with the tabular F value given in Appendix E. The results indicate that all interactions were significant, implying that there are differences in the treatment combinations. To appropriately interpret large interactions, we need to supplement the analysis with a detailed examination of the nature of the interactions. In Chapter 7 we will discuss mean comparisons that provide added information in interpreting results.

Finally, we will compute the coefficient of variation (*cv*) for the four error mean squares. Because of the inadequate degrees of freedom associated with factor *B*, we have not calculated the value of *cv(b)*. The *cv* value for *a*, *c*, and *d* error mean squares are computed as shown:

$$cv(a) = \frac{\sqrt{\text{Error}(a)\,MS}}{\text{Grand Mean}} \times 100 \qquad (6.25)$$

$$= \frac{\sqrt{0.01}}{4.11} \times 100 = 2.43\%$$

$$cv(c) = \frac{\sqrt{\text{Error}(c)\,MS}}{\text{Grand Mean}} \times 100 \tag{6.27}$$

$$= \frac{\sqrt{0.09}}{4.11} \times 100 = 7.30\%$$

$$cv(d) = \frac{\sqrt{\text{Error}(d)\,MS}}{\text{Grand Mean}} \times 100 \tag{6.34}$$

$$= \frac{\sqrt{0.04}}{4.11} \times 100 = 4.87\%$$

As before, the value of *cv* indicates the degree of precision associated with the measurement of the effects of the different factors. *cv(a)* is a measure of the degree of precision associated with the vertical factor, and *cv(b)* shows the degree of precision associated with the horizontal factor. The degree of precision associated with the interaction effect between factor *A* and *B* is given by *cv(c)*. *cv(d)* captures the precision of all effects associated with the subplot factor.

6.4 Factorial Experiments in Fractional Replication

An advantage of using factorial experiments is that the experimenter is able to study several factors simultaneously. However, as the number of factors included in the experiment increases, so does the requirement for resources. This can be very costly and time consuming. Additionally, technical considerations may dictate against the use of a fully replicated design. Consider, for example, an experiment in which six or seven factors each at two levels are to be studied simultaneously. A full replicate for each case would require the use of 64 and 128 experimental units, respectively, whether or not confounding is used. In such a situation, it is worth considering if estimates of the main effects and/or interactions, with a given precision, can be obtained by using a fraction of the full replicate. Finney (1945), who described a half-replicate of a 4×2^4 agricultural experiment, proposed the use of *fractional replication* and outlined the methods of construction for 2^n and 3^n factorials.

The major advantage of using the design is that five or more factors, for example, can be included simultaneously in an exploratory experiment of a practicable size in which factors that affect response can be identified and separated from factors in which the level is unimportant. Another feature of the design is reduced block size. As the design does not require that all treatment combinations be tested, the experimenter is able to reduce block size and thus improve the homogeneity of the experimental unit.

The advantages of using fractional replication should be weighed against the loss of information from a reduced number of replications, especially when fewer factors are involved. Loss of information is not a major concern in experiments in which there are a large number of factors resulting in a large number of interactions. The larger the number of factors in an experiment, the greater flexibility the investigator may have in the choice of factor interactions to be sacrificed. Prior knowledge of factor effects and interactions is helpful in deciding which factor combinations are to be included and which are to be sacrificed. In practice, high-order interactions such as four- or five-factor interactions are sacrificed. As suggested by Gomez and Gomez (1984, p. 168) in almost all cases, it is essential that the experimenter test a set of treatments that allows for estimation of all main effects and two-factor interactions. Another concern with use of fractional replication is the misinterpretation of the results, a condition which is less of a problem in fully replicated designs. In Subsection 6.4.1 we will discuss reasons for the problems of misinterpretation.

6.4.1   Aliases and Defining Contrasts

In using fractional factorial designs, care must be taken in the interpretation of results that are affected by *aliases*. Aliases refer to two factorial effects that are represented by the same comparison. Another term that needs clarification is *defining contrasts*. A defining contrast refers to a contrast that is used to split the factorial into two half-replicates. To have a better understanding of these two terms, consider a 2^3 factorial experiment in which the following four combinations — a, b, c, and abc are tested. These four treatment combinations are only half of the complete replicate. We can estimate the main effects and interactions as contrasts, using the coefficients given in Table 6.4.1.1.

Note that in Table 6.4.1.1 all of the treatment combinations for contrast ABC carry a + sign, indicating that this contrast cannot be estimated at all from the selected treatments. This is called a defining contrast. You should note that when using a half-replicate we lose one factorial effect (ABC) entirely, and each of the main effects is mixed with one of the two factor interactions.

When two factorial effects are represented by the same comparison, we refer to them as contrasts that are equivalent to each other, as is the case with $A = BC$, $B = AC$, and $C = AB$. These are called aliases. This means that the contrast that estimates A, for example, is the same as the contrast that estimates BC. Such equivalencies in contrasts make it difficult to distinguish

TABLE 6.4.1.1

Coefficients for Factorial Effects Using Four Treatments
from a 2^3 Factorial Experiment

	Treatment Combination			
Contrast	*a*	*b*	*c*	*abc*
A	+	−	−	+
B	−	+	−	+
AB	−	−	+	+
C	−	−	+	+
AC	−	+	−	+
BC	+	−	−	+
ABC	+	+	+	+

Note: + = defining contrasts. For example, for the main effect A,
the yields for units that contain a is added while we subtract
yields for units that do not, i.e., A = (abc) + (a) − (b) − (c).

whether the contrast is affected by *A*, or *BC*, or by a mixture of both. In fractional replication, every contrast has at least one other contrast as an alias. Difficulty in appropriately identifying the alias to which an effect is to be attributed often causes the problems alluded to earlier of misinterpretation of results.

To minimize the risk of misinterpreting results, we must know the aliases of the factorial effects we are interested in. To find aliases, use is made of the following rule:

> In the 2^n system, the alias of any factorial contrast is its generalized interaction with the defining contrast.

Using this rule, the alias for *A* in the defining contrast *ABC* is the interaction between *A* and *ABC*, which is *AABC* or *A²BC*. When interpreting a generalized interaction, squared terms are canceled (Cochran and Cox, 1957). Thus, *A²BC* is read as *BC*, and the alias of *B* is *AB²C* or *AC*.

To determine which factorial effect one should use as the defining contrast in a given situation, the following rule should be followed:

> In order to get a fractional replicate, do not use as the defining contrast any factorial effect that must be estimated from an experiment run on that fractional replicate.

Following this rule, the experimenter should select as the defining contrast the factorial effect that has the largest number of interactions. Higher-order interactions are negligible and thus can be sacrificed.

Fractional factorial designs are widely used and can be constructed to fit most factorial experiments. However, the procedure for construction is complex and beyond the scope of this book. In the following example, we have presented a simple analysis of fractional replication. The reader is referred to Cochran and Cox (1957), Finney (1945), Kempthorne (1952), and Brownlee, Kelly, and Loraine (1948) for further elaborations on — and more complex analyses of — fractional factorial designs. Appendix G presents several basic plans, each of which identifies the defining contrasts and the estimable factors that can be used for exploratory research. The following sections present the randomization and layout of a fractional factorial design and its analysis.

Randomization and layout: The randomization procedure requires several steps as described:

Step 1: Using Appendix G, select a basic plan that meets the needs of the experiment in terms of factors and levels of the factors to be tested. Assume that we have an experiment in which six factors each at two levels are used. Thus, from Appendix G, we select Plan 6, which meets the requirement for a 2^6 factorial experiment to be conducted in a half-replicate. The selected plan has the following treatment numbers assigned to each block as shown in Table 6.4.1.2.

Step 2: Randomly assign the block arrangement in the basic plan to the blocks in the field. This should be done in cases in which there is more than one replication per block, to allow for appropriate randomization. Suppose we decide to have two replications, each with two blocks. This means we will have 32 experimental plots with each block having 16 plots in them. Follow any of the randomization schemes in Chapter 5 to assign the block numbers of the basic plan to the blocks in the field.

Step 3: Randomly reassign the treatment in each block arrangement of the basic plan to the experimental plot separately and independently. This means

TABLE 6.4.1.2

A Half-Replicate of a 2^6 Factorial

Treatment Number	1	2	3	4	5	6	7	8
	9	10	11	12	13	14	15	16
Block I	(1)	ab	ac	bc	ae	af	ad	bd
	abef	ef	de	df	bf	be	ce	cf
Block II	acde	acdf	abdf	acef	cd	abcd	abcf	abce
	bcdf	bcde	bcef	abde	abcdef	cdef	bdef	adef

that the 16 treatment numbers in each of the four blocks (two in each replication) are randomly assigned to each block separately and independently. The results of this reassignment are shown in Table 6.4.1.3.

Based on this reassignment of the treatment numbers in the basic plan, we will assign treatment number 2 of the basic plan to plot 1 of block 1 of replication I, and treatment number 5 of the basic plan to plot 2 of block 1

TABLE 6.4.1.3

Randomly Reassigned Basic Plan Treatments to the
Experimental Plot

Basic Plan Treatment Number	Assigned Field Plot Number			
	Rep. I		Rep. II	
	Block 1	Block 2	Block 1	Block 2
1	2	8	10	6
2	5	16	12	9
3	11	6	9	10
4	10	2	16	14
5	7	9	7	1
6	3	15	13	11
7	16	1	8	5
8	4	13	6	2
9	1	5	14	16
10	6	11	4	3
11	14	7	1	12
12	8	4	15	4
13	12	10	2	13
14	9	12	3	8
15	13	3	11	15
16	15	14	5	7

of replication I. This is done for each plot in each block in each replication as shown in Figure 6.4.1.1.

Analysis of variance: In performing the analysis, there are several methods to compute the sum of squares for main effects and two-factor interactions. For simplicity, let us again assume we have a 2^6 factorial experiment with a half-replicate. To compute the sum of squares, we may construct 15 two-way tables for every pair of variables. Another approach would be to write down the combination of signs for each of the effects that are to be estimated. For this procedure, consult Cochran and Cox (1957). The quickest method of analyzing the fractional replication of a 2^n factorial is through an adaptation of Yates' algorithm.

Such an adaptation proceeds as follows:

1. The fractional replication design that contains 2^{n-k} combinations has a complete replication of n-k factors. Arrange the n-k factors in Yates' format, which ignores the k of the factors. A factor that is ignored is placed inside brackets. This is a reminder that the ignored factor has no part in the systematic order. For the 2^6 factorial experiment in which we have six factors (*ABCDEF*) each at two levels, there are 32 treatment combinations with a complete replication of the first five factors. This means that the last factor, or factor *F*, will be ignored in the arrangement of treatment combinations.

Block 1		Block 2			Block 1		Block 2	
1	2	1	2		1	2	1	2
ab	ae	abce	adef		ef	df	abcd	bcdf
3	4	3	4		3	4	3	4
de	ef	abcd	acdf		abef	cf	bcde	cdef
5	6	5	6		5	6	5	6
ad	ac	bcdf	bdef		ad	bf	acde	bcef
7	8	7	8		7	8	7	8
cf	bc	acde	abcdef		bd	af	cd	acdf
9	10	9	10		9	10	9	10
(1)	af	cd	bcef		be	bc	adef	abdf
11	12	11	12		11	12	11	12
be	bd	abcf	acef		(1)	ce	abde	acef
13	14	13	14		13	14	13	14
df	abef	bcde	abde		ab	ac	abcdef	abce
15	16	15	16		15	16	15	16
bf	ce	abdf	cdef		de	ae	bdef	abcf

Replication I   Replication II

FIGURE 6.4.1.1
Sample layout of a 2^6 factorial experiment with half of the treatments arranged in two blocks of 16 plots each.

2. Apply Yates' method as if the experiment were a complete replication of a 2^{n-k} factorial set. In our example, this would be a complete replicate of a 2^5 experiment.

3. Reintroduce the ignored factor (F) in the "Factorial Effect Identification column" as subsequently shown in Table 6.4.1.5. By doing so, each effect will have one or more aliases involving the ignored factor (F).

4. Finally, select the main effect and low-order interactions from the alias set to be identified with the contrast. For our example, we have used $ABCDEF$ as the defining contrast, hence the alias for A is $BCDEF$, and the alias for B is $ACDEF$, and so on. When we reintroduce the ignored factor, we have to identify its main and interaction effects for estimation purposes. The main effect of the ignored factor

is *F*, and its interactions are *AF, BF, CF, DF*, and *EF*. The aliases for these effects are *F = ABCDE, AF = BCDE, BF = ACDE, CF = ABDE, DF = ABCE*, and *EF = ABCD*. The identification assigned to each of the 32 factorial effects appears under the column labeled "Contrasts" in Table 6.4.1.5.

The following example is used to illustrate the analysis of variance for a fractional factorial design experiment.

Example 6.4.1

An animal scientist using a 2^6 factorial experiment in a half replicate is interested in knowing the impact of several factors on the daily weight gain of grazing beef steers. Specifically, the scientist is interested in knowing whether the three-factor interactions are significant. Factors *A, B, C, D, E*, and *F* are known to have an effect on the performance of the animal. The experiment was conducted with two replications (to allow for the estimation of the three-factor interactions) each with 32 treatment combinations. Each replication was further divided into two blocks each containing 16 plots. The daily weight gain data gathered from the experiment are given in Table

TABLE 6.4.1.4

Daily Weight Gain of Grazing Beef Steers from a 2^6 Factorial Experiment Conducted as a Half-Replicate in Blocks of 16 Units with Two Replications

Treatment Combination	Block 1			Treatment Combination	Block 2		
	Rep. I	Rep. II	Total		Rep. I	Rep. II	Total
(1)	2.13	2.18	4.31	ad	2.20	2.50	4.70
de	2.17	2.12	4.29	ae	2.35	2.24	4.59
df	2.40	2.11	4.51	af	2.34	2.34	4.68
ef	2.11	2.19	4.30	bd	1.50	1.89	3.39
ab	2.50	2.56	5.06	be	2.38	2.54	4.92
ac	2.33	2.43	4.76	bf	2.19	2.21	4.40
bc	2.15	2.30	4.45	cd	1.99	2.10	4.09
abde	1.99	2.00	3.99	ce	2.43	2.65	5.08
abdf	2.50	2.34	4.84	cf	2.00	2.10	4.10
abef	2.67	2.51	5.18	adef	1.50	2.35	3.85
acde	2.12	2.25	4.37	bdef	2.44	2.18	4.62
acdf	2.57	2.41	4.98	cdef	1.17	1.82	2.99
acef	2.00	2.28	4.28	abcd	2.67	2.41	5.08
bcde	2.82	2.69	5.51	abce	2.45	2.39	4.84
bcdf	2.16	2.19	4.35	abcf	1.90	1.85	3.75
bcef	1.56	1.89	3.45	abcdef	2.16	2.05	4.21
Total (RB)	36.18	36.45			33.67	35.62	

6.4.1.4.

SOLUTION

Step 1: Determine the number of factors, each at two levels, that is equal to the number of treatment combinations to be tested. As was stated in the previous section, in a fractional replication design of 2^{n-k} combinations, there is a complete replication of $n-k$ factors. The number of treatment combination (t) equals 2^n, ignoring the k factors. As the number of treatment combinations to be tested in the present example is 32, then $n = 5$ factors. This means that the complete replication contains the first five factors (*ABCDE*) and ignores the last factor (*F*).

Step 2: Apply Yates' method as if the experiment were a complete replicate of a 2^5 factorial experiment, and arrange the treatments such that when treatment (1) is included in the set of treatments, it always is listed as the first treatment; then comes those treatments with fewer number of letters. For example, *a* comes before *ab*, and *ac* comes before *ad* and so on. When arranging the treatments in this fashion, omit the treatment identification letters corresponding to the ignored factor. In our example, the treatment identification letter *af* comes before *ab* because factor *F* is ignored and plays no role in this arrangement process. As a reminder, those treatments that involve factor *f* are placed in brackets as shown in Table 6.4.1.5.

Step 3: Determine the degrees of freedom associated with each source of variation as given by the basic plan. The basic plan identifies the block, the main effect, and the two- and three-factor interactions. Because this experiment is performed with replications to allow for the measurement of the three-factor interactions, we must account for the variability from the replications, and the block x replication interaction. Thus, the following sources of variation and degrees of freedom are associated with this experiment:

Effects	d.f.
Replication	= 1
Block	= 1
Block × replication	= 1
Main	= 6
2-factor	= 15
3-factor	= 9
Error	= 30
Total	= 63

Step 4: Calculate the replication total (R), the block total (B), replication × block totals (RB) as shown in Table 6.4.1.4, and the grand total (G) as follows:

Rep. I (R_1) = 36.18 + 33.67 = 69.85

Rep. II (R_2) = 36.45 + 35.62 = 72.07

Block 1 (B_1) = 36.18 + 36.45 = 72.63

Block 2 (B_2) = 33.67 + 35.62 = 69.29

G = 69.85 + 72.07 = 141.92

Step 5: Calculate the correction factor and the total sum of squares as:

TABLE 6.4.1.5

Yates' Method of Computing the Sum of Squares for a 2^6 Factorial Experiment Conducted as a Half-Replicate in Blocks of 16 Units with Two Replications Using the Daily Weight Gain Data

Treatment Combination	t_0	t_1	t_2	t_3	t_4	t_5	Factorial Effect Identification Contrast	Alias
(1)	4.31	8.99	18.45	35.51	71.45	141.92	(G)	(G)
a(f)	4.68	9.46	17.06	35.94	70.47	4.40	A	A
b(f)	4.40	8.86	17.44	36.64	4.25	2.16	B	B
ab	5.06	8.20	18.50	33.83	0.15	−0.30	AB	AB
c(f)	4.10	9.21	18.99	0.99	− 0.81	−1.34	C	C
ac	4.76	8.23	17.65	3.26	2.97	2.24	AC	AC
bc	4.45	9.07	16.75	1.14	0.03	− 0.18	BC	BC
abc (f)	3.75	9.43	17.08	− 0.99	−0.33	−3.72	ABC	ABC[a]
d (f)	4.51	8.89	1.03	− 0.19	−0.33	−2.57	D	D
ad	4.70	10.10	− 0.04	− 0.62	−1.01	0.14	AD	AD
bd	3.39	9.36	1.64	0.14	1.05	2.26	BD	BD
abd (f)	4.84	8.29	1.62	2.83	1.05	− 2.48	ABD	ABD
cd	4.09	8.14	0.55	− 1.07	0.21	4.12	CD	CD
acd (f)	4.98	8.61	0.59	1.10	− 0.32	0.02	ACD	ACD
bcd (f)	4.35	7.36	−1.07	2.16	−3.07	6.64	BCD	BCD
abcd	5.08	9.72	0.08	−2.49	− 0.65	−4.86	ABCD	EF
e (f)	4.30	0.37	0.47	− 1.39	0.43	− 0.98	E	E
ae	4.59	0.66	− 0.66	1.06	− 3.00	− 4.10	AE	AE
be	4.92	0.66	− 0.98	− 1.34	2.27	3.78	BE	BE
abe (f)	5.18	− 0.70	0.36	0.33	− 2.13	− 0.36	ABE	ABE
ce	5.08	0.19	1.21	1.07	− 0.43	− 0.68	CE	CE
ace (f)	4.28	1.45	−1.07	− 0.02	2.69	0.14	ACE	ACE
bce (f)	3.45	0.89	0.47	0.04	2.17	−0.53	BCE	BCE
abce	4.84	0.73	2.36	1.15	−4.65	2.42	ABCE	DF
de	4.29	0.29	0.29	− 1.13	2.45	−2.57	DE	DE
ade (f)	3.85	0.26	−1.36	1.34	1.67	−4.40	ADE	ADE
bde (f)	4.62	− 0.80	1.26	−2.28	−1.09	3.12	BDE	BDE
abde	3.99	1.39	− 0.16	1.89	1.11	− 6.82	ABDE	CF
cde (f)	2.99	− 0.44	− 0.03	−1.65	2.47	− 0.78	CDE	CDE
acde	4.37	− 0.63	2.19	−1.42	4.17	2.20	ACDE	BF
bcde	5.51	1.38	0.19	2.22	0.23	1.70	BCDE	AF
abcde (f)	4.21	−1.30	−2.68	2.87	−v 5.09	− 5.32	ABCDE	F

[a] Block.

$$C = \frac{G^2}{rt} \tag{6.35}$$

where

C = Correction factor

G = Grand total

r = Number of replications

t = Total number of treatments tested

Thus,

$$C = \frac{(141.92)^2}{2 \times 32}$$

$$= 314.71$$

$$\text{Total } SS = \sum X^2 - C \tag{6.2}$$

$$= [(2.13)^2 + (2.17)^2 + \ldots + (2.05)^2] - 314.71$$

$$= 320.56 - 314.71$$

$$= 5.85$$

Step 6: Calculate the replication, block, and block × replication sum of squares as:

$$\text{Replication } SS = \frac{\sum R^2}{t} - C \tag{6.35}$$

$$= \frac{(69.85)^2 + (72.07)^2}{32} - 314.71$$

$$= 314.78 - 314.71$$

$$= 0.07$$

$$\text{Block } SS = \frac{\sum B^2}{t} - C \tag{6.36}$$

$$= \frac{(72.63)^2 + (69.29)^2}{32} - 314.71$$

$$= 314.88 - 314.71$$

$$= 0.17$$

$$\text{Block x Rep. } SS = \frac{\sum (RB)^2}{t/b} - (C + \text{Rep. } SS + \text{Block } SS) \tag{6.37}$$

where

b = number of blocks in each replication.

Therefore, the block x replication sum of square is:

$$= \frac{(36.18)^2 + (36.45)^2 + (33.67)^2 + (35.62)^2}{32/2} - (314.71 + 0.07 + 0.17)$$

$$= 315.00 - 314.95$$

$$= 0.05$$

Step 7: Calculate the *t* factorial effect totals by following the procedures outlined in the following text.

1. Place the original 32 treatment combinations as the first set of *t* treatments or t_0 in column 2 of Table 6.4.1.5, following Yates' arrangement.

2. Divide the 32 t_0 treatment combinations into half (*t*/2), forming 16 successive pairs. Then place the sum of the values of successive pairs (t_1) into column 3 of Table 6.4.1.5. This task provides the first half of the t_1 values. To compute the values for the second half of t_1, the value of the first treatment in each of the 16 successive pairs is subtracted from the value of the second treatment combination, respectively.

   Computation of the values for the first and second half of t_1 is illustrated in the following text.

   1st pair = 4.31 + 4.68 = 8.99
   2nd pair = 4.40 + 5.06 = 9.46
   3rd pair = 4.10 + 4.76 = 8.86
   .

   .

   .

   16th pair = 5.51 + 4.21 = 9.72

   The values for the second half of column 3 are computed as:

   9th pair = 4.68 − 4.31 = 0.37
   10th pair = 5.06 − 4.40 = 0.66
   11th pair = 4.76 − 4.10 = 0.66

.

.

.

16th pair = 4.21 − 5.51 = − 1.30

3. Compute the values for t_2 following the same procedures outlined in the previous step, except now we use t_2 instead of t_0 in our computation. Apply the same process to compute t_3, t_4, and t_5, each time using the t values of the previous step. With n factors, the process is continued until we reach column (n).

Step 8: Label each of the factorial effect totals computed in the last step with the letters of the corresponding treatments, and place them under the heading "Contrast" as shown in column 8 of Table 6.4.1.5. Note that the first value of column 7 or t_5 is the grand total (G). The second value refers to the treatment combination *a* (*f*). Given that factor *F* was ignored in arranging the treatment combinations, this value is therefore assigned to the *A* main effect. The assignment of the ignored factor was discussed earlier. Proceed with the identification of each of 32 specific factorial effects.

Step 9: Identify the aliases for the factorial effects and place them under the heading "Alias" in column 9 of Table 6.4.1.5. As was explained earlier, when we reintroduce the ignored factor we have to identify its main and interaction effects for estimation purposes. The main effect of the ignored factor is *F*, and its interactions are *AF*, *BF*, *CF*, *DF*, and *EF*. The aliases for these effects are *F* = *ABCDE*, *AF* = *BCDE*, *BF* = *ACDE*, *CF* = *ABDE*, *DF* = *ABCE*, and *EF* = *ABCD*. As we used blocking in this experiment, *ABC* = *DEF* is lost by confounding.

Step 10: Calculate the sum of square for the main effect, the two-factor interactions, the three-factor interactions, and the error, using the values given in column 7 of Table 6.4.1.5, in the following manner:

$$\text{Main effect } SS = \frac{(A)^2 + (B)^2 + (C)^2 + (D)^2 + (E)^2 + (F)^2}{(r)(2)^k} \tag{6.38}$$

$$= (4.40)^2 + (2.16)^2 + (-1.34)^2 + (-2.57)^2 + (-098)^2 + (-5.32)^2 / (2)(32)$$

$$= 61.69/64$$

$$= 0.96$$

$$\text{Two-factor interaction } SS = \left[(AB)^2 + (AC)^2 + \dots + (BF)^2 + (AF)^2 \right] / (r)(2)^k \tag{6.39}$$

$$= \left[(-0.30)^2 + (2.24)^2 + \ldots + (2.20)^2 + (1.70)^2 \right] / (2)(32)$$

$$= 149.13/64$$

$$= 2.33$$

Three-factor interaction $SS = \left[(ABD)^2 + (ACD)^2 + \ldots + (BDE)^2 + (CDE)^2 \right]$ (6.40)

$$/ (r)(2)^k$$

$$= 80.37/(2)(32)$$

$$= 1.26$$

To compute the error sum of squares, we subtract all the sum of squares from the total sum of squares computed in Step 5. This is shown in the following text.

Error SS = Total SS – (the sum of all other SS) (6.41)
$$= 5.85 - (0.07 + 0.17 + 0.05 + 0.96 + 2.33 + 1.26)$$
$$= 5.85 - 4.84$$
$$= 1.01$$

Step 11: Calculate the mean square for each source of variation as:

Replication MS = Replication $SS/1$
$$= 0.07/1$$
$$= 0.07$$

Block MS = Block $SS/1$
$$= 0.17/1$$
$$= 0.17$$

Block × Replication MS = Block × Rep. $SS/1$
$$= 0.05/1$$
$$= 0.05$$

Main effect MS = Main effect $SS/6$
$$= 0.96/6$$
$$= 0.16$$

2-factor interaction MS = 2-factor interaction $SS/15$
$$= 2.33/15$$

$$= 0.16$$

3-factor interaction MS = 3-factor interaction $SS/9$
$$= 1.26/9$$
$$= 0.14$$

Error MS = Error $SS/30$
$$= 1.01/30$$
$$= 0.03$$

Step 12: Compute the F value for each source of variation as shown in Table 6.4.1.6.

Step 13: Compare the computed F of Table 6.4.1.6 with the tabular F value given in Appendix E. The results indicate that the main effects and all interactions were highly significant. Similarly, the results show that the block effect was significant at the 5% level, implying that there are differences in the treatment combinations among blocks. It should be noted once again that without the use of replications in the experiment, it would not have been possible to estimate the three-factor interactions. Therefore, in experiments in which three-factor interactions may not be the primary concern of the experimenter, the experiment could be conducted without replications, following the same procedures outlined earlier.

In summary, there are several points that should be considered when using fractional replication. First, the experimenter would have to make a choice regarding the number of levels for factors to be included in the experiment. Generally, use is made of two levels, with the exception of the cases in which fitting a second-degree response surface for a quantitative factor is desired. Second, experimenters often desire to know the direction and amount of the main effect of a factor and whether there is interaction between two factors.

TABLE 6.4.1.6

Analysis of Variance of Daily Weight Gain Data from a Fractional Replication Design

Source of Variation	Degree of Freedom	Sum of Squares	Mean Square	F
Replication	1	0.07	0.07	2.33[ns]
Block	1	0.17	0.17	5.67[*]
Block × replication	1	0.05	0.05	1.67[ns]
Main effect	6	0.96	0.16	5.33[**]
2-factor interaction	15	2.33	0.16	5.33[**]
3-factor interaction	9	1.26	0.14	4.67[**]
Error	30	1.01	0.03	
Total	63	5.85		

Note: ns = nonsignificant; [*] = significant at 5% level; [**] = significant at 1% level.

Under such circumstances, use of a two-level design that estimates the main effect free of a two-factor or a three-factor interaction is warranted. Third, when use is made of smaller designs, estimation of the residual standard deviation is quite difficult — and at times impossible — from the observations. In such conditions, knowledge of a prior estimate is desirable. This is not a problem with larger designs where an estimate can be found if three-factor or higher-order interactions are negligible. Fourth, doubtful results may be obtained because of the presence of aliases. When there is difficulty in discerning some important effects, further investigation is a desirable alternative. In some circumstances it may prove helpful to conduct an experiment of similar size to the initial one, perhaps converting a $1/16$th-replicate into a $1/8$th-replicate to provide for a more favorable alias structure and, in particular, separating out the pair or pairs of contrasts that have given rise to interpretation difficulties. Finally, fractional replication designs allow for the inclusion of many factors that an experimenter may deem appropriate, based on prior knowledge and theoretical considerations. Such flexibility in the design is the hallmark of fractional replication designs.

References and Suggested Readings

Brownlee, K.A., Kelly, B.K., and Loraine, P.K. 1948. Fractional replication arrangement for factorial experiments with factors at two levels. *Biometrika* 35: 268–276.

Cochran, W.G. and Cox, G.M. 1957. *Experimental Designs*. 2nd ed. New York: John Wiley & Sons. chap. 6A.

Finney, D.J. 1945. The fractional replication of factorial arrangements. *Ann. Eugen.* 12: 291–301.

Fiske, W.F., Rekaya, R., and Weigel, K.A. 2003. Assessment of environmental descriptors for studying genotype by environment interaction. *Livest. Prod. Sci.* 82: 223–231.

Gomez, K.A. and Gomez, A.A. 1984. *Statistical Procedures for Agricultural Research.* 2nd ed. New York: John Wiley & Sons. chap. 4.

Hayes, B.J. Carrick, M., Bowman, P., and Goddard, M.E. 2003. Genotype x environment interaction for milk production of daughters of Australian dairy sires from test-day records. *J. Dairy Sci.* 86: 3736–3744.

Kempthorne, O. 1952. *The Design and Analysis of Experiments*. New York: John Wiley & Sons.

Kempthorne, O. and Tischer, R.G. 1953. An example of the use of fractional replication. *Biometrics* 9: 295–303.

Petersen, R.G. 1985. *Design and Analysis of Experiments,* New York: Marcel Dekker. chap. 8.

Exercises

1. A crop scientist is interested in using a split-split-plot design experiment in which there are three factors of interest. He wishes to assign the four planting dates (P_1, P_2, P_3, P_4) to the main plot arranged in randomized complete blocks (I, II, III, and IV). Subplots are to be fumigated (F_1) and nonfumigated (F_2) for mite control. The four harvest dates (H_1, H_2, H_3, H_4) are to be assigned to the sub-subplots. Show how you would lay out this experiment in the field.

2. Certain varieties of cotton such as Acala SJ-2 show potassium deficiency even when the soils, by test, do not show any deficiency. This problem is severe on soils with high levels of Verticillium wilt. In a $2 \times 2 \times 3$ split-split-plot design experiment to determine whether potassium deficiency in cotton is disease induced, the researchers have assigned the fertilizers (NPK) to the main plot, the treatments to the subplots, and the depth of soil as a sub-subplot factor. They have collected soil samples from two depths in fumigated and nonfumigated plots and recorded the following data.

Effect of Soil Fumigation on Levels of N, P, and K at Two Sampling Depths

	Parts per Million								
	N			P			K		
Treatment	Rep. I	Rep. II	Rep. III	Rep. I	Rep. II	Rep. III	Rep. I	Rep. II	Rep. III
				0–12 in. depth					
Fumigated	30.2	30.3	31.0	24.2	24.5	23.8	92.0	94.0	93.3
Nonfumigated	68.0	67.8	68.2	19.0	19.3	19.5	104.8	105.1	104.6
				12–24 in. depth					
Fumigated	28.2	30.3	28.5	20.2	21.5	20.8	89.0	88.4	88.7
Nonfumigated	30.2.	29.8	30.1	17.9	18.3	18.5	93.8	93.0	92.9

(a) Do a main-plot analysis.

(b) Do a subplot analysis.

(c) Do a sub-subplot analysis.

(d) Compute the coefficient of variation.

3. An environmental horticulturist is interested in finding out whether (1) stress-adapted landscapes save water, (2) whether irrigation equal to 15% or less reference evapotranspiration (ET_0) can be applied to established shrubs and ground cover without any drought-related injury. This is a three-factor experiment designed to

test the effect of three irrigation regimes (no irrigation, 12.0 in., and 24.0 in. of water) and two different irrigation methods (drip and furrow) on the growth of shrubs and ground covers such as Xylosma, Oleander, Cotoneaster, Juniper, Ice Plant, and Hedera. The experiment is strip-split-plot design replicated three times. The data collected at the end of a 2-yr period are given in the following table.

Growth of Shrubs and Ground Cover as a Function of Irrigation Water Received from April to August

Plantings	Irrigation Method	Inches of Water Applied	Growth in Inches		
			Rep. I	Rep. II	Rep. III
Xylosma	Drip	0.0	8.0	8.4	9.5
		12.0	19.5	20.1	20.2
		24.0	30.6	31.0	31.4
	Furrow	0.0	6.0	5.4	5.8
		12.0	12.8	16.9	17.4
		24.0	28.2	27.6	29.4
Oleander	Drip	0.0	18.0	19.4	19.5
		12.0	39.5	40.1	40.3
		24.0	60.6	59.0	61.4
	Furrow	0.0	16.0	15.4	15.7
		12.0	22.8	36.9	37.4
		24.0	48.2	47.6	49.4
Cotoneaster	Drip	0.0	6.0	6.4	6.5
		12.0	35.5	31.1	30.6
		24.0	40.6	41.0	41.3
	Furrow	0.0	4.0	4.4	4.8
		12.0	19.8	16.9	18.4
		24.0	25.2	27.6	29.5
Juniper	Drip	0.0	12.0	12.4	12.7
		12.0	10.5	10.1	10.2
		24.0	20.6	18.0	19.3
	Furrow	0.0	10.0	11.1	10.8
		12.0	9.8	6.9	7.4
		24.0	13.2	14.8	15.4
Ice Plant	Drip	0.0	22.0	18.8	19.5
		12.0	26.5	28.1	27.2
		24.0	33.6	33.0	32.4
	Furrow	0.0	16.0	15.4	15.8
		12.0	22.5	26.9	25.4
		24.0	28.8	29.6	29.9
Hedera	Drip	0.0	16.0	17.4	18.5
		12.0	20.5	22.1	20.7
		24.0	40.6	41.0	41.4
	Furrow	0.0	12.0	13.2	14.6
		12.0	15.8	14.9	14.6
		24.0	28.2	27.6	29.4

(a) Perform the analysis of variance.

(b) What conclusions can you draw from the analysis?

(c) Compute the coefficient of variation for the factors.

4. Irrigation specialists have found that subsurface drip irrigation has a number of potential advantages over conventional surface irrigation. However, the depth at which a subsurface irrigation system is placed has implications for seed germination. The researcher is interested in the depth of burial of the subsurface drip tape, the depth of seed placement, and irrigation frequency reference evapotranspiration (ET_0). Using a strip-split-plot design experiment with melon seeds, the following data were collected when the experiment was replicated four times:

Number of Emerged Seedlings out of 100 Seeds Planted 14 d after Initial Irrigation

Planting Depth (in.)	Burial Depth of Irrigation Tape	Amount of Water Applied	Number of Emerged Seedlings			
			Rep. I	Rep. II	Rep. III	Rep. IV
0.5	6	0.25 ET_0	35	32	34	36
1.5			53	54	52	56
2.5			1	0	0	2
0.5	9		40	38	41	36
1.5			38	37	35	38
2.5			0	0	1	1
0.5	12		9	10	12	8
1.5			30	32	28	30
2.5			0	0	0	0
0.5	6	0.50 ET_0	58	60	62	58
1.5			45	47	48	52
2.5			6	4	5	3
0.5	9		40	38	42	36
1.5			58	57	55	58
2.5			3	1	5	4
0.5	12		30	30	32	28
1.5			60	52	58	60
2.5			5	6	3	4
0.5	6	0.75 ET_0	50	50	52	46
1.5			55	57	58	52
2.5			1	1	2	2
0.5	9		48	48	44	46
1.5			45	47	42	40
2.5			4	3	1	2
0.5	12		28	30	31	26
1.5			36	34	38	32
2.5			2	2	4	2

(a) What do you conclude from the analysis?

(b) What do the coefficients of variation in this problem show?

5. In a fractional factorial design experiment (half-replicate), a crop scientist is interested in knowing the impact of five different fertilizers each at two levels on yield of corn. The scientist wishes to have

two blocks in each of the two replications. Show how you would lay out the experiment for the scientist.

6. Suppose that in the previous example, the data collected from the experiment are as shown below.

Corn Yield from a 2^5 Factorial Experiment Conducted as a Half-Replicate with Two Blocks Each Containing Two Replicates

| | Grain Yield (bu/ac) | | | Grain Yield (bu/ac) | |
| | Block 1 | | | Block 2 | |
Treatment	Rep. I	Rep. II	Treatment	Rep. I	Rep. II
(1)	132	125	ac	112	114
ae	151	145	ad	122	132
ab	136	138	bc	145	148
be	144	142	bd	161	158
acde	171	162	de	145	138
cd	154	159	ce	162	168
bcde	143	138	abde	144	140
abcd	132	136	abce	155	159

(a) Perform the analysis of variance.

(b) Which one of the main effects and interactions were highly significant?

7

Comparisons of Treatment Means

7.1 Introduction

Researchers have at their disposal various methods to compare means of treatments from experiments. Deciding which treatment means are significantly different is called *mean separation*. In conducting experiments to determine which treatment mean is significant, it is essential to keep in mind that the best method is one that meets the objectives of the experiment, is simple, and does not require assumptions that are difficult to meet. Given these criteria, researchers can design an appropriate experiment that compares treatment means and provides answers to their queries.

Researchers often use two or more factors in a factorial experiment to determine the average effect of the factors over a wide range of conditions. For example, a horticulturist may be interested in knowing how a variety of beans responds to the different rates of fertilizer application. Depending on what question the investigator is trying to answer, the experiment can take several forms. For instance, if the horticulturist is interested in knowing whether there is any response to the fertilizer, the question can easily be answered by comparing the mean yield from the fertilized fields with the mean yield from the control plots. If, on the other hand, the horticulturist wants to know if there are differences between the different rates of fertilizer applications, the mean yield of each treatment (rate of application) can be compared with the others to determine the best among them.

Additional factors can be introduced in this simple mean comparison example to illustrate the diversity of the mean comparisons. For instance, if the horticulturist decides to introduce an added criterion, such as band or broadcast application of fertilizer, in conducting the experiment, the question that the experimenter is asking is whether there is a difference in yield if fertilizer is applied through band application or broadcast. A comparison of means between those fields with band and broadcast applications should provide the answer.

The preceding comparisons are classified as either *paired* or *group comparisons*. Paired comparisons are the simplest and are used extensively in agricultural research. Group comparisons are made when treatments are classified according to some common characteristics. Because planned group comparisons allow for a more precise mean separation than multiple comparison tests, the *F*-test is a much more meaningful test among the groups. Once it is determined that a significant *F* exists, we then ask which of the mean values are significantly different. The ordinary *t*-test for the difference between means is applied to every pair of means. To save time in such a comparison, the least significant difference is computed. This will be explained in the following section.

7.2   Comparisons of Paired Means

As mentioned earlier, agricultural researchers often use this simple method of comparing treatment means. Paired comparisons can be carried out either as *planned* or *unplanned*. When planned, the researcher identifies, prior to conducting the experiment, a specific pair of treatments to be compared. The simplest comparison of this kind usually involves comparing each of the treatments with the control treatment. In the unplanned comparison, on the other hand, the researcher attempts to compare every possible pair of treatment means instead of choosing a particular set of treatments to be compared.

The procedures used in making a paired comparison are the least significant difference (LSD) test and Duncan's multiple range test (DMRT). When using a planned paired comparison, LSD is more appropriate, whereas in an unplanned paired comparison, use should be made of the DMRT.

7.2.1   Least Significant Difference Test

The least significance test is a form of *t*-test that is used to determine whether the difference between two treatment means is significant at a prescribed level of α. The LSD, although a suitable test for such a comparison, should not be used for comparing all possible means — especially when the *F*-test is not significant. Its indiscriminate use can lead to misleading interpretation of results. Appropriately, LSD should be used for planned comparisons when the number of comparisons is not large. As the number of comparisons increases, the chance of spuriously significant results increases. This means that the researcher will observe some differences that are significant, but not at the level of significance chosen. It has been pointed out by Cochran and Cox (1957) that if several *t*-tests have been performed, the probability that at least one of these is apparently significant is greater than 0.05. If the *t*-tests are independent, this probability is .23 for 5 tests, .40 for 10 tests, and .64 for 20 tests.

What this signifies is that comparison between means further apart than any two in an array will be made at lower levels of significance. Therefore, it is wise not to use LSD for comparing all possible pairs of means. Gomez and Gomez (1984) suggested that such comparisons should be limited to less than six for the results to be meaningful.

As an LSD test is similar to a *t*-test, the test for the statistical significance of the difference between two means follows the same reasoning. Suppose we have two treatments *i* and *j*, and are interested in using a *t*-test to compare the difference between their means; we would have:

$$t = \frac{\bar{d} - \mu_{\bar{d}}}{s_{\bar{d}}} \qquad (7.1)$$

where

$\bar{d} = \bar{X}_1 - \bar{X}_2$ or the difference between two means
$\mu_{\bar{d}}$ = population of mean differences where the mean is 0
$s_{\bar{d}}$ = the standard error of the mean difference

To perform the *t*-test using LSD, we would simply replace $\bar{d}$ with LSD and $\mu_{\bar{d}}$ with 0 (as it is assumed that the mean from a population of mean differences is 0), and our equation becomes:

$$t_\alpha = \frac{LSD - 0}{s_{\bar{d}}} \qquad (7.2)$$

or

$$t_\alpha = \frac{LSD}{s_{\bar{d}}} \qquad (7.3)$$

Using Equation 7.3 to solve for LSD, we will have:

$$LSD = \left(t_\alpha\right)\left(s_{\bar{d}}\right) \qquad (7.4)$$

We can use Equation 7.4 to test the LSD at a prescribed level of significance. The level of significance α serves as the boundary between significant and nonsignificant differences between any pair of treatment means. Once the level of significance attached to the test is identified, Equation 7.4 can be written as:

$$LSD_\alpha = \left(t_\alpha\right)\left(s_{\bar{d}}\right) \qquad (7.5)$$

where

t_α = the tabular t value for a level of significance () with
n = error degrees of freedom

$$s_{\bar{d}} = \sqrt{(s_1^2 / r_1) + (s_2^2 / r_2)}$$

In computing the LSD at a prescribed level of significance α, the two treatments are considered to be significantly different if their difference exceeds the computed LSD value. On the other hand, if the difference between the treatments is less than the computed LSD, the results are not significantly different. In using the LSD procedure, it should be noted that the appropriate standard error of estimate ($s_{\bar{d}}$) is used. Depending on the design of the experiment, and the types of means compared, the standard error of estimate to be used varies. The steps in computing the LSD are illustrated with the following examples using different experimental designs, and comparing different types of treatment means.

Complete block design: In a complete block design in which the experiment may be a completely randomized design (CRD), a randomized complete block (RCB), or a Latin square in which only one error term is involved, the standard error is computed as:

$$s_{\bar{d}} = \sqrt{\frac{2s^2}{r}} \qquad (7.6)$$

where

s^2 = the mean square error (MSE) in the analysis of variance
r = the number of replications common to both treatments in the pair

On the other hand, if the treatments have different numbers of replications, then the standard error of estimate is computed as:

$$s_{\bar{d}} = \sqrt{\left(\frac{s^2}{r_1}\right) + \left(\frac{s^2}{r_2}\right)} \qquad (7.7)$$

We will illustrate the case with equal and unequal numbers of replications, using the data from previous examples.

Equal replications: The data from a completely randomized experiment involving seven treatments with a control (Example 4.2.1.2) are used to determine if one or more of the seven vitamin supplementation treatments is better than the control treatment. The steps in performing the LSD test are given in the following text.

TABLE 7.2.1.1

Mean Weight Gain Differences between the Seven Treatments and the Control Treatment

Treatment	Treatment Mean (kg)	Difference from Control (kg)
A	4.94	0.64[ns]
B	5.02	0.72[ns]
C	6.58	2.28**
D	5.24	0.94[ns]
E	4.44	0.14[ns]
F	6.54	2.24**
G	5.18	0.88[ns]
H (Control)	4.30	—

Note: [ns] = nonsignificant, ** = significant at the 1% level.

Step 1: Compute the mean difference between the control and each of the seven treatments as shown in Table 7.2.1.1.

Step 2: Compute the LSD value at the 5% and 1% levels of significance. Because the MSE for this example was 0.76 with 32 error degrees of freedom and five replications, the tabular t value from Appendix I is 2.041 for the 5% level and 2.741 for the 1% level of significance. Thus, we have:

$$LSD_{.05} = (t_{.05})\sqrt{\frac{2s^2}{r}}$$

$$= 2.041\sqrt{\frac{2(0.61)}{5}}$$

$$= 1.00$$

$$LSD_{.01} = 2.741\sqrt{\frac{2(0.61)}{5}} = 1.35$$

Step 3: Make a comparison of each of the mean differences computed in Step 1 with the computed LSD values in Step 2. It appears that only treatments C and F are significant, implying that these two treatments show increased weight gains, whereas the other five treatments were not significantly different from the control treatment.

Unequal replications: Researchers often are faced with situations in which, for unforeseen reasons, elements of the experimental unit are destroyed or lost in the process. In such conditions, the researcher may not have equal numbers of replications in the experiment. When we have unequal numbers of replications in an experiment, the appropriate comparison is the planned-pair comparisons. Using the data from an unequal replication experiment (Example 4.2.1.3 in Chapter 4), in which response to nitrogen fertilization on different maize hybrids were studied in a completely randomized design,

TABLE 7.2.1.2

Mean Yield Differences between the Five Hybrids (Treatments) and the Control Treatment

Treatment	Number of Replications	Mean Yield bu/ac	Difference from Control (bu/ac)	LSD Values 5%	LSD Values 1%
P3747	3	154.0	12.0^{ns}	114	160
P3732	3	145.7	3.7^{ns}	114	160
Mo17 × A634	2	149.0	7.0^{ns}	128	179
LH 74 × LH51	2	166.5	24.5^{ns}	128	179
CP18 × LH54	2	140.5	1.5^{ns}	128	179
Control	3	142.0	—	—	—

Note: ^{ns} = nonsignificant.

the agronomist wishes to determine if any of the five hybrids (treatments) performed better than the control treatment.

Step 1: Compute the mean difference between each of the five treatments and the control as shown in column 4 of Table 7.2.1.2.

Step 2: Compute the LSD value at the 5% and 1% level of significance. As some treatments have two replications and others have three, we must compute two sets of the LSD values. Because the MSE for this example was 4966.72 (see Table 4.2.1.6) with 9 error degrees of freedom, the tabular *t* value from Appendix I is 2.262 for the 5% level and 3.250 for the 1% level of significance. Thus, the LSD values are computed as:

A. Treatments with three replications compared to the control:

$$LSD_{.05} = 2.262\sqrt{\frac{2(4,966.72)}{3}}$$

$$=130\ bu/ac$$

$$LSD_{.01} = 3.250\sqrt{\frac{2(4,966.72)}{3}}$$

$$=187\ bu/ac$$

B. Treatments with two replications compared with control: as the number of replications of the control mean is 3 and each of the treatments has only two treatments, we use Equation 7.5 and substitute Equation 7.7 for the standard error of estimate in it. The LSD values are computed as follows:

$$LSD_\alpha = t_\alpha \sqrt{\left(\frac{s^2}{r_i}\right) + \left(\frac{s^2}{r_j}\right)} \tag{7.8}$$

$$LSD_{.05} = 2.262\sqrt{\left(\frac{4,966.72}{2}\right) + \left(\frac{4,966.72}{3}\right)}$$

$$=145 \text{ bu/ac}$$

$$LSD_{.01} = 3.250\sqrt{\left(\frac{4,966.72}{2}\right) + \left(\frac{4,966.72}{3}\right)}$$

$$=209 \text{ bu/ac}$$

Step 3: Make a comparison of each of the mean differences computed in Step 1 with the computed LSD values in Step 2. Indicate by an appropriate asterisk or the nonsignificant (ns) sign whether the treatments are different from the control. From our analysis, it appears that none of the treatments are significantly better than the control treatment.

Balanced lattice design: Recall from Chapter 4 (Section 4.3.1) that in these designs every pair of treatments occurs once in the same incomplete block. Hence, the degree of precision in comparing all pairs of treatments is the same. In the analysis of variance, we computed the total, the treatment, and the replication sum of squares in the usual manner. However, the sum of squares for blocks within replications was adjusted for treatment effects. As we apply the LSD test to the data from a balanced lattice design, we must use the adjusted treatment means in computing the mean difference, and the effective MSE to compute the standard error of the mean difference.

To illustrate the LSD test, we use the data from a 3×3 balanced lattice design experiment in which each treatment is replicated four times. The researcher has gathered the following data on alfalfa forage yield. He wishes to know if there are any differences in yield between the eight treatments and the control. In performing the analysis of variance, the researcher has found the effective MSE with 16 degrees of freedom to be 2.1. The following adjusted mean data are reported and are presented in Table 7.2.1.3. The steps in performing the LSD test are given below.

Step 1: Compute the mean difference between the control and the eight treatments, as shown in column 3 of Table 7.2.1.3.

Step 2: Calculate the LSD value at level of significance as shown below:

$$LSD\alpha = t_\alpha \sqrt{\frac{2 \text{ (effective error MS)}}{r}} \tag{7.9}$$

TABLE 7.2.1.3

Adjusted Mean Yield of Alfalfa Forage from First Harvest in Oklahoma

Treatment Number	Adjusted Mean (t/ac)	Difference from Control (t/ac)
1	2.68	1.18^{ns}
2	2.61	1.11^{ns}
3	2.42	0.92^{ns}
4	1.84	0.64^{ns}
5	1.97	0.47^{ns}
6	2.58	1.08^{ns}
7	1.79	0.29^{ns}
8	2.62	1.12^{ns}
9 (Control)	1.50	—

Note: ^{ns} = nonsignificant.

As the tabular t values from Appendix I with 16 degrees of freedom are 2.120 for the 5% and 2.921 for the 1% levels of significance, respectively, we have:

$$LSD_{.05} = 2.120\sqrt{\frac{2(2.1)}{4}}$$

$$= 2.17 \text{ t/ac}$$

$$LSD_{.01} = 2.921\sqrt{\frac{2(2.1)}{4}} = 2.99 \text{ t/ac}$$

Step 3: Compare the computed LSD values at the different levels of significance with the mean difference of each pair computed in Step 2. It appears that none of the treatments are significantly better than the control treatment.

Partially balanced lattice design: Performing the LSD test on the data from a partially balanced lattice design experiment is similar to the balanced lattice design, in that the computation of the mean difference requires the use of the adjusted mean and the effective MSE is used to compute the standard error of the mean difference. As we apply the LSD test to a partially balanced lattice experiment, we must remember there are two effective error mean squares to be considered. The first (MS_1) involves treatments that were tested in the same incomplete block, and the second (MS_2) deals with the treatment pairs that never appeared in the same incomplete block. Thus, the LSD test for each would be:

$$LSD_\alpha = t_\alpha\sqrt{\frac{2 \text{ (effective error } MS_1)}{r}} \tag{7.10}$$

$$LSD_\alpha = t_\alpha \sqrt{\frac{2 \text{ (effective error } MS_2)}{r}} \qquad (7.11)$$

For the purpose of illustrating the procedure for testing the LSD, we will take the data from Table 4.3.2.3 and assume that treatment number 5 is the control treatment, and that the researcher wishes to determine if any of the hybrid corns is significantly better than the control.

Before we start performing the LSD test, we have to adjust the data that were presented in Table 4.3.2.3. First, to compute the adjusted treatment means, we use the following formula:

$$\mu_a = \frac{T_a}{k+1} \qquad (7.12)$$

where

μ_a = adjusted treatment mean
T_a = adjusted treatment total
$k + 1$ = the number of replications (r)

As Equation 7.12 calls for the adjusted treatment totals, we must convert the treatment totals given in Table 4.3.2.3 as follows:

$$T_a = T + \mu \sum C_b \qquad (7.13)$$

where T is the treatment total, and C_b is the difference between the sum of treatment totals for all treatments appearing in a block and the product of the number of replications and block totals ($C_b = \Sigma T - nB_T$). The C_b value for our example was computed and presented in Table 4.3.2.4. The adjusted treatment totals are computed using Equation 7.13, and the results are presented in Table 7.2.1.4. With this information, we are ready to perform the LSD test. The step-by-step procedures are outlined in the following text.

Step 1: Calculate the mean difference between the control treatment (treatment number 15) and each of the 24 hybrid corns. The mean difference is shown in column 5 of Table 7.2.1.4.

Step 2: Calculate the LSD values, using the two effective error means MS_1 and MS_2. These values were computed in Chapter 5 as $MS_1 = 50.8$, and $MS_2 = 51.5$. The tabular t values from Appendix I with $n = 56$ degrees of freedom are 2.004 for 5% and 2.397 for 1% levels of significance, respectively. Thus, the LSD for comparing two corn hybrids tested in the same incomplete block at these two levels are

TABLE 7.2.1.4

Comparison of the Adjusted Treatment Means with the Control Treatment

Treatment Number	Treatment Total (T)	Adjusted Treatment Total (T_a)	Adjusted Treatment Mean (μ_a)	Difference from Control
1	119	119.29	29.82	18.33**
2	72	72.29	17.82	6.33ns
3	121	121.29	30.32	18.83**
4	66	66.29	16.57	5.08ns
5	82	82.29	20.57	9.08ns
6	71	70.62	17.66	6.17ns
7	105	104.62	26.16	14.67**
8	70	69.62	17.41	5.92ns
9	87	86.62	21.66	10.17*
10	92	91.62	22.91	11.42*
11	83	82.97	20.74	9.25ns
12	112	111.97	27.99	16.50**
13	84	83.97	20.99	9.50ns
14	60	59.77	14.99	3.50ns
15	46	45.97	11.49	—
16	112	111.56	27.89	16.40**
17	83	82.56	20.64	9.15ns
18	84	83.56	20.89	9.40ns
19	90	89.56	22.39	10.90*
20	121	120.56	30.14	18.65**
21	80	80.56	20.14	8.65ns
22	87	87.56	21.89	10.40*
23	104	104.56	26.14	14.65**
24	49	49.56	12.39	0.90ns
25	106	106.56	26.64	15.15**

Note: Treatment numbers for treatments tested in the same incomplete block with the control are shown in bold; ns = nonsignificant; * = significant at 1% level; ** = significant at 5% level.

$$LSD_1 = 2.004\sqrt{\frac{2(50.8)}{4}} = 10.10 \text{ bu/ac}$$

$$LSD_1 = 2.397\sqrt{\frac{2(50.8)}{4}} = 12.08 \text{ bu/ac}$$

For those corn hybrids that were not tested together in the same incomplete block, the LSD values are:

$$LSD_2 = 2.004\sqrt{\frac{2(51.5)}{4}} = 10.17 \text{ bu/ac}$$

$$LSD_2 = 2.397\sqrt{\frac{2(51.5)}{4}} = 12.16 \text{ bu/ac}$$

Step 3: Compare each of the mean differences computed in Step 1 with the calculated LSD values. We use the LSD_1 for those pairs that were tested together in the same incomplete block, and the LSD_2 for those pairs of treatments that were not tested in the same incomplete block. The results of all pairs of comparison with the control treatment are shown with the appropriate signs to indicate their significance and nonsignificance in column 5 of Table 7.2.1.4.

Applying the LSD test to the next set of designs that involve two or more factors is as simple as those illustrated with the single factor. We will illustrate the application of the LSD test only for the split-plot design, as the procedures can be generalized to the strip-plot and the split-split-plot design easily. There are only two methodological adjustments that are necessary. One relates to the use of the appropriate standard error of the mean difference ($s_{\bar{d}}$), and the other is the use of the weighted t values in the calculation of the LSD values. Depending on the type of paired comparison, the calculation of the standard error of the mean difference varies. When the standard error of the mean involves more than one error mean square, the appropriate t value to use will be the weighted t value. The condition that calls for the use of the weighted t values is explained in the following section when we discuss the split-plot design.

Split-plot design: Recall from Chapter 6 that in a split-plot design, the researcher is able to work with two variable factors in an experiment. The design allows the investigator to increase the precision in estimating certain effects while sacrificing certain other effects. Usually, the precision of the treatment assigned to the main plots is sacrificed for the increased precision of the average treatment effect assigned to the subplots. With this randomization of the treatments within a block, there are two error terms that the researcher will work with: the main-plot error and the subplot error. As one would expect, the main-plot error is usually larger as a result of the increased variability from widely spaced main plots, whereas the subplot error is smaller because of the reduced variability from closely spaced subplots within the main plot.

So in performing the LSD test on the data from a split-plot design, the researcher will work with four mean comparisons, as we have two factors and two error terms. The four mean comparisons and the standard error for the mean difference $s_{\bar{d}}$ are given in Table 7.2.1.5.

In the formulas shown in Table 7.2.1.5, E_i or EMS_i refers to the MSE of the ith factor, a is the number of main-plot treatments, and b is the number of subplot treatments. Whenever the standard error of the mean difference involves two error terms (as is the case in comparison D), we use the following equation to compute the weighted t values:

TABLE 7.2.1.5

Standard Error of the Mean Difference for the Four Types of Paired Comparisons in a Split-Plot Design

	Paired Comparison	$s_{\bar{d}}$
A.	Two main-plot means averaged over all subplot treatments averaged over all vertical treatments	$\sqrt{\dfrac{2E_a}{rb}}$
B.	Two subplot means averaged over all main-plot treatments, or 2 vertical means averaged over all horizontal treatments	$\sqrt{\dfrac{2E_b}{ra}}$
C.	Two subplot means in the same main-plot treatment	$\sqrt{\dfrac{2E_b}{r}}$
D.	Two main-plot means in the same or different subplot treatments	$\sqrt{\dfrac{2\left[(b-1)\,E_b+E_a\right]}{rb}}$

$$t_w = \frac{(b-1)\,EMS_b t_b + EMS_a t_a}{(b-1)\,EMS_b + EMS_a} \tag{7.14}$$

To illustrate the LSD test for the split-plot design, let us use the data from Example 5.5.1, and compare the between-hybrid means with the same nitrogen rate. The steps are:

Step 1: Calculate the standard error of the mean difference. As two subplot means within the same main plot are compared in this example, the formula to calculate the standard error of the mean difference is

$$s_{\bar{d}} = \sqrt{\frac{2\,EMS_b}{r}} \tag{7.15}$$

For our example, the error mean square (*b*), which is 29.99, is taken from Table 5.5.4 and is substituted in Equation 7.15 to get:

$$s_{\bar{d}} = \sqrt{\frac{2(29.99)}{2}}$$

$$= 5.48 \text{ bu/ac}$$

Step 2: Calculate the LSD value at level of significance, using Equation 7.5 as shown in the following text.

$$LSD_\alpha = \left(t_\alpha\right)\left(s_{\bar{d}}\right) \tag{7.5}$$

The tabular t values from Appendix I with $n = 16$ error b degrees of freedom are 2.120 for 5% and 2.921 for 1% levels of significance, respectively. Thus, the LSD test is:

$$LSD_{.05} = \left(2.120\right)\left(5.48\right) = 11.62$$

$$LSD_{.01} = \left(2.921\right)\left(5.48\right) = 16.01$$

Step 3: Compare the difference between two hybrid means at the same nitrogen fertilization rate with the LSD values. To do this, we need to construct a two-way table of means that shows the mean yield of the hybrids over the two replications for each rate of fertilizer applied. This is shown in Table 7.2.1.6.

We are now able to compare the difference between any two hybrid means at the same rate of nitrogen fertilization and the LSD value. The difference between the mean yield of hybrid P3747 and P3732 is 137.5 − 132.5 = 5.0. This yield difference is smaller than the LSD value of 11.62, implying the nonsignificance of the yield difference. On the other hand, the mean yield difference between hybrids P3732 and Mo17 × A64, as well as A632 × LH38, is 137.5 − 122.5 = 15. This yield difference is significant at the 5% level. The mean yield difference for the different hybrids at the same rate can be compared with the LSD values similarly, and a decision can be arrived at regarding the significance of the difference.

The procedure for performing the LSD test for the strip-plot and the split-split-plot is similar to the split-plot, except that the formulas for the standard error of the mean difference are different for the different comparisons. The reader is referred to Gomez and Gomez (1984), which provides a detailed

TABLE 7.2.1.6

Mean Yields of Five Corn Hybrids Tested with Four Nitrogen Rates in a Split-plot Design Experiment

Nitrogen Rate (lb/ac)	Mean Yield from Two Replications (bu/ac)				
	P3747	P3732	Mo17 × A64	A632 × LH38	LH74 × LH51
0	132.5	137.5	122.5	122.5	130.0
70	160.0	155.0	147.5	145.0	175.0
140	180.0	170.0	160.0	167.5	177.5
210	175.0	182.5	162.5	155.0	185.0

list of the formulas for the standard error of the mean difference for the strip-plot and the split-split-plot designs.

7.2.2   Duncan's Multiple-Range Test (DMRT)

Among the several multiple-range tests, Duncan's test is the most widely used and misused in agriculture. The test is especially useful when the number of all possible pairs of treatment means compared is large and when several unrelated treatments are included in an experiment. However, its inappropriate use has been pointed out in the literature extensively (Maindonald and Cox, 1984; Bryon-Jones and Finney, 1983; and Morse and Thompson, 1981).

The principle behind the test is to take an ordered set of treatment means and test each group of two, three, or more successive means to determine whether the spread between the highest and the lowest of each group is significantly large in terms of the experimental error mean square, at the same time allowing for the number of treatments being tested in the group. Unlike other multiple comparison tests such as the Fisher's Significance Difference (FSD) test, the DMRT does not require that the F-test be significant before making pairwise comparisons. Instead, the DMRT replaces the F-test by a range test of the difference between the smallest mean in the set of k means. If this range is not significant, then none of the pairwise comparisons can be significant in terms of the DMRT.

As was stated in the beginning of Section 7.2.1, when several t-tests have been performed and the t-tests are independent, the probability that one of these tests is significant decreases as the number of tests performed increases. This, of course, presents problems in reporting the results of a test. The same problem surfaces when several confidence intervals are computed. Duncan's multiple-range test tries to provide some level of protection against statements made in the summary of an experiment. Consider for a moment that a specified pair of true means is identical. As such, the combined F- and t-tests give a protection level of at least 95% that the means will be declared indistinguishable. Suppose now we have three identical true means. The protection level for these three means depends on the position of the fourth mean. If the fourth mean happens to be far removed from the three identical means in the array, the level of protection is about 87% (Cochran and Cox, 1957). Depending on how many means are compared and how far they are from each other, the level of protection varies. Duncan's multiple-range test specifies the level of protection that guards against erroneously finding significant results. The level of protection that has been specified by Duncan is 95% for sets of two means, 90.25% or $(0.95)^2$ for sets of three means, and 85.7% or $(0.95)^3$ for sets of four means, and so on. Even though there is a reduced level of protection in this test, the sensitivity of the test is greater than the Newman-Keuls test, which keeps all levels at 95% in detecting real differences.

To apply Duncan's multiple-range test for comparing all pairs of means involving equal sample sizes, we arrange the k treatment averages in ascending (or descending) order, and compute the standard error of a mean as:

$$s_{\bar{d}} = \sqrt{\frac{2s^2}{r}} \tag{7.6}$$

In the case of unequal sample size (r), we replace the r in Equation 7.16 with the harmonic mean r_h of the (r_i), where

$$r_h = \frac{k}{\displaystyle\sum_{i=1}^{k}(1/r_i)} \tag{7.16}$$

In Equation 7.16 if $r_1 = r_2 = \ldots = r_k$, then $r_h = r$. Once we have computed the standard error of the mean, we then obtain from Appendix H the values of $r_\alpha(p,f)$, for $p = 2, 3, \ldots, k$, where α is the significance level, p is the number of means for range being tested, and f is the number of degrees of freedom for error. Convert these ranges into a set of $k-1$ least significant ranges (R_p) for $p = 2, 3, \ldots, k$, by calculating

$$R_p = r_a(p,f)s_{\bar{d}} \quad \text{for } p = 2, 3, \ldots, k \tag{7.17}$$

Now we are ready to test the observed differences between means by taking the difference between the largest and smallest means and comparing it with the least significant range R_p. Next, the difference of the largest and the second smallest is computed and compared with the least significant range R_{k-1}. The comparisons are continued until all means have been compared with the largest mean. A pair of means is considered significantly different if the observed difference is greater than the corresponding least significant range. It should be kept in mind, however, that no differences between a pair of means is considered significant if the two means involved fall between two other means that do not differ significantly. What this means is that, for example, if among six means in an array (A B C D E F), we find A to be significantly different from B, and B not to be significantly different from E, then we draw a line joining B and E as shown below.

A <u>B C D E</u> F

In the above situation it is not necessary to test C and D against E, as they fall within two other means that are not significantly different. Our next set

TABLE 7.2.2.1

Rank Order of the Treatment Means from an Animal Weight-Gain
Experiment Involving Eight Treatments and Five Replications

Treatment	Treatment Mean (kg/animal)	Rank
C	6.58	1
F	6.54	2
D	5.24	3
G	5.18	4
B	5.02	5
A	4.94	6
E	4.44	7
H (Control)	4.30	8

of means to be compared will be D and F. If the difference is not significant, D and F will be joined by a line as shown below.

$$\text{A} \quad \text{B} \quad \text{C} \quad \underline{\text{D} \quad \text{E} \quad \text{F}}$$

The steps in performing a Duncan's multiple-range test are given in the following text, using the data from a previous example (Example 4.2.1.2), in which the mean weight gain from vitamin supplementation of the eight treatments with five replications were recorded as 4.94, 5.02, 6.58, 5.24, 4.44, 6.54, 5.18, and 4.30 kg. The MSE was computed as 0.61.

Step 1: Rank the treatment means in a descending order as shown in Table 7.2.2.1.

Step 2: Using Equation 7.6 compute the standard error of the mean as:

$$s_{\bar{d}} = \sqrt{\frac{2(0.61)}{5}} = 0.49 \text{ kg}$$

Step 3: From Appendix H, determine the $r_\alpha(p,f)$ for =.05 and 32 error degrees of freedom. Thus, we have:

$$r_{.05}(2,32) = 2.88$$

$$r_{.05}(3,32) = 3.03$$

$$r_{.05}(4,32) = 3.11$$

$$r_{.05}(5,32) = 3.19$$

$$r_{.05}(6,32) = 3.24$$

$$r_{.05}(7,32) = 3.28$$

Step 4: Using Equation 7.17, determine the $(k-1)$ values of the shortest significant range (R_p) as:

$$R_2 = r_{.05}(2,32)s_{\bar{d}} = (2.88)(0.49) = 1.41$$

$$R_3 = r_{.05}(3,32)s_{\bar{d}} = (3.03)(0.49) = 1.48$$

$$R_4 = r_{.05}(4,32)s_{\bar{d}} = (3.11)(0.49) = 1.52$$

$$R_5 = r_{.05}(5,32)s_{\bar{d}} = (3.19)(0.49) = 1.56$$

$$R_6 = r_{.05}(6,32)s_{\bar{d}} = (3.24)(0.49) = 1.59$$

$$R_7 = r_{.05}(7,32)s_{\bar{d}} = (3.28)(0.49) = 1.61$$

Step 5: Compare the largest mean with the smallest, using the information provided in Table 7.2.2.1. If the difference between these means and the R_p value is equal to or greater, the means are significantly different. Repeat this process of comparison for the remaining treatment means as shown in the following text.

C vs. H: $6.58 - 4.30 = 2.28 > 1.61$ (R_7)

C vs. E: $6.58 - 4.44 = 2.14 > 1.59$ (R_6)

C vs. A: $6.58 - 4.94 = 1.64 > 1.56$ (R_5)

C vs. B: $6.58 - 5.02 = 1.56 > 1.52$ (R_4)

C vs. G: $6.58 - 5.18 = 1.40 < 1.48$ (R_3)

C vs. D: $6.58 - 5.24 = 1.34 < 1.41$ (R_2)

F vs. H: $6.54 - 4.30 = 2.24 > 1.59$ (R_6)

F vs. E: $6.54 - 4.44 = 2.10 > 1.56$ (R_5)

F vs. A: $6.54 - 4.94 = 1.60 > 1.52$ (R_4)

F vs. B: $6.54 - 5.02 = 1.52 > 1.48$ (R_3)

F vs. G: $6.54 - 5.18 = 1.36 < 1.41$ (R_2)

D vs. H: $5.24 - 4.30 = 0.90 < 1.56$ (R_5)

Step 6: Indicate statistical significance by lines notation as shown in the following text.

C	F	D	G	B	A	E	H
6.58	6.54	5.24	5.18	5.02	4.94	4.44	4.30

You will notice that in the preceding presentation there is no significant difference (at the 5% level) between treatments C and G, as well as between treatments D and H. Because treatments F and D are in a subset with non-significant range, it is not necessary to test F treatment against D and G treatments. Similarly, as there is no significant difference between the D and H treatments, it is not necessary to test all those treatments such as G, B, A, and E that are a subset of a nonsignificant range.

We can also use the alphabet notation to show the nonsignificance of the means compared. When using the alphabet notation to present the results, significant differences can be shown even if the means are not arrayed. Alphabet letters such as a, b, c, etc., are used to show that the means compared are not significantly different at the 5% level. For the present example we have:

C	F	D	G	B	A	E	H
6.58	6.54	5.24	5.18	5.02	4.94	4.44	4.30

a

b

7.3   Comparisons of Grouped Means

In the preceding section we discussed two of the most widely used pairwise comparisons in agriculture. When an experiment involves $k \geq 3$ means and if we test all pairwise differences among means, we actually have $c = k(k-1)/2$ comparisons or tests of significance to perform. In making these group comparisons, we are simply partitioning the treatment sum of squares into meaningful component comparisons that meet the experimental objectives. That is, the researcher must choose from a number of comparisons those sets of group comparisons that provide answers to the research questions. Once the desired group comparisons have been selected, the researcher then can compute the sum of squares for the components of each group and perform an F-test, as was the case in the analysis of variance. The components may be orthogonal class comparisons or trend comparisons. In the following sections, we shall discuss these comparisons using *orthogonal coefficients*.

7.3.1   Orthogonal Class Comparisons

A researcher, in making two comparisons such as d_i and d_j on the same set of k treatments, would have an *orthogonal* or *independent* comparison if the sum of the products of the corresponding coefficients for d_i and d_j is equal to 0, that is, if

$$\sum_{i}^{k} a_{ki}a_{kj} = a_{1i}a_{1j} + a_{2i}a_{2j} + \ldots + a_{ki}a_{kj} = 0 \tag{7.18}$$

Equation 7.18 states that i and j are distributed independently of each other. This independence is relevant to the partitioning of the treatment sum of squares. If the researcher wishes to divide the treatment sum of squares into the contribution from a comparison d_1 and the remainder, and wishes to further subdivide the remainder, comparisons must be chosen that are orthogonal to d_1. In this context, after removing the contribution of d_2, the next comparison d_3 must be orthogonal to both d_1 and d_2 so on.

A group of comparisons are said to be *mutually orthogonal* if the sum of the products of the coefficients for all possible pairs of comparisons is equal to 0. In an experiment with k treatments, we can construct only $k-1$ single degree of freedom mutually orthogonal contrasts. The sum of their sum of squares equals the treatment sum of squares as given in the following text.

$$\text{Treatment } SS = SS(M_1) + SS(M_2) + \ldots + SS(M_{k-1}) \tag{7.19}$$

where M_1, M_2, and M_{k-1}, are mutually orthogonal single *d.f.* contrasts. Each of the sum of squares on the right-hand side of Equation 7.19 are also a mean square with one *d.f.* Thus, in performing the F-test, each of the treatment mean squares is divided by the error mean square.

To show how we construct a table of comparison coefficients, let us take three of the many comparisons that might be made on a set of $k = 4$ treatment

TABLE 7.3.1.1

Three Comparisons on $k = 4$ Treatment Means with Values of Coefficients for Each Comparison and Notation for the Coefficients

Comparisons	Values of Coefficients				Notation for Coefficients				$\sum a_{.i}^2$
	$\bar{X}_{1.}$	$\bar{X}_{2.}$	$\bar{X}_{3.}$	$\bar{X}_{4.}$	$\bar{X}_{1.}$	$\bar{X}_{2.}$	$\bar{X}_{3.}$	$\bar{X}_{4.}$	
d_1	1	−1	0	0	a_{11}	a_{21}	a_{31}	a_{41}	2
d_2	0	0	−1	1	a_{12}	a_{22}	a_{32}	a_{42}	2
d_3	1/2	1/2	−1/2	−1/2	a_{13}	a_{23}	a_{33}	a_{43}	1

means. Table 7.3.1.1 shows the comparisons and values of the coefficients and the notation for the coefficients.

Comparisons such as those shown in Table 7.3.1.1 are linear functions of the treatment means. Any linear function of the treatment means such as the one in Equation 7.20 is called a *comparison* if at least two of the coefficients are not equal to 0 and if the sum of the coefficients is equal to 0. That is, if $\sum a_{.i} = 0$.

$$d_i = a_{1i}\bar{X}_{1.} + a_{2i}\bar{X}_{2.} + ... + a_{ki}\bar{X}_{k.} \tag{7.20}$$

In Table 7.3.1.1, the means of the treatments are given as $\bar{X}_{1.}$; $\bar{X}_{2.}$; $\bar{X}_{3.}$; and $\bar{X}_{4.}$. The numbers given in each row of the table are called the *coefficients* of the treatment means, and the value of a with the appropriate subscript shown in the right-hand side of the table represent these coefficients. For example, the first subscript refers to a particular treatment mean, which is to be multiplied by the coefficient, and the second subscript indicates a particular comparison. So by multiplying the treatment means by the coefficients in the first row, $\bar{X}_{1.}(1)$, $\bar{X}_{2.}(-1)$, $\bar{X}_{3.}(0)$, and $\bar{X}_{4.}(0)$ we obtain the following comparison:

$$d_1 = \bar{X}_{1.} - \bar{X}_{2.} \tag{7.21}$$

To obtain comparison d_2, we multiply the treatment means by the coefficients in the second row, and we have:

$$d_2 = \bar{X}_{4.} - \bar{X}_{3.} \tag{7.22}$$

As we multiply the treatment means by the coefficients in the last row, we have the comparison d_3, which is written as:

$$d_3 = \frac{1}{2}\left(\bar{X}_{1.} + \bar{X}_{2.}\right) - \frac{1}{2}\left(\bar{X}_{3.} + \bar{X}_{4.}\right) \tag{7.23}$$

Equation 7.23 states that comparison d_3 is the difference between the average of the means for treatments 1 and 2 and the average of the means for treatments 3 and 4. It should be noted that the first two comparisons are pairwise comparisons, but the third is not.

In constructing a table of comparison coefficients such as the one presented in Table 7.3.1.1, the following simple rules have been used:

1. When comparing groups of equal size, assign coefficients of +1 to one group and –1 to members of the other group. It does not matter which group is assigned a + or a.

2. When comparing groups that contain different numbers of treatments, assign to the first group coefficients equal to the number of treatments in the second group, and the number of treatments in the first to the second group, but with opposite signs. For example, if we have five treatments such that the first three treatments are to be compared with the last two treatments, then the coefficients would be +2, +2, +2, –3, –3. For simplicity, the coefficients can be reduced to the smallest possible integer. For example, if we are comparing a group of four treatments with a group of two, the coefficients can be written as:

$$+2, +2, +2, +2, -4, -4$$

or in its reduced form:

$$+1, +1, +1, +1, -2, -2$$

3. To find the interaction coefficients, multiply the corresponding coefficients of the main effects.

Our discussion so far has dealt with mean comparisons. We can also use treatment sums for our comparisons. To do this, let D_i represent the difference obtained by multiplying each of the treatment sums by the coefficients; then we have

$$D_i = a_{1i}\sum X_{1.} + a_{2i}\sum X_{2.} + \ldots + a_{ki}\sum X_{k.} \tag{7.24}$$

To compute the sum of squares for a comparison coefficient, we use the following equation:

$$SS = \frac{D_i^2}{r \sum a_{.i}^2} \tag{7.25}$$

where
$a_{.i}$ = comparison coefficient
D_i = as defined above
r = number of replications

Equation 7.25 is a component of the treatment sum of squares with one degree of freedom. As such, it is also a mean square and therefore can be written as:

$$MS_{D_i} = \frac{D_i^2}{r \sum a_{.i}^2} \tag{7.26}$$

To test the significance of a comparison on the treatment sum, we have

$$F = \frac{MS_{D_i}}{\text{Error } MS} \tag{7.27}$$

The *d.f.* associated with the numerator is 1, while the denominator *d.f.* is equal to those associated with the error mean square.

As it is possible to analyze the treatment sum of squares into more than one set of orthogonal comparisons, the particular set of comparisons to be made should be determined by experimental interests. The testing of orthogonal comparisons for significance should be used only if the comparisons have been planned in advance.

The following example will show how a class comparison is made using the orthogonal contrasts.

Example 7.3.1.1

In determining the yield difference due to the application of P and K on kale, a horticulturist has conducted a randomized complete block design experiment with four replications and four treatments. She is particularly interested in the following questions: (1) Is there a difference in yield between the fields applied with P and those receiving K? (2) Are there yield differences between the fields receiving P and K separately and those receiving P and K together? (3) Is there a response to fertilization at all? In response to these questions,

TABLE 7.3.1.2

Yield Data, Treatment Total, and Treatment Mean from a Randomized
Complete Block Design with Four Replications and Four Treatments

Treatment	Kale Yield (t/ac)				Treatment Total (T)	Treatment Mean
	Rep. I	Rep. II	Rep. III	Rep. IV		
Control	21.2	20.5	19.8	19.6	81.1	20.28
P added	24.6	25.3	26.8	25.9	102.6	25.65
K added	23.2	22.5	25.4	21.4	92.5	23.13
P and K added	28.7	29.8	27.2	30.1	115.8	28.95
Rep. total (R)	97.7	98.1	99.2	97.0		
Grand total					392.0	
Grand mean						24.5

TABLE 7.3.1.3

Analysis of Variance of Kale Yield Data in a Randomized Complete Block
Design with Four Treatments and Four Replications

Source of Variation	Degrees of Freedom	Sum of Squares	Mean Square	F
Treatment	3	163.46	54.49	28.38**
Replication	3	0.64	0.21	
Error	9	17.28	1.92	
Total	15	181.38		

** = significant at 1% level.

the horticulturist has designed the following set of treatments: (1) Control,
(2) P added to the plots, (3) K added to the plots, and (4) both P and K added
to the plots. The yield data have been collected by the horticulturist and are
shown in Table 7.3.1.2.

SOLUTION

To answer these questions, we first perform an analysis of variance for a
randomized complete block design following the steps outlined in Section
4.2.2. The results are shown in Table 7.3.1.3.

In calculating the sum of squares for treatment components using the
comparison coefficients, we follow these steps:

Step 1: Construct a set of orthogonal contrasts among the treatments,
following the rules stated earlier. This is shown in Table 7.3.1.4.

Step 2: Sum the squares for each comparison. Thus, we have:

$$\sum a_{.1}^2 = (3)^2 + (-1)^2 + (-1)^2 + (-1)^2 = 12$$

$$\sum a_{.2}^2 = (+1)^2 + (-1)^2 = 2$$

TABLE 7.3.1.4

A Set of Orthogonal Contrasts among the Treatments

	Treatment			
Contrast	Control	P Applied	K Applied	P and K Applied
Response to fertilizer	3	−1	−1	−1
P vs. K	0	+1	−1	0
P and K vs. PK together	0	−1	−1	+2

TABLE 7.3.1.5

A Set of Orthogonal Contrasts among the Treatments

	Control	P Applied	K Applied	P and K Applied	Value of
Contrast	81.1	102.6	92.5	115.8	D_i
Response to fertilizer	3	−1	−1	−1	67.60
P vs. K	0	+1	−1	0	10.10
P and K vs. PK together	0	−1	−1	+2	36.50

$$\sum a_{.3}^2 = (-1)^2 + (-1)^2 + (+2)^2 = 6$$

Step 3: Compute the coefficients (D_i) for each of the three comparisons, using either the treatment means or the treatment sums. We have used the treatment sums in our computation, and the results are presented in Table 7.3.1.5. To compute D_1, for example, Equation 7.24 was used and we get:

$$81.1(3) + 102.6(-1) + 92.5(-1) + 115.8(-1) = -67.60$$

Step 4: Compute the sum of squares for the treatment components, using Equation 7.25:

$$SS_{Fertilizer} = \frac{(-67.60)^2}{4(12)} = 95.20$$

$$SS_P = \frac{(10.10)^2}{4(2)} = 12.75$$

$$SS_K = \frac{(36.50)^2}{4(6)} = 55.51$$

As was pointed out earlier, the preceding sums of squares are also mean squares because we have only a single degree of freedom associated with each. Thus, in computing the F value, each of the preceding sum of squares is divided by the MSE, as shown in the next step. Note that the sum of all the mean squares is equal to the treatment sum of squares given in Table 7.3.1.3. That is,

$$MS_{D_1} + MS_{D_2} + MS_{D_3} = 95.20 + 12.75 + 55.51 = 163.46$$

Step 5: Perform the F-test using Equation 7.27 and the data from Table 7.3.1.3 as follows:

$$F_1 = \frac{95.20}{1.92} = 49.58$$

$$F_2 = \frac{12.75}{1.92} = 6.64$$

$$F_3 = \frac{55.51}{1.92} = 28.91$$

Step 6: Compare the computed F values with the tabular F given in Appendix E with = 0.05, and = 0.01 for one and nine degrees of freedom. The tabular F for 5% and 1% levels of significance are 5.12 and 10.56, respectively. Thus, we would conclude that D_1, D_2, and D_3 are significant. Based on these results, we have an answer to each of the questions posed when the experiment was planned.

7.3.2 Trend Comparisons

Our discussions so far have dealt with mean comparisons that focus on specific treatments. In some experiments treatments may consist of different values of an ordered variable that covers the whole range of treatments. For example, we may test the response to different increments of a fertilizer or test different groups of animals after increasing the dosage of a drug. In such experimental situations a wider range of treatment responses is observed. Thus, it is not appropriate to use specific mean comparisons.

In experiments in which the treatments consist of different values of an ordered variable, and if it can be assumed that the differences between the values are uniform (equal), then the researcher may be interested in determining whether the treatment means (or totals) are functionally related to the different values of the treatment variable: that is, what the nature of the

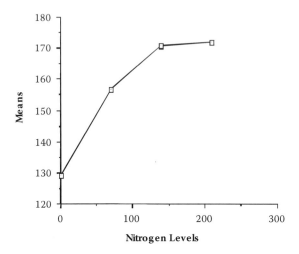

FIGURE 7.3.2.1
Treatment means for each of the four levels of fertilizer applied.

response of the experimental unit is to the varying levels of a treatment. Specifically, the researcher may want to examine whether the treatment means are linearly related to the treatment variable, and if so, whether there is a significant curvature in the trend of the means.

To illustrate the computational procedures for a trend or a factorial comparison, we have used the data from Example 5.5.1, in which corn hybrids received (0, 70, 140, and 210 lb/ac of nitrogen). Note that the treatments have an equal interval of 70 lb/ac of nitrogen fertilizer. The ordered treatment means were 129, 156.5, 171.2, and 172.0. Figure 7.3.2.1 shows the treatment means graphed against the level of nitrogen applied. The trend of the mean does not appear to be linear.

In order to determine whether the linear component of the trend of the sums is statistically significant and whether the treatment sums (or means) deviate significantly from linearity, use is made of the table of coefficient for orthogonal polynomials given in Appendix J. The steps in computing the sum of squares and performing the F-test are given in the following text.

Step 1: Determine the coefficients for the linear, quadratic, and cubic component for $k = 4$ treatments from Appendix J. For the present example the coefficients are given in Table 7.3.2.1.

TABLE 7.3.2.1

Coefficients for the Linear, Quadratic, and Cubic Components for $k = 4$ Treatments

Comparison	Treatment Totals			
	1,290	1,565	1,710	1,720
Linear	−3	−1	1	3
Quadratic	1	−1	−1	1
Cubic	−1	3	−3	1

Step 2: Sum the squares for each comparison given in Table 7.3.2.1 as shown:

$$\sum a_{.L}^2 = (-3)^2 + (-1)^2 + (+1)^2 + (+3)^2 = 20$$

$$\sum a_{.Q}^2 = (+1)^2 + (-1)^2 + (-1)^2 + (+1)^2 = 4$$

$$\sum a_{.C}^2 = (-1)^2 + (+3)^2 + (-3)^2 + (+1)^2 = 20$$

Step 3: Compute the coefficients for each of the three comparisons, linear (L), quadratic (Q), and cubic (C), by multiplying the treatment sums by the appropriate coefficient:

$$L = (-3)(1{,}290) + (-1)(1{,}565) + (1)(1{,}710) + (3)(1{,}720) = 1{,}435$$

$$Q = (1)(1{,}290) + (-1)(1{,}565) + (-1)(1{,}710) + (1)(1{,}720) = -265$$

$$C = (-1)(1{,}290) + (3)(1{,}565) + (-3)(1{,}710) + (1)(1{,}720) = -5$$

Step 4: Compute the sum of squares (mean squares) for the treatment components, using Equation 7.25:

$$MS_L = SS_L = \frac{(1{,}435)^2}{10(20)} = 10{,}296.13$$

$$MS_Q = SS_Q = \frac{(-265)^2}{10(4)} = 1{,}755.63$$

$$MS_C = SS_C = \frac{(-5)^2}{10(20)} = 0.13$$

As a check, note that the sum of squares for the treatment components add up to the treatment sum of squares as given in Table 5.5.4. As the components are an orthogonal set, they must equal the sum of squares partitioned.

Step 5: Perform the test of significance for the linear, quadratic, and cubic comparisons as shown:

$$F = \frac{MS_{\text{L}}}{\text{Error } MS} = \frac{10,296.13}{93.96} = 109.58$$

$$F = \frac{MS_{\text{Q}}}{\text{Error } MS} = \frac{1,755.63}{93.96} = 18.68$$

$$F = \frac{MS_{\text{C}}}{\text{Error } MS} = \frac{0.13}{93.96} < 1$$

Step 6: Compare the *F* values with the tabular *F*'s from Appendix E at the prescribed level of significance. This is shown in Table 7.3.2.2.

The results of Table 7.3.2.2 show that there is a significant linear trend in the treatment sums. This means that the relationship between yield and nitrogen rate of application is linear within the range of rates tested. The results also indicate that there is significant curvature in the trend of the sums. From Table 7.3.2.2 we note that beyond the quadratic component, the results are nonsignificant. Thus, we conclude that the yield is a quadratic function of the nitrogen rates. The interpretation of the preceding results should be done with caution. Recall from earlier chapters that when the number of error degrees of freedom is not adequate, as is the case in the present example, the test of significance is not valid.

The same procedures applied in partitioning the treatment components can also be applied to make between-group comparison on the means of the three hybrids.

TABLE 7.3.2.2

Analysis of Variance of Hybrid × N Response Experiment Partitioned into Linear, Quadratic, and Cubic Components of Response Curve

Source of Variation	Degrees of Freedom	Sum of Squares	Mean Square	F	Required F 5%	1%
Replication	1	4,100.63	4,100.63			
Nitrogen (A)	3	12,051.88	4,017.29	42.76**	9.28	29.46
Linear	(1)	10,296.13	10,296.13	109.58**	10.13	34.12
Quadratic	(1)	1,755.63	1,755.63	18.68*		
Cubic	(1)	0.13	0.13	<1		
Error (a)	3	281.87	93.96			
Hybrid (B)	4	2,466.25	822.10	27.41**		
Nitrogen x Hybrid (A × B)	12	863.75	71.98	2.40ns		
Error (b)	16	479.97	29.99			
Total	39	20,244.38				

ns = Nonsignificant; ** = significant at 1% level; * = significant at 5% level.

References and Suggested Readings

Bryon-Jones, J. and Finney, D.J. 1983. On an error in Instruction to Authors. *Hort. Sci.* 18: 279–282.

Cochran, W.G. and Cox, G.M. 1957. *Experimental Designs.* 2nd ed. New York: John Wiley & Sons. Chap. 6A.

Cramer, S.G. and Swanson, M.R. 1973. An evaluation of ten pairwise multiple comparison procedures by Monte Carlo methods. *J. Am. Stat. Assoc.* 68: 66–74.

Chew, V. 1976. Comparing treatment means: a compendium. *Hort. Sci.* 11: 348–357.

Duncan, D.B. 1955. Multiple range and multiple F tests. *Biometrics* 11: 1–42.

Gomez, K.A. and Gomez, A.A. 1984. *Statistical Procedures for Agricultural Research.* 2nd ed. New York: John Wiley & Sons. Chap. 5.

Maindonald, J.H. and Cox, N.R. 1984. Use of statistical evidence in some recent issues of DSIR agricultural journals. *N.Z. J. Agric.* Vol. 27: 597–610.

Mead, R. 1988. *The Design of Experiments: Statistical Principles for Practical Applications.* Cambridge: Cambridge University Press. Chap. 4 and Chap. 12.

Miller, R.G., Jr. 1977. Developments in multiple comparisons. 1966–1976. *J. Am. Stat. Assoc.* 72: 779–788.

Morse, P.M. and Thompson B.K. 1981. Presentation of experimental results. *Can. J. Plant Sci.* 61: 799–802.

Peterson, R.G. 1977. Use and misuse of multiple comparison procedures. *Agron. J.* 69: 205–208.

Wyman, J.A., Chapman, R.K., and Longridge, J.L. 1982. Project Report: Vegetable Crops Entomology Field Research. University of Wisconsin. Madison.

Exercises

1. In a completely randomized field experiment, measurements were made to study the response of field-grown cassava (*Manihot esculenta* Crantz) to changes in the application of a fertilizer. The objective of the research was to find out if there were any differences in the amount of the dry matter produced under five different fertilizer regimes. The mean yield from the experiment with four replications were as follows:

Mean Yield of Dry Matter Production (t/ha) of Cassava as a Result of Five Different Fertilizer Applications

Treatment	Treatment Mean (t/ha)
Control	2.14
50 kg/ha	2.54
100 kg/ha	2.68
150 kg/ha	3.81
200 kg/ha	4.42

(a) Compute the LSD value at 5% and 1% levels of significance.

(b) What conclusions can you draw from the comparisons?

2. Suppose that in the previous example the treatment means were computed from the four replications where some data were lost as a result of a natural disaster, as shown in the following table.

Mean Yield of Dry Matter Production (t/ha) of Cassava as a Result of Five Different Fertilizer Applications

Treatment	Rep. I	Rep. II	Rep. III	Rep. IV	Treatment Mean
Control	2.20	—	2.25	2.01	2.15
50 kg/ha	2.40	2.56	2.66	2.52	2.54
100 kg/ha	—	2.68	2.79	—	2.74
150 kg/ha	3.00	3.56	4.00	4.66	3.81
200 kg/ha	3.50	4.98	5.00	4.20	4.42

(a) What type of a comparison of means would be appropriate?

(b) Compute the LSD at the 5% and 1% levels of significance.

(c) Compare each of the mean differences.

3. In a balanced lattice design experiment in which each treatment was replicated four times, a natural scientist studied the impact of chlorpyrifos on ash borer infestation. The ash borer represents serious threat to ash and olive trees. The treatments were applied during the moth flight season. The scientist was interested in knowing if there are differences in the number of attacks on the ash tree as a result of the treatments. In performing the analysis of variance, the scientist found the effective MSE with 16 degrees of freedom to be 1.8.

Treatment Number	Adjusted Mean Number of Attack Sites
1	3
2	4
3	2
4	3
5	1
6	4
7	2
8	5
9 (Control)	8

(a) Compute the LSD for the 5% and 1% levels of significance.

(b) What conclusions can be drawn about the effectiveness of the treatments as compared with the control?

4. To make a mean comparison on the data from a 4×4 partially balanced triple lattice design experiment, an animal scientist recorded the following data on the role of progestrone in stimulating sexual receptivity in estrogen-treated gilts. The experiment involved 16 ovariectomized gilts treated with estradil benzoate (EB), and a control group that did not receive estradil benzoate (treatment number 16). After EB treatment, gilts were moved to an evaluation pen in which boars were brought in. Gilts remained in the evaluation pen for 5 min, during which time the number of mounts attempted by the boar were recorded. The treatment numbers appear in parentheses.

Incomplete Block Number	Mounts, Number/5 min			
	Rep. I			
1	7(01)	5(02)	4(03)	4(04)
2	5(05)	2(06)	1(07)	3(08)
3	4(09)	3(10)	4(11)	5(12)
4	3(13)	3(14)	2(15)	1(16)
	Rep. II			
1	6(01)	6(05)	6(09)	2(13)
2	4(02)	2(06)	3(10)	3(14)
3	3(03)	3(07)	1(11)	5(15)
4	1(04)	4(08)	3(12)	2(16)
	Rep. III			
1	7(01)	5(06)	4(11)	2(16)
2	4(05)	4(02)	3(15)	6(12)
3	1(09)	2(14)	2(03)	4(08)
4	5(13)	3(10)	4(07)	2(04)

(a) How many effective error mean squares are involved in this experiment?

(b) Before performing the LSD test, do you need to compute the adjusted treatment means?

(c) Compute the mean difference between the control treatment and each of the 15 treatments.

5. To study the impact of four insecticides on the omnivorous looper, a split-plot design experiment was conducted. The insecticide treatments were assigned to the subplots, whereas the concentration rate

of active ingredients was assigned to the main plots. The results of the experiment are shown in the following table.

Average Number of Larvae Found per Tree 14 d Posttreatment from an Experiment Replicated Three Times

Active Ingredient	Dylox 80 SP	Krycide 8 F	Lannate L	Orthene 75 SP	Control
1 lb/ac	9.66	10.33	3.00	2.33	17.33
2 lb/ac	7.33	5.00	2.66	2.33	20.00
4 lb/ac	3.66	5.33	2.00	1.33	16.33

(a) Perform an LSD test on the data.

(b) What conclusions can be drawn from the analysis?

6. Research has shown that aldicarb is the most preferred pesticide used by potato farmers. In a study conducted in Wisconsin, researchers used a variety of pest management strategies including six different chemicals: aldicarb (Temik), phorate (Thimet), disulfoton (Disyston), carbofuran (Furadan), oxamyl (Vydate), and terbufos (Counter) on Atlantic variety potatoes. The data on the yield of potato per acre are given in the following table. The MSE and error degrees of freedom for the experiment were 20.96 and 39, respectively.

Mean Yield of Atlantic Variety Potatoes Treated with Soil-Applied Systemic Insecticide

Treatment (lb ai/ac)	Mean Yield (cwt/ac)
Temik 15G (3) with fertilizer at planting	327.4
Temik 15G (3) topdress at emergence	369.8
Temik 15G (2) topdress at emergence	360.5
Disyston 15G (3) with fertilizer at planting	318.4
Disyston 15G (3) topdress at emergence	307.4
Disyston 15G (3+3) with fertilizer at planting and topdress at emergence	322.0
Thimet 20G (3) with fertilizer at planting	319.7
Thimet 20G (3) with fertilizer at planting, foliar spray	317.5
Furadan 15G (3) with fertilizer at planting	329.3
Vydate 10G (3) with fertilizer at planting	296.9
Vydate 10G (3) topdress at emergence	315.7
Counter 15G (3) with fertilizer at planting	317.8
Untreated	239.6

Source: Data adapted from Wyman, J.A., Chapman, R.K., and Longridge, J.L. 1982. Project Report: Vegetable Crop Entomology Field Research. University of Wisconsin, Madison.

(a) Perform a Duncan's Multiple-Range Test on the data.

7. In an attempt to provide adequate moisture for germination, cotton farmers in California irrigate their fields before planting. This pre-

season irrigation has been identified as a major contributor to drainage problems. To remedy this problem a study was conducted to determine if plastic mulch could be used effectively to replace preseason irrigation of cotton fields. Three different plastic mulches (white, black) were used in this experiment on Acala cotton. The experiment was replicated four times. Plant height was used as a criterion in determining the differences between the mulches and the control. Perform a grouped mean comparison on the data.

Height (Inches) of Acala Cotton Under Different Treatment Conditions

Treatment	Rep. I	Rep. II	Rep. III	Rep. IV
Control	10.0	8.5	9.2	11.0
White plastic	20.3	19.5	21.0	19.8
Black plastic	24.5	23.8	23.6	22.4

(a) Is there a difference in the height of cotton plants from fields with white mulch and black mulch?

(b) Is there a difference between the height of plants receiving white mulch and the control?

(c) Is there a difference between the height of plants receiving black mulch and the control?

8. To determine whether application of fertilizers produce more range forage in drought than normal years, a study was conduct by scientist in Colorado. The results of the experiment produced the following oven dry weight of range forage.

Range Forage Yield according to Rate of Nitrogen Application in Normal and Five Drought Years

| Replication | Precipitation (in.) | Fertilizer Treatment (lb/ac) | | | | |
		0	30	60	90	120
I.	Normal (16.5)	3,520	3,600	3,750	4,150	4,390
	Drought (8.5)	2,650	3,211	3,476	4,485	4,863
	Drought (10.5)	2,652	3,356	4,291	6,310	6,050
	Drought (6.0)	1,432	2,397	2,513	3,200	3,252
	Drought (8.0)	1,451	2,498	2,568	2,734	3,682
II.	Normal (16.5)	3,319	3,560	3,552	4,230	4,410
	Drought (8.5)	2,578	3,276	3,399	4,345	4,674
	Drought (10.5)	2,602	3,416	4,300	5,810	5,980
	Drought (6.0)	1,392	2,278	2,567	3,660	3,000
	Drought (8.0)	1,400	2,399	2,510	2,932	3,862
III.	Normal (16.5)	3,423	3,545	3,759	4,252	4,270
	Drought (8.5)	2,760	3,200	3,498	4,375	4,542
	Drought (10.5)	2,752	3,452	4,312	6,110	6,000
	Drought (6.0)	1,487	2,377	2,542	3,189	3,342
	Drought (8.0)	1,493	2,456	2,575	2,831	3,732

(a) Perform a trend or a factorial comparison on the data.

9. An animal scientist is comparing two diets for fattening beef cattle. Fifteen pairs of animals were selected from an existing herd. Each pair of animals was selected based on their heredity factors as closely as possible. The paired animals were randomly allocated one to each diet. The following table shows the weight gain after a 90-d test period.

Pair	Diet A	Diet B	Difference
1	320	380	−60
2	420	400	20
3	480	450	30
4	395	380	15
5	415	425	−10
6	430	425	5
7	382	395	−13
8	400	405	−5
9	382	365	17
10	415	395	20
11	398	365	33
12	345	320	25
13	415	442	−27
14	300	295	5
15	325	314	11
Mean	388.13	382.4	5.7
SD	47.8	45.1	24.2

(a) Compute the standard error of the mean difference.

(b) Test for a difference between Diets A and B using a paired t-test at $\alpha = 0.05$.

(c) Construct a 95% confidence interval for μ_d.

8

Sample Designs Over Time

8.1 Terminology and Concepts

Our discussions so far have centered around those experimental designs that are conducted in one period in time and on a single site. Care must be taken in the generalization of the results of such experiments. No matter how well such experiments are conducted, their results differ from those obtained from the same experiments conducted over a longer period of time (several seasons or several years) or at several sites, or both. In this chapter we will address how such experiments are conducted and the unique conditions associated with these experiments.

Agricultural researchers are often interested in the applicability of their findings over a wide range of physical and environmental conditions. For example, they would be interested in the applicability of their recommendations from a corn experiment conducted in Nebraska not only to the neighboring states, but also to other corn-producing regions around the world. To accomplish such an objective, the experiment on a single-site would have to emulate all the physical and climatic conditions. This is very difficult, as there are certain factors that are not easily controlled. To modify such factors as rainfall, sunshine, or the inherent fertility of the soil is outside the control of the experimenter. Given this dilemma, researchers would have to rely on experiments conducted at several sites or over a number of periods of time, or both, so that they can generalize the results of their experiments over a wide range of climatic and physical conditions.

To analyze the results of a series of experiments conducted at different sites and times requires great care. Such experiments and their analyses are usually carried out in stages. The preliminary stage entails conducting individual experiments at different sites. The data obtained from such experiments are then combined to see if there are differences among the treatments from the different experiments. At this stage the experimenter is able to determine if a particular treatment consistently performs well in comparison to others. Furthermore, it is possible to determine whether certain treatments do well in certain circumstances. The second stage of the analysis can be used for estimation and comparison of the average effects of treatments over the whole series of experiments.

Multisite experiments, although helpful in overcoming the shortfalls of the single-site experiments, present difficulties that must be recognized at the outset. One such problem is that treatments x places interactions may not be homogeneous. The other is the inequality of error variance at different sites. When the treatments x places interactions are not homogeneous, it implies that some sites will give a better estimate of the difference than will others. Under such conditions the difference is estimated overall by weighting the difference at each site by the reciprocal of its variance. The reason for such weighting is to give greater weight to the sites that provide a better determination, on account of either higher replication or lower error variance.

In cases where the interactions are large, interpreting the results of the combined analysis is not too difficult. A large F ratio for interactions permits the researcher to be confident in the statistical significance of the experiment. Additionally, when interactions are large, the F-test of the treatment mean square against the interaction mean square is little affected by the inequality in the error variance.

When the researcher suspects that there is interaction between treatments and places, as well as the presence of unequal error at different sites, it is best to work with specified contrasts between treatments. Under such circumstances it is advisable to test any component of the treatment sum of squares against its own interaction with places (Snedecor and Cochran, 1967). This means that the treatments mean square has to be subdivided into sets of comparisons; then, the interactions mean squares for each set are computed and tested separately (Snedecor and Cochran, 1967; Pearce, 1983).

To check for the inequality of the error variance, Bartlett's test could be used. If the test shows that variances are heterogeneous, the validity of using the F-test of treatments x places interactions is in question. When comparisons are made over a subset of places, the pooled error for these places should be used instead of the overall pooled error (Yates and Cochran, 1938; Snedecor and Cochran, 1967).

In summary, experiments that are conducted at several sites, with equal care, design and structure, are easily combined for analysis. The results of such an analysis are efficient and theoretically sound. However, care should be taken in the generalization of the results, especially from the ordinary test of significance. As was pointed out, experiments conducted at several sites are prone to heterogeneous error and interaction variances.

8.1.1   Representing Time Spans

Agricultural field experiments are often repeated over a number of seasons and for a number of years. This is done to determine if the subject under investigation performs differently at different time periods. In conducting a time-span experiment, the aim of the researcher is to obtain data (from successive years) that are regarded as independent. To accomplish this objective, researchers should select a new site and use a new randomization for the experiment. The experimental arrangement, however, need not be uniform throughout the whole series of experiments, in order to obtain data considered independent. In such experiments, the interaction between treatment and seasons are more important than the treatment and year.

The statistical analysis that we should use with the data from a series of experiments depends on the objectives of the study. However, it is safe to say that the analysis in the preliminary stages tend to be the same. In this chapter, we will present an introductory account of these procedures. In the next two sections, procedural steps for the combined analysis of experiments over the years and over the seasons are presented.

8.2 Analysis of Experiments over Years

Researchers, in conducting experiments over the years, are interested in determining how treatments respond to the different environmental conditions that normally occur from year to year or over a number of years. Because researchers are not able to predict what environmental conditions may prevail, years are generally considered as a random variable in such experiments. We will illustrate the combined analysis for experiments that are performed over several years, by using only the data from an experiment that was conducted over a 2-yr period. In the following example, we will illustrate the step-by-step procedure in analyzing the data from such an experiment.

Example 8.2.1

Field experiments to evaluate potential differences in yield response of five corn hybrids widely grown in the northern "Corn Belt" were conducted during 1990 and 1991. A randomized complete block design with four replications was used for the study. The experiments were conducted on separate but adjacent sites, where corn had been grown since 1985. The 1991 experimental plots were randomized differently compared to those in 1990. Soil fertility conditions, planting date, plant density were all similar in both years. The following results in bu/ac were obtained from the experiments. Perform the combined analysis of variance on the data.

| Hybrid | 1990 Yield (bu/ac) | | | | | 1991 Yield (bu/ac) | | | | |
	Rep. I	Rep. II	Rep. III	Rep. IV	Total	Rep. I	Rep. II	Rep. III	Rep. IV	Total
P3747	110	135	120	120	485	115	120	113	117	465
P3732	150	150	148	160	608	140	155	138	140	573
A630 × LH2	113	120	135	116	484	118	129	136	138	521
LH70 × LH3	120	130	138	120	508	130	142	150	149	571
P3742	122	126	132	130	510	128	143	110	139	520
Total	615	661	673	646	2595	631	689	647	683	2650

TABLE 8.2.1

Results of the Analyses of Variance for the Individual Years Using
an RCB Design with Four Replications and Five Corn Hybrids

Source of Variation	Degree of Freedom	Sum of Squares	Mean Square	F
1990				
Replication	3	376.95	125.65	
Hybrid	4	2626.00	656.50	13.07**
Error	12	602.80	50.23	
1991				
Replication	3	471.00	157.00	
Hybrid	4	1984.00	496.00	6.64**
Error	12	896.00	74.67	

** = Significant at 1% level.

SOLUTION

To perform the combined analysis, we first need to perform a separate analysis of variance for each year in the experiment. The procedure for computing the sums of squares is shown below, using the 1990 data. The sum of squares for the year 1991 is computed similarly. The results of the analysis are shown in Table 8.2.1. The steps in performing the individual analysis of variance and the combined analysis are given below.

Step 1: Compute the correction factor and the various sums of squares as:

$$C = \frac{G^2}{rt} \tag{8.1}$$

$$= \frac{(2,595)^2}{4 \times 5} = \frac{6,734,025}{20}$$

$$= 336,701.25$$

$$\text{Total } SS = \sum_{j=1}^{t} \sum_{i=1}^{r} X_{ij}^2 - C \tag{8.2}$$

$$\text{Total } SS = \left[(110)^2 + (150)^2 + ... + (130)^2 \right] - 336,701.25$$

$$= 340,307 - 336,701.25$$

$$= 3,605.75$$

$$\text{Replication } SS = \frac{\sum R_j^2}{t} - C \tag{8.3}$$

$$= \frac{\left(615\right)^2 + \left(661\right)^2 + \left(673\right)^2 + \left(646\right)^2}{5} - 336,701.25$$

$$= 337078.20 - 336,701.25$$

$$= 376.95$$

$$\text{Treatment } SS = \frac{\sum T_i^2}{r} - C \tag{8.4}$$

$$= \frac{\left(485\right)^2 + \left(608\right)^2 + \left(484\right)^2 + \left(508\right)^2 + \left(510\right)^2}{4} - 336,701.25$$

$$= 339,327.25 - 336,701.25$$

$$= 2,626$$

$$\text{Error } SS = \text{Total } SS - \text{Rep. } SS - \text{Treatment } SS \tag{8.5}$$

$$= 3,605.75 - 376.95 - 2,626$$

$$= 602.8$$

Step 2: Compute the mean square for each source of variation as follows:

$$\text{Replication } MS = \frac{\text{Replication } SS}{r-1} \tag{8.6}$$

$$= \frac{376.95}{3} = 125.65$$

$$\text{Treatment } MS = \frac{\text{Treatment } SS}{t-1} \tag{8.7}$$

$$= \frac{2,626}{4} = 656.50$$

$$\text{Error } MS = \frac{\text{Error } SS}{(r-1)(t-1)} \tag{8.8}$$

$$= \frac{602.80}{12} = 50.23$$

Step 3: Compute the F value for testing the treatment difference as:

$$F = \frac{\text{Treatment } MS}{\text{Error } MS} \tag{8.9}$$

$$= \frac{656.50}{50.23} = 13.07$$

To perform the combined analysis of variance, we use the data given in Table 8.2.1 and proceed as follows.

Step 4: Determine the degrees of freedom associated with each source of variation. Because in the present case year (Y) is considered a random variable, the error term is represented by the year × hybrid interaction MS. Thus, we have:

Year (Y) *d.f.* $= y - 1 = 1$
Reps. within year $= y (r - 1) = 6$
Hybrid $= t - 1 = 4$
Year × hybrid $= (y - 1)(t - 1) = 4$
Pooled error $= y (r - 1)(t - 1) = 24$
Total *d.f.* $= yrt - 1 = 39$

Step 5: Compute the sum of squares for the replications within the year and the pooled error as follows:

$$\text{Rep. within year } SS = \sum_{i-1}^{y} \left(\text{Rep. } SS \right)_i \tag{8.10}$$

$$\text{Pooled error } SS = \sum_{i-1}^{y} \left(\text{Error } SS \right)_i \tag{8.11}$$

Thus, we have:

$$\text{Reps. within year } SS = 376.95 + 471.00 = 847.95$$

$$\text{Pooled error } SS = 602.80 + 896.00 = 1{,}498.80$$

Step 6: Calculate the correction factor as:

$$C = \frac{\left(\sum\limits_{i=1}^{y} G_i\right)^2}{yrt} \tag{8.12}$$

$$C = \frac{(2{,}595 + 2{,}650)^2}{40} = 687{,}750.63$$

Step 7: Compute the sum of squares for the year, treatment, and the Year × Treatment interaction as follows:

$$\text{Year } SS = \sum\limits_{i=1}^{y} \frac{G_i^2}{rt} - C \tag{8.13}$$

$$\text{Year } SS = \frac{(2{,}595)^2 + (2{,}650)^2}{4(5)} - 687{,}750.63$$

$$= 687826.25 - 687{,}750.63$$

$$= 75.62$$

$$\text{Treatment (Hybrid) } SS = \sum\limits_{j=1}^{t} \frac{T_j^2}{yr} - C \tag{8.14}$$

$$= \frac{\left[(950)^2 + (1{,}181)^2 + (1{,}005)^2 + (1{,}079)^2 + (1{,}030)^2\right]}{2(4)} - 687{,}750.63$$

$$= 5{,}532{,}427/8 - 687{,}750.63$$

$$= 3802.75$$

$$\text{Year} \times \text{Treatment } SS = \sum_{i=1}^{y} \sum_{j=1}^{t} \frac{(YT)_{ij}^2}{r} - (C + \text{Year } SS + \text{Treatment } SS) \quad (8.15)$$

where

$(YT)_{ij}$ = the total of the *j*th treatment in the *i*th year

$$\text{Year} \times \text{Treatment } SS = \frac{\left[(485)^2 + (608)^2 + \dots + (520)^2\right]}{4}$$

$$- \left(687,750.63 + 75.62 + 3,802.75\right)$$

$$= 2,769,745/4 - 691,629$$

$$= 692436.25 - 691,629$$

$$= 807.25$$

Step 8: Compute the mean square for each source of variation as shown as follows:

$$\text{Year } MS = \frac{\text{Year } SS}{y-1} = \frac{75.62}{1} = 75.62$$

$$\text{Rep. within year } MS = \frac{\text{Rep. within year } SS}{y(r-1)} = \frac{847.95}{6} = 141.33$$

$$\text{Treatment}(\text{hybrid}) MS = \frac{\text{Treatment } SS}{t-1} = \frac{3,802.75}{4} = 950.69$$

$$\text{Year} \times \text{Treatment } MS = \frac{\text{Year} \times \text{Treatment } SS}{(y-1)(t-1)} = \frac{807.25}{1(4)} = 201.81$$

$$\text{Pooled error } MS = \frac{\text{Pooled error } SS}{y(r-1)(t-1)} = \frac{1,498.80}{2(3)(4)} = 62.45$$

Step 9: Perform the test of homogeneity of error variance, using the data from the individual analysis of variance given in Table 8.2.1. This test could be performed either through the application of the chi-square or the *F*-test. We have used the *F*-test to determine the homogeneity as follows:

$$F = \frac{\text{Larger error } MS}{\text{Smaller error } MS} \tag{8.16}$$

$$F = \frac{74.67}{50.23} = 1.49$$

In comparing the computed *F* value with the corresponding *F* value (2.69 at the 5% level) found in Appendix E for 12 degrees of freedom, we cannot reject the hypothesis of homogeneous error variances. Had we rejected the homogeneity of the error variances, then we would need to partition the pooled error sum of squares into components corresponding to the Year × Treatment sum of squares. To compute the *F* value for each component of the Year × Treatment interaction, we use the corresponding component in the pooled error as its error term. Because, in the present case, we cannot reject the hypothesis of homogeneous error variance, we proceed in computing the *F* value for testing the significance of the various effects.

Step 10: Compute the *F* value for each source of variation. As was mentioned earlier, when data are combined over years, the error term is the Year × Treatment interaction mean square. Thus the *F* value for the replications within year, treatment, and the Year × Treatment interaction is computed as follows and shown in Table 8.2.2.

$$F \text{ (Treatment)} = \frac{\text{Treatment } MS}{\text{Year} \times \text{Treatment } MS} \tag{8.17}$$

$$= \frac{950.69}{201.81} = 4.71$$

TABLE 8.2.2

Combined Analysis of Variance of a Randomized Complete Block Experiment over 2 Yr

Source of Variation	Degree of Freedom	Sum of Squares	Mean Square	F
Year	1	75.62	75.62	a
Reps. within year	6	847.95	141.33	
Treatment (hybrid)	4	3,802.75	950.69	4.71[ns]
Year × Treatment	4	807.25	201.81	3.23*
Pooled error	24	1,498.80	62.45	
Total	39	7,032.37		

Note: a = As the *d.f.* is not adequate, a valid test of significance cannot be performed; [ns] = nonsignificant, * = significant at 5% level.

$$F \text{ (Year} \times \text{Treatment)} = \frac{\text{Year} \times \text{Treatment } MS}{\text{Pooled error } MS} \qquad (8.18)$$

$$= \frac{201.81}{62.45} = 3.23$$

The results of the F-tests show that the hybrid effect is nonsignificant, whereas the interaction effect between the hybrid and year is significant. When the interaction between year and treatment (hybrid) is significant, it is important to determine the relative size of such interaction with the average effect of the treatments.

If the interaction is large relative to the average effect, and the ranking of treatments (hybrids) changes over years — one treatment performs better than another some years and does not do well in other years — then we need to examine the nature of the interaction. However, if the interaction effect is small and ranking of the treatments (hybrids) over the years is stable — hybrid A will do better than hybrid B in all years — then the interaction could be ignored.

As can be seen in Table 8.2.2, the interaction for the present example is large, and the average effect of the hybrid is not significant. We need to further examine the mean difference between years for the five hybrids. Table 8.2.3 shows the mean yield of the five corn hybrids tested in 1990 and 1991.

In examining the mean difference between years for each of the five corn hybrids, we make the following observations:

1. With the exception of the first two hybrids, P3747 and P3732, which had lower yield in the second year of the experiment, all the other hybrids had higher yields in the second year than in the first.

2. Our analysis shows that hybrid P3732 gave the highest average yield (147.63 bu/ac) over the 2-yr period; however, its performance was superior only in the first year of the experiment.

TABLE 8.2.3

Mean Yield of 5 Corn Hybrids Tested in 1990 and 1991

Hybrid	Mean Yield (bu/ac)		Average	Difference
	1990	1991		
P3747	121.25	116.25	118.75	–5.00
P3732	152.00	143.25	147.63	–8.75
A630 × LH2	121.00	130.25	125.63	9.25
LH70 × LH3	127.00	142.75	134.88	15.75
P3742	127.50	130.00	128.75	2.50
Average	129.75	132.50	156.25	2.75

3. Hybrid P3742 shows some consistency in its performance over the 2 yr.

Based on these results, the experimenter is well advised to conduct further studies on the hybrids with special attention to P3742 and P3732.

8.3   Analysis of Experiments over Seasons

The object of the combined analysis of experiments over seasons is to provide the experimenter with information on the nature of the interaction between treatment and seasons. Treatment may be any factor of production. Once it is determined that the interaction between the treatment and season is significant, then such information is used to make recommendations on the use of treatments for different seasons. For example, if data from a study on a new hybrid (treatment) show that it performs better in spring planting rather than winter, then the recommendation for its use will undoubtedly favor spring planting.

   In the previous section, we mentioned that because researchers cannot predict environmental conditions such as rainfall or sunshine from one year to the next, year was used as an *independent variable* in the analysis. In conducting seasonal analysis, planting seasons within a year are distinct periods and therefore can be considered as a *fixed variable*. This fact makes it possible to determine the superiority of a treatment for a specific season.

   The following example will illustrate the step-by-step procedures for analyzing the data from an experiment conducted over two seasons.

Example 8.3.1
Field studies were conducted to determine the influence of planting season using five nitrogen rates on wheat. A randomized complete block design with four replications was used in this experiment. The data in Table 8.3.1 show the yield in bushels per acre for winter and spring planting. The step-by-step procedures are outlined as follows.

SOLUTION

   To perform a combined analysis, we first need to conduct a separate analysis of variance for each planting season in the experiment. The procedure for performing the individual analysis of variance is similar to that of Example 8.2.1 and need not be repeated. Table 8.3.2 shows the result of the analysis, which is based on a randomized complete block design.

TABLE 8.3.1

Yield Data from Spring and Winter Planting of Wheat

Nitrogen Rate, lb/ac	Yield (bu/ac)				Total	Mean
	Rep. I	Rep. II	Rep. III	Rep. IV		
Spring Planting						
0	27.8	24.6	28.2	26.9	107.5	26.88
50	30.0	29.2	30.1	28.9	118.2	29.55
100	29.9	28.3	29.7	30.0	117.9	29.48
150	31.4	32.0	31.7	31.8	126.9	31.73
200	30.8	31.3	29.9	32.0	124.0	31.00
250	30.5	31.2	33.0	31.8	126.5	31.63
Total	180.4	176.6	182.6	181.4	721.0	
Winter Planting						
0	25.1	24.9	26.2	24.2	100.4	25.10
50	24.4	29.2	28.1	26.9	108.6	27.15
100	30.4	26.8	28.2	29.5	114.9	28.73
150	30.3	34.3	32.1	36.2	132.9	33.23
200	31.5	33.6	35.8	32.9	133.8	33.45
250	34.2	35.4	33.6	31.2	134.4	33.60
Total	175.9	184.2	184.0	180.9	725.0	

TABLE 8.3.2

Results of the Analyses of Variance for the Individual Seasons Using an RCB Design with Five Nitrogen Application Rates and Four Replications

Source of Variation	Degree of Freedom	Sum of Squares	Mean Square	F
Spring Planting Season				
Replication	3	3.37	1.12	
Nitrogen	5	67.40	13.48	15.32**
Error	15	13.25	0.88	
Winter Planting Season				
Replication	3	7.50	2.50	
Nitrogen	5	275.05	55.01	15.41**
Error	15	53.51	3.57	

** = Significant at 1% level.

Step 1: Determine the degrees of freedom associated with each source of variation based on the RCB design used in the experiment. The following sources of variation are identified:

Season (S) $d.f. = s - 1 = 1$
Reps. within season $= s(r - 1) = 6$
Treatment (Nitrogen) $= t - 1 = 5$
Season x Nitrogen $= (s - 1)(t - 1) = 5$
Pooled error $= s(r - 1)(t - 1) = 30$
Total $d.f. = srt - 1 = 47$

Step 2: Compute the sum of squares for the replications within the season and the pooled error as follows:

$$\text{Rep. within season } SS = \sum_{i=1}^{s} \left(\text{Rep.}SS\right)_i \tag{8.19}$$

$$\text{Pooled error } SS = \sum_{i=1}^{s} \left(\text{Error } SS\right)_i \tag{8.20}$$

Thus we have:

Reps. within season $SS = 3.37 + 7.50 = 10.87$

Pooled error $SS = 13.25 + 53.51 = 66.76$

Step 3: Calculate the correction factor as:

$$C = \frac{\left(\sum_{i=1}^{s} G_i\right)^2}{srt} \tag{8.21}$$

$$C = \frac{\left(721.0 + 725.0\right)^2}{48} = 43,560.75$$

Step 4: Compute the sum of squares for the season, the nitrogen, and the Season × Nitrogen interaction as follows:

$$\text{Season } SS = \sum_{i=1}^{s} \frac{G_i^2}{rt} - C \tag{8.22}$$

$$\text{Season } SS = \frac{(721)^2 + (725)^2}{4(6)} - 43,560.75$$

$$= 43,561.08 - 43,560.75$$

$$= 0.33$$

$$\text{Nitrogen } SS = \sum_{j=1}^{t} \frac{T_j^2}{sr} - C \tag{8.23}$$

$$= \frac{\left[(207.9)^2 + (226.8)^2 + (232.8)^2 + (259.8)^2 + (257.8)^2 + (260.9)^2\right]}{2(4)} - 43,560.57$$

$$= 43,860.27 - 43,560.75$$

$$= 299.52$$

$$\text{Season } \times \text{ Treatment } SS = \sum_{i=1}^{s} \sum_{j=1}^{t} \frac{(st)_{ij}^2}{r} - (C + \text{Season } SS + \text{Treatment } SS) \tag{8.24}$$

where

$(st)_{ij}$ = the total of the jth treatment in the ith season

$$\text{Season } \times \text{ Treatment } SS = \frac{\left[(107.5)^2 + (118.2)^2 + \ldots + (134.4)^2\right]}{4}$$

$$- (43,560.75 + 0.33 + 299.52)$$

$$= 43,903.53 - 43,860.60$$

$$= 42.93$$

Step 5: Compute the mean square for each source of variation as shown below:

$$\text{Season } MS = \frac{\text{Season } SS}{s-1} = \frac{0.33}{1} = 0.33$$

$$\text{Rep. within season } MS = \frac{\text{Rep. within season } SS}{s(r-1)} = \frac{10.87}{6} = 1.81$$

$$\text{Nitrogen } MS = \frac{\text{Nitrogen } SS}{t-1} = \frac{299.52}{5} = 59.90$$

$$\text{Season} \times \text{Nitrogen } MS = \frac{\text{Season} \times \text{Nitrogen } SS}{(s-1)(t-1)} = \frac{42.93}{1(5)} = 8.59$$

$$\text{Pooled error } MS = \frac{\text{Pooled error } SS}{s(r-1)(t-1)} = \frac{66.76}{2(3)(5)} = 2.23$$

Step 6: Compare the computed F value for each source of variation shown in Table 8.3.3 with the table F value in Appendix E.

The tabular F value for 5 and 30 degrees of freedom at the 1% level of significance is 3.70. The computed F value for the nitrogen treatment and the Season × Nitrogen interaction is significant. This implies that wheat yield is responsive to the application of nitrogen, but there is a difference in response between the spring and winter planting seasons.

Step 7: Perform the test of homogeneity of error variance using the data from the individual analysis of variance given in Table 8.3.2. As was pointed

TABLE 8.3.3

Combined Analysis of Variance of a Randomized Complete Block Experiment over Two Planting Seasons

Source of Variation	Degree of Freedom	Sum of Squares	Mean Square	F
Season	1	0.33	0.33	*a*
Reps. within season	6	10.87	1.81	<1
Nitrogen	5	299.52	59.90	26.86**
Season × Nitrogen	5	42.93	8.59	3.85**
Pooled error	30	66.76	2.23	
Total	47	420.41		

Note: *a* = As the *d.f.* is not adequate, a valid test of significance cannot be performed; ** = significant at 1% level.

out in Section 8.2, this test could be performed either through the application of the chi-square or the F-test. We have used the F-test to determine the homogeneity as follows:

$$F = \frac{\text{Larger error } MS}{\text{Smaller error } MS} \tag{8.16}$$

$$F = \frac{3.57}{0.88} = 4.06$$

In comparing the computed F value with the corresponding F value (3.52 at the 1% level) found in Appendix E for 15 degrees of freedom, we reject the hypothesis of homogeneous error variances over seasons. Thus, we need to partition the Season × Nitrogen interaction into a set of orthogonal contrasts. Such a partition will shed light on why the relative performance of nitrogen applications differed over the spring and winter planting seasons.

Step 8: Construct a set of mutually orthogonal contrasts on one of the factors. Because nitrogen is highly significant, the most natural set of contrasts that can explain the nature of the interaction between nitrogen and season will be the orthogonal polynomials on nitrogen. This means that the Season × Nitrogen sum of squares needs to be partitioned into Season × Nitrogen$_{\text{linear}}$, Season × Nitrogen$_{\text{quadratic}}$, and so on. However, for simplicity of analysis we will consider only the linear and quadratic orthogonal polynomial contrasts. Because the nitrogen rates tested in this experiment have equal intervals and $t = 6$, the single-degree of freedom contrast coefficients are obtained directly from Appendix J and are given as follows:

	Orthogonal Polynomial Coefficient	
Nitrogen Rates (lb/ac)	Linear	Quadratic
0	−5	+5
50	−3	−1
100	−1	−4
150	+1	−4
200	+3	−1
250	+5	+5

If the treatment intervals were not equal, the polynomial coefficients have to be derived. Gomez and Gomez (1984) provide the derivation of the coefficients when treatment intervals are not equal.

Step 9: Calculate the sum of squares for each single *d.f.* contrast, using the following equations:

$$\text{Nitrogen}_{\text{Linear}}SS = \frac{\left[(-5)(N_1)+(-3)(N_2)+(-1)(N_3)+(1)(N_4)+(3)(N_5)+(5)(N_6)\right]^2}{(r)(a)\left[(-5)^2+(-3)^2+(-1)^2+(1)^2+(3)^2+(5)^2\right]}$$

$$\tag{8.25}$$

$$\text{Nitrogen}_{\text{Quadratic}} SS = \frac{\left[(5)(N_1)+(-1)(N_2)+(-4)(N_3)+(-4)(N_4)+(-1)(N_5)+(5)(N_6)\right]^2}{(r)(a)\left[(5)^2+(-1)^2+(-4)^2+(-4)^2+(-1)^2+(5)^2\right]}$$

$$(8.26)$$

where

$N_1 \dots N_6$ = Nitrogen totals
r = number of replications
a = number of seasons

To compute the sum of squares, we need to construct the Season × Nitrogen table of totals, as shown in Table 8.3.4.

$\text{Nitrogen}_{\text{Linear}} SS$

$$= \frac{\left[(-5)(207.9)+(-3)(226.8)+(-1)(232.8)+(1)(259.8)+(3)(257.8)+(5)(260.9)\right]^2}{(4)(2)\left[25+1+16+16+1+25\right]}$$

$$= \frac{(385)^2}{560} = 264.69$$

$\text{Nitrogen}_{\text{Quadratic}} SS$

$$= \frac{\left[(5)(207.9)+(-1)(226.8)+(-4)(232.8)+(-4)(259.8)+(-1)(257.8)+(5)(260.9)\right]^2}{(4)(2)\left[25+1+16+16+1+25\right]}$$

$$= 18.33$$

TABLE 8.3.4

The Season × Nitrogen Table of Yield Totals Computed from Data in Table 8.3.1

Nitrogen	Yield Total				Nitrogen Total
	Rep. I	Rep. II	Rep. III	Rep. IV	
0	52.9	49.5	54.4	51.1	207.9
50	54.4	58.4	58.2	55.8	226.8
100	60.3	55.1	57.9	59.5	232.8
150	61.7	66.3	63.8	68.0	259.8
200	62.3	64.9	65.7	64.9	257.8
250	64.7	66.6	66.6	63.0	260.9

TABLE 8.3.5

Treatment Total and the Linear and Quadratic Sum of Squares Computed from Data in Table 8.3.1

			Treatment Total				Sum of Squares	
Season	0	50	100	150	200	250	Linear	Quadratic
Spring	107.5	118.2	117.9	126.9	124.0	126.5	52.64	9.44
Winter	100.4	108.6	114.9	132.9	133.8	134.4	248.16	12.69

Step 10: Following the procedures of Step 9, compute the sum of squares for each contrast based on the nitrogen totals at each season. This is shown as follows, and the results are presented in Table 8.3.5.

$$\text{Nitrogen}_{\text{Linear}} SS = \frac{\left[(-5)(N_1)+(-3)(N_2)+(-1)(N_3)+(1)(N_4)+(3)(N_5)+(5)(N_6)\right]^2}{(r)\left[(-5)^2+(-3)^2+(-1)^2+(1)^2+(3)^2+(5)^2\right]}$$

$$= \frac{\left[(-5)(107.5)+(-3)(118.2)+(-1)(117.9)+(1)(126.9)+(3)(124.0)+(5)(126.5)\right]^2}{(4)\left[25+9+1+1+9+25\right]}$$

$$= 52.64 \tag{8.27}$$

and the quadratic sum of square is:

$$\text{Nitrogen}_{\text{Quadratic}}\ SS = \frac{\left[(5)(N_1)+(-1)(N_2)+(-4)(N_3)+(-4)(N_4)+(-1)(N_5)+(5)(N_6)\right]^2}{(r)\left[(-5)^2+(-1)^2+(-4)^2+(-4)^2+(-1)^2+(5)^2\right]}$$

$$= \frac{\left[(5)(107.5)+(-1)(118.2)+(-4)(117.9)+(-4)(126.9)+(-1)(124.0)+(5)(126.5)\right]^2}{(4)\left[25+1+16+16+1+25\right]}$$

$$= 9.44 \tag{8.28}$$

Step 11: Calculate the component of the Nitrogen × Season interaction corresponding to the set of mutually orthogonal contrasts as follows:

$$\text{Nitrogen}_i \times \text{Season } SS = \sum_{j=1}^{s}\left(\text{Nitrogen}_i SS\right)_j - \text{Nitrogen}_i SS \tag{8.29}$$

The subscript i is used to distinguish between the linear and quadratic contrasts, and subscript j corresponds to the two seasons. Thus, we have:

$$\text{Nitrogen}_{\text{Linear}} \times \text{Season } SS = (52.64 + 248.16) - 264.69$$

$$= 36.11$$

$$\text{Nitrogen}_{\text{Quadratic}} \times \text{Season } SS = (9.44 + 12.69) - 18.33$$

$$= 3.80$$

Step 12: Enter all the values of the analysis in an ANOVA table as shown in Table 8.3.6. The analysis shows that only the linear component of the sum of squares varied significantly with season. This implies that the Season × Nitrogen interaction is mainly due to the difference in the linear part of the yield responses to nitrogen rates of the different seasons. We also note that both the linear and the quadratic component of the Season × Nitrogen interaction are significant. This implies that yield initially increased at lower rates, tapered off at a maximum, and finally decreased at higher rates.

The results indicate that different nitrogen rates need to be used for the spring and winter seasons. The average yield response to nitrogen fertilizer between seasons is important in making accurate economic fertilizer recommendations that are environmentally sound. Experiments such as this serve as the basis for making recommendations on the actual rates of nitrogen

TABLE 8.3.6

Analysis of Variance of a Partitioned Treatment Sum of Squares of a Randomized Complete Block Experiment

Source of Variation	Degree of Freedom	Sum of Squares	Mean Square	F
Season	1	0.33	0.33	a
Reps. within season	6	10.87	1.81	<1
Nitrogen	(5)	(299.52)	59.90	26.86**
Nitrogen$_L$	1	264.69	264.69	59.34**
Nitrogen$_Q$	1	18.33	18.33	8.21**
Nitrogen$_{Res.}$	3	16.50	5.50	2.47^{ns}
Nitrogen x Season	(5)	(42.93)	8.59	3.85**
Nitrogen$_L$ × Season	1	36.11	36.11	16.19**
Nitrogen$_Q$ × Season	1	3.80	3.80	1.70^{ns}
Nitrogen$_{Res}$ × Season	3	3.02	1.01	1.00^{ns}
Pooled error	30	66.76	2.23	
Total	47			

Note: a = As the *d.f.* is not adequate, a valid test of significance cannot be performed; ^{ns} = nonsignificant, ** = significant at 1% level.

fertilizer to be used. Predictions based on multiyear and multilocation studies provide the researcher with information on changes in nitrogen fertilizer requirements for different seasons.

If the test of homogeneity is not significant, the combined analysis of the experiment would have been completed at this stage and the results reported as shown in Table 8.3.6. However, if the test of homogeneity is significant, then the appropriate test of significance requires that the error term be the component of the pooled error and not the pooled error itself. The following steps show how this is accomplished.

Step 13: Partition the pooled error sum of squares into components corresponding to those of the Season × Nitrogen sum of squares as:

$$\text{Rep. within Season} \times \text{Nitrogen}_{\text{Lin}} = \sum_{i=1}^{s}\left[\sum_{j=1}^{r}\left(\text{Nitrogen}_{\text{Lin}}SS\right)_{ji} - \left(\text{Nitrogen}_{\text{Lin}}SS\right)_{i}\right]$$

(8.30)

$$\text{Rep. within Season} \times \text{Nitrogen}_{\text{Quad}} = \sum_{i=1}^{s}\left[\sum_{j=1}^{r}\left(\text{Nitrogen}_{\text{Quad}}SS\right)_{ji} - \left(\text{Nitrogen}_{\text{Quad}}SS\right)_{i}\right]$$

(8.31)

To determine what the values are for the left-hand side of the above equations, we first need to compute the sum of squares for the right-hand side. To do this, we need to compute the $(\text{Nitrogen}_{\text{Lin}}SS)_{ji}$, which is the linear component of the Nitrogen SS computed from the jth replication in the ith season. The $\text{Nitrogen}_{\text{Lin}}SS_i$ is the corresponding component that is computed from the totals over all replications, as shown in Table 8.3.7.

$$\text{Nitrogen}_{\text{Linear}}SS = \frac{\left[(-5)(N_1)+(-3)(N_2)+(-1)(N_3)+(1)(N_4)+(3)(N_5)+(5)(N_6)\right]^2}{\left[(-5)^2+(-3)^2+(-1)^2+(1)^2+(3)^2+(5)^2\right]}$$ (8.32)

$$= \frac{\left[(-5)(27.8)+(-3)(30.0)+(-1)(29.9)+(1)(31.4)+(3)(30.8)+(5)(30.5)\right]^2}{\left[25+9+1+1+9+25\right]}$$

$$= \frac{\left[-258.9+276.3\right]^2}{70} = \frac{302.76}{70}$$

$$= 4.33$$

TABLE 8.3.7

Treatment Totals and the Linear and Quadratic Sum of Squares for the Pooled Error *SS*

Replication Number	Treatment Total						Sum of Squares	
	0	50	100	150	200	250	Linear	Quadratic
				Spring				
I	27.8	30.0	29.9	31.4	30.8	30.5	4.33	2.50
II	24.6	29.2	28.3	32.0	31.3	31.2	26.41	6.13
III	28.2	30.1	29.7	31.7	29.9	33.0	9.22	0.002
IV	26.9	28.9	30.0	31.8	32.0	31.8	18.11	2.54
Total	107.5	118.2	117.9	126.9	124.0	126.5	52.64	9.44
				Winter				
I	25.1	24.4	30.4	30.3	31.5	34.2	63.56	0.06
II	24.9	29.2	26.8	34.3	33.6	35.4	76.55	0.39
III	26.2	28.1	28.2	32.1	35.8	33.6	58.51	0.44
IV	24.2	26.9	29.5	36.2	32.9	31.2	50.92	24.75
Total	100.4	108.6	114.9	132.9	133.8	134.4	248.16	12.69

Similarly, the quadratic sum of squares for nitrogen is computed as:

$\text{Nitrogen}_{\text{Quadratic}} \; SS$

$$= \frac{\left[(5)(27.8) + (-1)(30.0) + (-4)(29.9) + (-4)(31.4) + (-1)(30.8) + (5)(30.5) \right]^2}{\left[(5)^2 + (-1)^2 + (-4)^2 + (-4)^2 + (-1)^2 + (5)^2 \right]}$$

$$= \frac{\left[-306 + 291.5 \right]^2}{84} = \frac{210.25}{84}$$

$$= 2.50$$

Now we are ready to compute the components of the pooled error sum of squares as:

$\text{Rep. within Season} \times \text{Nitrogen}_{\text{Lin}} =$

$$\left[(4.33 + 26.41 + 9.22 + 18.11) - 52.64 \right] + \left[(63.56 + 76.55 + 58.51 + 50.92) - 248.16 \right]$$

$$= 6.81$$

$\text{Rep. within Season} \times \text{Nitrogen}_{\text{Quad}} =$

$$\left[(2.50 + 6.13 + 0.002 + 2.54) - 9.44 \right] + \left[(0.06 + 0.39 + 0.44 + 24.75) - 12.69 \right]$$

$$= 14.68$$

Rep. within Season $\times$ Nitrogen$_{Res.}$= Pooled error –Reps. within Season $\times$ Nitrogen$_{Lin}SS$ –Reps. within Season $\times$ Nitrogen$_{Quad}SS$

$$= 66.76 - (6.81 + 14.68)$$
$$= 66.76 - 21.49$$
$$= 45.27$$

The results summarized in Table 8.3.8 show each source of variation of the pooled error.

Step 14: Test the significance of the Season $\times$ Nitrogen interaction, using the three components of the pooled error, as follows:

$$F_{\text{Nitrogen } Lin \times \text{ Season}} = \frac{\text{Nitrogen}_{Lin} \times \text{Season } MS}{\text{Reps. within Season} \times \text{Nitrogen}_{Lin} MS} \tag{8.33}$$

$$F_{\text{Nitrogen } Quad \times \text{ Season}} = \frac{\text{Nitrogen}_{Quad} \times \text{Season } MS}{\text{Reps. within Season} \times \text{Nitrogen}_{Quad} MS} \tag{8.34}$$

$$F_{\text{Nitrogen } Res. \times \text{ Season}} = \frac{\text{Nitrogen}_{Res.} \times \text{Season } MS}{\text{Reps. within Season} \times \text{Nitrogen}_{Res.} MS} \tag{8.35}$$

Thus, we have:

$$F_{\text{Nitrogen } Lin \times \text{ Season}} = \frac{36.11}{1.14} = 31.68$$

$$F_{\text{Nitrogen } Quad \times \text{ Season}} = \frac{3.80}{2.45} = 1.55$$

$$F_{\text{Nitrogen } Res. \times \text{ Season}} = \frac{3.02}{2.52} = 1.20$$

Again, we note that only the linear component of the Nitrogen $\times$ Season is significant (F values from Appendix E for the 5% and 1% levels of significance are 5.99 and 13.74, respectively). The interpretation of the results

TABLE 8.3.8

Components of the Pooled Error Sum of Squares

Source of Variation	Degree of Freedom	Sum of Squares	Mean Square
Pooled Error	(30)	(66.76)	2.23
Reps. within Season $\times$ Nitrogen$_L$	6	6.81	1.14
Reps. within Season $\times$ Nitrogen$_Q$	6	14.68	2.45
Reps. within Season $\times$ Nitrogen$_{Res.}$	18	45.27	2.52

remains the same as given in Step 12. The only difference in the analysis is that now we have used the appropriate error term in the *F*-test of significance.

References and Suggested Reading

Cochran, G.W. and Cox, G.M. 1957. *Experimental Designs*. New York: John Wiley & Sons.

Cochran, G.W. 1954. The combination of estimates from different experiments. *Biometrics* 10: 101–129.

Fisher, R.A. 1971. *The Design of Experiments*. 9th ed. New York: Hafner Press.

Gomez, K.A. and Gomez, A.A. 1984. *Statistical Procedures for Agricultural Research*. New York: John Wiley & Sons.

Pearce, S.E. 1983. *The Agricultural Field Experiment: A Statistical Examination of Theory and Practice*. Chichester, England: John Wiley & Sons. Chap. 9.

Snedecor, G.W. and Cochran, W.G. 1967. *Statistical Methods*. 6th ed. Ames, IA: Iowa State University Press. Chap. 12.

Yates, F. and Cochran, W.G. 1938. The analysis of groups of experiments. *J. Agric. Sci.* 28: 556–580.

Exercises

1. Agricultural engineers were interested in evaluating the impact of limited capacity sprinkler irrigation systems on soybean yields. A 3-yr field study was conducted using a randomized complete block experimental design. The irrigation treatments included a nonirrigated (NI) check, irrigation scheduled by soil moisture depletion (SCH), irrigation which began no earlier than the flowering stage (FL), and irrigation which began no earlier than pod elongation (POD). Soybean yields as influenced by irrigation treatments were recorded as follows.

Soybean Yield (bu/ac) as Influenced by Irrigation Treatments

Year	Irrigation Treatment	Rep. I	Rep. II	Rep. III	Rep. IV
1990	NI	37	34	35	35
	POD	50	51	50	49
	FL	53	48	50	51
	SCH	46	48	47	47
1991	NI	32	29	25	30
	POD	34	33	30	35
	FL	43	42	40	38
	SCH	49	48	45	43
1992	NI	30	31	33	32
	POD	40	41	40	38
	FL	43	47	50	45
	SCH	43	47	48	46

(a) Perform a combined analysis over the years.

(b) Perform a test of homogeneity of variance. Can the hypothesis of homogeneous variances be accepted?

(c) What can be said about the nature of the interaction effect between the irrigation treatments and years?

2. As more attention is being paid to the issue of soil conservation, a number of practices are suggested as viable methods. Conventional and no-tillage systems are two approaches being tested growing different crops. In a study in Nebraska, the performances of commercial corn hybrids were investigated using the conventional (CT) and no-tillage (NT) systems. The other objective of the study was to determine if hybrid × tillage system interactions exits for grain yield. The experimental design was a randomized complete block in a split-plot arrangement with four replications. The data collected are shown as follows.

Grain Yield (bu/ac) under Conventional and No-Tillage Systems for 10 Corn Hybrids

Year	Hybrid	CT Replications				NT Replications			
		I	II	III	IV	I	II	III	IV
2002	Pioneer 3737	135.4	132.0	133.9	136.1	125.2	122.4	126.0	122.8
	Pioneer 3744	100.9	101.8	102.0	105.3	110.0	109.6	106.4	108.5
	Funk G-4312	105.0	104.9	102.1	106.3	111.3	110.2	110.9	111.0
	Funk G-4342	122.5	121.3	122.9	122.0	110.9	110.2	111.2	110.5
	DeKalb 484	128.2	127.6	125.9	126.3	123.9	124.0	124.1	124.2
	DeKalb 524	126.8	127.2	126.9	128.0	133.1	132.5	131.9	132.3
	Cargil 842	138.1	137.6	137.2	135.7	120.0	123.3	124.7	120.5
2003	Pioneer 3737	133.4	132.0	133.9	131.1	118.2	126.4	123.0	123.7
	Pioneer 3744	117.9	115.8	112.0	115.5	120.0	129.7	126.3	128.4
	Funk G-4312	109.0	108.9	105.3	108.7	108.3	106.4	109.9	107.0
	Funk G-4342	104.5	101.3	106.9	107.4	120.9	122.2	121.2	120.8
	DeKalb 484	124.8	126.6	125.3	124.9	113.8	114.0	114.1	114.2
	DeKalb 524	136.9	137.2	136.9	140.0	132.1	136.5	131.3	135.8
	Cargil 842	108.9	107.6	107.2	105.8	118.3	119.3	120.7	121.4

(a) Perform a combined analysis over the years.

(b) Perform a test of homogeneity of variance. Can the hypothesis of homogeneous variances be accepted?

(c) What can be said about the nature of the interaction effect between the hybrid and tillage interaction?

3. In an attempt to determine the effect of harvest management on forage yield from "Nitro," a cultivar of alfalfa (*Medicago sativa* L.), a randomized complete block experiment with four replicates was conducted. The treatments consisted of the following four harvest management systems:

H1: No harvest during the growing season

H2: Two harvests at bud and herbage regrowth harvested in the fall

H3: Three harvests at bud and herbage regrowth harvested in the fall

H4: Two harvests at first flower and herbage regrowth harvested in the fall

The data on forage yield collected for the summer and fall are given as follows.

Effect of Harvest Management on Forage Yield (t/ac)

Harvest Management	Rep. I	Rep. II	Rep. III	Rep. IV
Summer				
H1	—	—	—	—
H2	1.9	2.0	2.1	2.0
H3	3.0	3.1	3.0	2.8
H4	2.5	2.4	2.8	2.3
Fall				
H1	0.5	2.3	1.4	2.1
H2	0.7	0.9	0.8	0.7
H3	0.4	0.6	0.2	0.1
H4	0.3	0.5	0.3	0.4

(a) Perform a combined analysis over the seasons.

(b) Perform a test of homogeneity of variance. Can the hypothesis of homogeneous variances be accepted?

(c) What can be said about the nature of interaction effect between the seasons and harvest management practices?

4. As grazing of hay fields during some part of the year is used as an alternative to harvested hay for beef cattle, an animal scientist compared four management systems to determine the yield of forage produced during the spring and fall. The data gathered from the

randomized complete block experiment with five replications are shown as follows.

Total Seasonal Herbage Dry Matter Harvested as Influenced by Management (lb/ac)

Management	Rep. I	Rep. II	Rep. III	Rep. IV	Rep. V
			Spring		
Hay only	4015	4020	4000	3995	4005
Spring and fall grazing	3756	3776	3789	3886	3554
Spring grazing	3300	3312	3310	3345	3305
Fall grazing	4583	4539	4601	4520	4495
			Fall		
Hay only	2334	2486	2300	2319	2398
Spring and fall grazing	3300	3198	3264	3310	3315
Spring grazing	2404	2400	2445	2487	2404
Fall grazing	3360	3354	3329	3352	3390

(a) Perform a combined analysis over the seasons.

(b) Perform a test of homogeneity of variance. Can the hypothesis of homogeneous variances be accepted?

(c) What can be said about the nature of the interaction effect between the seasons and the management systems?

9

Regression and Correlation Analysis

9.1 Bivariate Relationships

In many agricultural research problems you will be faced with a situation in which you are interested in the relationship between two different random variables X and Y. Such a relationship is known as a *bivariate relationship*. For example, an animal scientist may be interested in the relationship between the amount of total digestible nutrient (TDN) consumed and the average daily weight gain of the animal. An agricultural economist may be interested in the bivariate relationship between the appraised value of farm real estate, X, and its sale price, Y; or an agronomist may use a crop–weather model to analyze the effects of weather (X) on the yield of a crop (Y).

To determine if one variable is a predictor of another variable, we use the bivariate modeling technique. The simplest model for relating a variable Y to a single variable X is a straight line. This is referred to as a *linear* relationship. *Simple linear regression* is a technique used to judge whether a relationship exists between Y and X. Furthermore, the technique is used to estimate the mean value of Y and to predict a future value of Y for a given value of X.

In simple regression analysis, we are interested in describing the pattern of the functional nature of the relationship between two variables. This is accomplished by estimating an equation called the *regression equation*. The variable to be estimated in the regression equation is called the *dependent variable* and is plotted on the vertical (or Y) axis. The variable used as the predictor of Y and that exerts influence in explaining the variation in the dependent variable is called the *independent variable*. This variable is plotted on the horizontal (or X) axis.

The linear relationship between the two variables Y and X is expressed by the general equation for a straight line as:

$$Y = a + bX \tag{9.1}$$

where

    Y = dependent variable
    a = regression constant, or the Y intercept
    b = regression coefficient, or the slope of the regression line
    X = independent variable

Other linear equations in which the parameters occur in a linear fashion, although the relationship between X and Y is definitely not linear but quadratic or cubic, are:

$$Y = a + bX + cX^2 \tag{9.2}$$

$$Y = a + bX + cX^2 + dX^3 \tag{9.3}$$

The graphical presentations of the three models are shown in Figure 9.1.1.

In correlation analysis, on the other hand, we are simply interested in the magnitude or closeness of the relationship between two variables. Regression and correlation analyses work together; one asks if there is any relationship between two variables and the other seeks to provide an answer to how close this relationship is.

In the following sections of this chapter we will discuss the techniques for estimating a regression equation, the standard error, and coefficients of determination and correlation. The concepts developed in Section 9.1 are directly applicable to our discussion of multiple regression in Section 9.4.

9.2 Regression Analysis

As was pointed out in the previous section, simple linear regression analysis is concerned with the relationship between two variables. Researchers use prior knowledge and past research as a basis for selecting the independent variables that are helpful in predicting the values of the dependent variable.

Once we have determined that there is a logical relationship between two variables, we can portray the relationship between the variables through a *scatter diagram*. A scatter diagram is a graph of plotted points, each of which represents an observed pair of values of the dependent and independent variables. The scatter diagram serves two purposes: (1) it provides for a visual presentation of the relationship between two variables, and (2) it aids in choosing the appropriate type of model for estimation.

Example 9.2.1 presents a set of data that is used to illustrate how a scatter diagram is helpful in determining the presence or lack of linear relationship between the dependent and the independent variables.

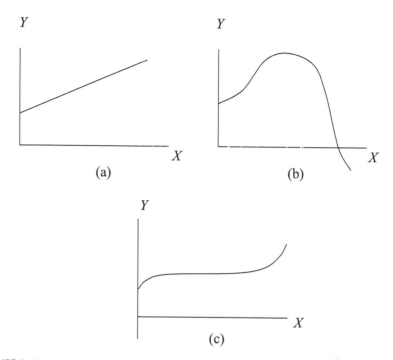

FIGURE 9.1.1
Polynomial relationships: (a) linear, (b) quadratic, and (c) cubic.

Example 9.2.1

An animal scientist interested in the milk yield of a lactating ewe has measured the milk yield (kg/d) by weighing the lamb before and after suckling. The following observations were obtained at different time intervals. The scatter diagram is constructed as described in the text that follows.

Day	Yield
10	1.78
14	1.66
18	1.62
22	1.59
26	1.55
30	1.60
34	1.58
38	1.54
42	1.50
46	1.48
50	1.43
54	1.40
58	1.37
62	1.35
66	1.32

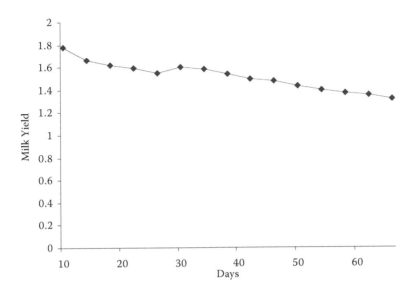

FIGURE 9.2.1
A scatter diagram of the milk yield of a lactating ewe at 4-d intervals.

SOLUTION

Following the standard convention of plotting the dependent variable along the Y axis and the independent variable along the X axis, we have the milk yield plotted along the Y axis and the day-intervals along the X axis. Figure 9.2.1 shows the scatter diagram for this problem.

An examination of Figure 9.2.1 shows that there is an *inverse*, or negative, relationship between the two variables. That is, as the X variable increases, the Y tends to decrease. If, on the other hand, Y increases as X increases, then there is a *direct*, or positive, relationship between the variables.

The scatter diagram is also used to determine if there is a *linear* or *curvilinear* relationship between variables. If a straight line can be used to describe the relationship between variables X and Y, there is a linear relationship. If the observed points in the scatter diagram fall along a curved line, there is a curvilinear relationship between the variables.

Figure 9.2.2 shows a number of different scatter diagrams depicting different relationships between variables. You will notice that scatter diagrams Figure 9.2.2(a) and Figure 9.2.2(b) illustrate a positive and negative linear relationship between two variables, respectively. Figure 9.2.2(c) and Figure 9.2.2(d) show positive and negative curvilinear relationships between variables X and Y, and Figure 9.2.2.(f) shows no relationship between variables.

Another curvilinear relationship is illustrated in Figure 9.2.2(e), in which X and Y rise at first, and then, as X increases, Y decreases. Such a relationship is observed in agricultural economics and business; for instance, a farmer's earned income tends to rise with the age of the farmer and then decline after the farmer retires.

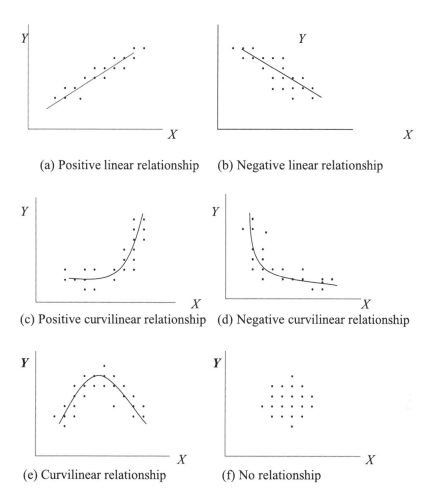

FIGURE 9.2.2
Examples of linear and curvilinear relationships found for scatter diagrams.

The linear regression equation: The mathematical equation of a line such as the one in the scatter diagram in Figure 9.2.1 that describes the relationship between two variables is called the *regression* or *estimating equation*. The regression equation has its origins in the pioneering work of Sir Francis Galton (1908), who fitted lines to scatter diagrams of data on the heights of fathers and sons. Galton found that the heights of the children of tall parents tended to regress toward the average height of the population. Galton referred to his equation as the regression equation.

The regression equation is determined by the use of a mathematical method referred to as the *least-squares method*. This method simply minimizes the sum of the squares of the vertical deviations about the line. Thus, the least-squares method is a best fit in the sense that $\sum\left(Y-\hat{Y}\right)^2$ is less than it would be for any other possible straight line. In this expression Y is the observed value and $\hat{Y}$ is the estimated value of a variable. Additionally, the least-squares regression line has the following property:

$$\sum\left(Y-\hat{Y}\right)=0 \tag{9.4}$$

This characteristic makes the total of positive and negative deviations equal to zero.

You should note that the linear regression equation, Equation 9.1, is just an estimate of the relationship between the two variables in the population that are given in Equation 9.5:

$$\mu_{y.x} = A + BX \tag{9.5}$$

where

$\mu_{y.x}$ = the mean of the Y variable for a given value of X

A and B = population parameters that must be estimated from sample data

The regression equation can be calculated by two methods. The first involves solving simultaneously two equations called the *normal equations*. They are:

$$\sum Y = na + b\sum X \tag{9.6}$$

$$\sum XY = a\sum X + b\sum X^2 \tag{9.7}$$

We use Equation 9.6 and Equation 9.7 to solve for a and b and obtain the estimating or regression equation.

The second method of arriving at a least-squares regression equation is by using the following computationally more convenient equations:

$$b = \frac{n\left(\sum XY\right)-\left(\sum X\right)\left(\sum Y\right)}{n\left(\sum X^2\right)-\left(\sum X\right)^2} \tag{9.8}$$

$$a = \frac{\sum Y}{n} - b\left(\frac{\sum X}{n}\right) = \bar{Y} - b\bar{X} \tag{9.9}$$

The following example will illustrate the computation of the regression equation, using either the normal equations or the shortcut formula given in Equation 9.8 and Equation 9.9.

Example 9.2.2

An agronomist is interested in the relationship between maize yield, Y, and the amount of fertilizer X applied. In order to determine whether there is a relationship, the agronomist has divided a field into nine plots of equal size in different localities in Iowa and has applied different amounts of fertilizer to each plot. The following yield data (in bushels) were recorded for the different amounts of fertilizer (in pounds) applied.

Yield Y	Fertilizer X
50	5
57	10
60	12
62	18
63	25
65	30
68	36
70	40
69	45
66	48

SOLUTION

To compute the regression equation, the data in Table 9.2.1 are used. We will substitute the appropriate values from Table 9.2.1 into Equation 9.6 and Equation 9.7 as follows:

$$\sum Y = na + b\sum X \tag{9.6}$$

$$\sum XY = a\sum X + b\sum X^2 \tag{9.7}$$

$$630 = 10a + 269b$$

$$17{,}702 = 269a + 9{,}343b$$

TABLE 9.2.1

Computation of Intermediate Values Needed
for Calculating the Regression Equation

Yield Y	Fertilizer X	Y^2	XY	X^2
50	5	2,500	250	25
57	10	3,249	570	100
60	12	3,600	720	144
62	18	3,844	1,116	324
63	25	3,969	1,575	625
65	30	4,225	1,950	900
68	36	4,624	2,448	1,296
70	40	4,900	2,800	1,600
69	45	4,761	3,105	2,025
66	48	4,356	3,168	2,304
630	269	40,028	17,702	9,343

To solve the preceding two equations for either of the unknowns a and b, we must eliminate one of the unknown coefficients. For example, to eliminate the unknown coefficient a in the equations, we multiply the first equation by 26.9. By doing so, the value of the a coefficient in the first equation is now equal to the value of a in the second equation. We then subtract the second from the first to obtain the value of b as shown:

$$16,947 = 269a + 7,236.1b$$

$$\underline{-17,702 = -269a - 9,343b}$$

$$-755 = -2,106.9b$$

$$b = 0.358$$

The value of b can be substituted in either Equation 9.6 or Equation 9.7 to solve for the value of a.

$$630 = 10a + 269(0.358)$$

$$630 = 10a + 96.302$$

$$533.69 = 10a$$

$$a = 53.369$$

The least-squares regression equation is:

$$\hat{Y} = 53.369 + 0.358X$$

Using the shortcut formula, you will obtain the following:

$$b = \frac{n\left(\sum XY\right) - \left(\sum X\right)\left(\sum Y\right)}{n\left(\sum X^2\right) - \left(\sum X\right)^2}$$

$$= \frac{10(17,702) - (269)(630)}{10(9,343) - (269)^2} \qquad (9.8)$$

$$= \frac{7,550}{21,069}$$

$$= 0.358$$

We will now substitute the value of b into Equation 9.9 to obtain the intercept of the line, or a, as follows:

$$a = \frac{630}{10} - 0.358\left(\frac{269}{10}\right)$$

$$= 63.0 - 0.358(26.9)$$

$$= 63.0 - 9.630$$

$$= 53.369$$

Hence, the least-squares regression line is:

$$\hat{Y} = 53.369 + 0.358X$$

In this equation, $a = 53.369$ and is an estimate of the Y intercept. The value of a means that the yield of maize from a plot of land with no fertilizer added is equal to 53.369 bushels. The b value indicates that the slope of the line is positive. This means that as fertilizer usage increases, so does the yield. The value of b implies that for each additional pound of fertilizer applied, the yield will increase by 0.358 bushels within the range of values observed.

This regression equation can also be used to estimate values of the dependent variable for given values of the independent variable. For example, if the agronomist wishes to estimate the yield from 42 lb of fertilizer, it can be calculated as shown:

$$\hat{Y} = 53.369 + 0.358X\,(42)$$

$$= 53.369 + 15.036$$

$$= 68.41 \text{ bushels of maize}$$

You must keep in mind that the sample regression equation should not be used for prediction outside the range of values of the independent variable given in a sample.

In order to graph the regression line, we need two points. Because we have determined only one ($X = 42$, $Y = 68.41$), we need one other point. The second point ($X = 10$, $Y = 56.95$) is shown, along with the original data, in Figure 9.2.3.

The estimated yield values of 56.95 and 68.41 should be treated as average values. This means that in the future, the average yield will vary from sample to sample due to the fertility of the soil, the temperature, the amount of water used during the growing season, and a host of other factors.

In the following section, we will examine a measure that helps us determine whether the estimate made from the regression equation is dependable.

The standard error of estimate ($S_{y.x}$): The regression equation is primarily used for estimation of the dependent variable, given values of the independent variable. Once we have estimated a regression equation, it is important to determine whether the estimate is dependable or not. Dependability is measured by the closeness of the relationship between the variables. If in a scatter diagram the points are scattered close to the regression line, there is a close

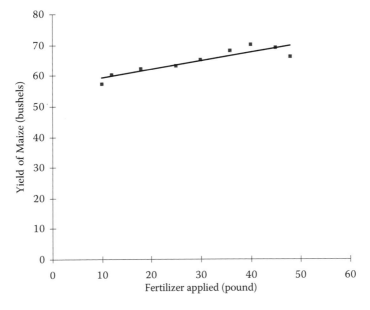

FIGURE 9.2.3
Regression line of the maize yield.

relationship between the variables. If, on the other hand, there is a great deal of dispersion of the points about the regression line, the estimate made from the regression equation is less reliable.

$S_{y.x}$ is used as a measure of the scatter or dispersion of the points about the regression line, just as one uses the standard deviation to measure the deviation of the individual observations about the mean of those values. The smaller $S_{y.x}$ is, the closer the estimate is likely to be to the ultimate value of the dependent variable. In the extreme case in which every point falls on the regression line, the vertical deviations are all 0; that is, $S_{y.x} = 0$. In such a situation, the regression line provides perfect predictions. On the other hand, when the points are highly dispersed, making the vertical deviations large ($S_{y.x}$ is large), the predictions of Y made from the regression line are subject to sampling error.

$S_{y.x}$ is computed by solving the following equation:

$$S_{y.x} = \sqrt{\frac{\sum \left(Y - \hat{Y}\right)^2}{n - 2}} \tag{9.10}$$

where

Y = the dependent variable
$\hat{Y}$ = the estimated value of the dependent variable
n = the sample size

The $n - 2$ value in the denominator represents the number of degrees of freedom around the fitted regression line. Generally, the denominator is $n\ k$, where k represents the number of constants in the regression equation. In the case of a simple linear regression, we lose two degrees of freedom when a and b are used as estimates of the constants in the population regression line. Note that Equation 9.10 requires a value of Y for each value of X. We must therefore compute the difference between each $\hat{Y}$ and the observed value of Y as shown in Table 9.2.2.

$S_{y.x}$ for Example 9.2.2 is calculated as follows:

$$S_{y.x} = \sqrt{\frac{\sum \left(Y - \hat{Y}\right)^2}{n - 2}}$$

$$= \sqrt{\frac{67.448}{8}} \tag{9.10}$$

$$= 2.90 \text{ bushels of maize}$$

TABLE 9.2.2

Computation of Intermediate Values Needed for
Calculating the Standard Error of Estimate

Yield Y	Fertilizer X	$\hat{Y}$	$Y - \hat{Y}$	$\left(Y - \hat{Y}\right)^2$
50	5	55.159	−5.159	26.615
57	10	56.949	0.051	0.003
60	12	57.665	2.335	5.452
62	18	59.813	2.187	4.783
63	25	62.319	0.681	0.464
65	30	64.109	0.891	0.794
68	36	66.257	1.743	3.038
70	40	67.689	2.311	5.341
69	45	69.479	−0.479	0.229
66	48	70.553	−4.553	20.729
630	269	630.000	0.0	67.488

This computational method requires a great deal of arithmetic, especially when large numbers of observations are involved. To minimize cumbersome arithmetic, the following shortcut formula is used in computing $S_{y.x}$.

$$S_{y.x} = \sqrt{\frac{\sum Y^2 - a\left(\sum Y\right) - b\left(\sum XY\right)}{n - 2}} \qquad (9.11)$$

All the values needed to compute $S_{y.x}$ are available from Table 9.2.1; the previously obtained values of a and b are used for the calculation:

$$S_{y.x} = \sqrt{\frac{40,028 - 53.369(630) - 0.358(17,702)}{8}}$$

$$= \sqrt{\frac{68.214}{8}}$$

$$= 2.92 \text{ bushels of maize}$$

The answers from the two approaches are similar, as expected. The minute difference in the two values of $S_{y.x}$ is due to rounding.

Because $S_{y.x}$ is theoretically similar to the standard deviation, there is also similarity in their interpretation. If the scatter about the regression line is normally distributed and we have a large sample, approximately 68% of the points in the scatter diagram will fall within 1 $S_{y.x}$ above and below the regression line; 95.4% of the points will fall within 2 $S_{y.x}$ above and below the regression line, and virtually all points above and below the regression line will fall within 3 $S_{y.x}$, as shown in Figure 9.2.4.

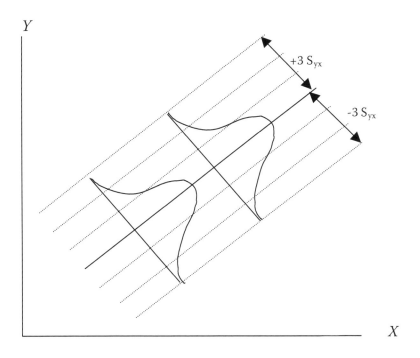

FIGURE 9.2.4
Illustration of the standard error of estimate about the estimated regression line.

Confidence interval estimate: We used the regression equation to estimate the value of Y given a value of X. An estimate of Y_1 was obtained by simply inserting a value for X_1 into the regression equation $Y_1 = a + bX_1$. The estimate of Y_1 is nothing more than a *point estimate*. To attach some confidence to this point estimate, we use $S_{y.x}$ to compute an interval estimate in which a probability value may be assigned to it. For a small sample, the interval estimate for an individual Y given a value of X, say X_0, is computed by the following equation:

$$Y_i = \hat{Y} \pm t\left(S_{y.x}\right)\sqrt{1 + \frac{1}{n} + \frac{\left(X_0 - \bar{X}\right)^2}{\sum X^2 - \dfrac{\left(\sum X\right)^2}{n}}} \qquad (9.12)$$

Earlier, the agronomist had estimated the yield of maize to be 68.41 bu when he applied 42 lb of fertilizer. How much confidence he has in this estimate depends on the probability value attached to it. To construct a 95% prediction interval, we use Equation 9.12 and the data from Table 9.2.2. The critical value of t for the 95% level of confidence is given as 2.306 in Appendix I. The computation of the prediction interval is as follows:

$$Y_i = 68.41 \pm 2.306(2.90)\sqrt{1 + \frac{1}{10} + \frac{(42 - 26.9)^2}{9,343 - \frac{(269)^2}{10}}}$$

$$Y_i = 68.41 \pm 7.35$$

or

$$61.72 < Y_i < 75.77 \text{ bu}$$

Hence, the prediction interval is from 61.72 bu to 75.77 bu. To probabilistically interpret this number, we would say that we are 95% confident that the single prediction interval constructed includes the true yield.

The prediction interval for an individual value of Y when using large samples is determined by the following expression:

$$\hat{Y} \pm z S_{y.x} \qquad (9.13)$$

9.3   Correlational Analysis

In regression analysis, we emphasized estimation of an equation that describes the relationship between two variables. In this section, we are interested in those measures that verify the degree of closeness or association between two variables and the strength of the relationship between them. We will examine two correlation measures: the *coefficient of determination* and the *coefficient of correlation*.

Before using these correlation measures, it is important to understand the assumptions of the two-variable correlation models. In correlation analysis, we make the assumption that both X and Y are random variables; furthermore, they are normally distributed. Also, the standard deviations of the Y's are equal for all values of X, and *vice versa*.

Coefficient of determination: As a measure of the closeness between variables, the coefficient of determination (r^2) can provide an answer as to how well the least-squares line fits the observed data. The relative variation of the Y values around the regression line and the corresponding variation around the mean of the Y variable can be used to explain the correlation that may exist between X and Y. Conceptually, Figure 9.3.1 illustrates three different deviations — namely, the total deviation, the explained deviation, and the unexplained deviation — that exist between a single point Y and the mean and the regression line.

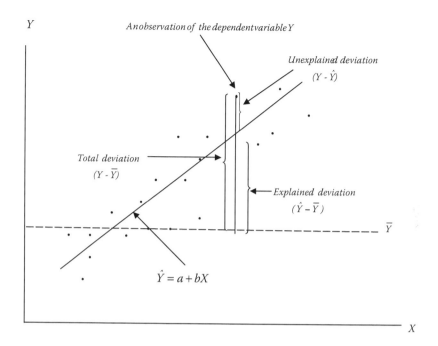

FIGURE 9.3.1
Total deviation, explained deviation, and unexplained deviation for one observed value of Y.

The vertical distance between the regression line and the $\overline{Y}$ line is the explained deviation, and the vertical distance of the observed Y from the regression line is the unexplained deviation. The unexplained deviation represents that portion of total deviation that was not explained by the regression line. The distance between Y and $\overline{Y}$ is called the *total deviation*. Stated differently, the total deviation is the sum of the explained and the unexplained deviations, that is,

Total deviation = explained deviation + unexplained deviation

$$Y - \overline{Y} = (\hat{Y} - \overline{Y}) + (Y - \hat{Y}) \qquad (9.14)$$

To transform Equation 9.14 into a measure of variability, we simply square each of the deviations and sum for all observations to obtain the squared deviations:

Total sum of squares = explained sum of squares + unexplained sum of squares

$$\sum \left(Y - \overline{Y}\right)^2 = \sum (\hat{Y} - \overline{Y})^2 + \sum (Y - \hat{Y})^2 \qquad (9.15)$$

The term on the left-hand side of Equation 9.15 is now the *total sum of squares*, which measures the dispersion of the observed values of the Y about their mean $\bar{Y}$. Similarly, on the right-hand side of Equation 9.15 we have the explained and unexplained sum of squares, respectively.

Given the above relationships, we can define the sample coefficient of determination as:

$$r^2 = 1 - \frac{\text{Unexplained sum of squares}}{\text{Total sum of squares}}$$

or

$$r^2 = \frac{\text{Explained sum of squares}}{\text{Total sum of squares}}$$

or

$$r^2 = \frac{\sum (\hat{Y} - \bar{Y})^2}{\sum (Y - \bar{Y})^2} \tag{9.16}$$

To compute r^2 using Equation 9.16, we need the values of the explained and total sum of squares, as computed in Table 9.3.1. Substituting these values into Equation 9.16 we get:

$$r^2 = \frac{269.97}{338} = 0.798$$

TABLE 9.3.1

Calculation of the Sum of Squares When $\bar{Y} = 63$

Yield Y	Fertilizer X	$\hat{Y}$	$(Y - \bar{Y})^2$	$(\hat{Y} - \bar{Y})^2$
50	5	55.159	169	61.48
57	10	56.949	36	36.61
60	12	57.665	9	28.46
62	18	59.813	1	10.15
63	25	62.319	0	0.46
65	30	64.109	4	1.22
68	36	66.257	25	10.60
70	40	67.689	49	21.98
69	45	69.479	36	41.97
66	48	70.553	9	57.04
630	269	630.000	338	269.97

As is apparent, computation of r^2 using Equation 9.16 is tedious, particularly with a large sample. To remedy the situation, we use a shortcut formula that utilizes the estimated regression coefficients and the intermediate values used to compute the regression equation. Thus, the shortcut formula for computing the sample coefficient of determination is:

$$r^2 = \frac{a \sum Y + b \sum XY - n\bar{Y}^2}{\sum Y^2 - n\bar{Y}^2} \qquad (9.17)$$

To calculate r^2 for the agronomist in Example 9.2.2, we have:

$$r^2 = \frac{53.369(630) + 0.358(17,702) - 10(63)^2}{40,028 - 10(63)^2}$$

$$= \frac{33,622.47 + 6,337.32 - 39,690}{40,028 - 39,690}$$

$$= \frac{269.79}{338}$$

$$= 0.798$$

The calculated r^2 (0.798) signifies that 79.8% of the total variation in the yield of maize Y can be explained by the relationship between the yield and the amount of fertilizer X applied. Because the greatest possible value that r can have is 1, the calculated r in this example implies a strong linear relationship between X and Y. Similar to the sample coefficient of determination, the population coefficient of determination ρ^2 is equal to the ratio of the explained sum of squares to the total sum of squares.

Sample correlation coefficient without regression analysis: Another parameter that measures the strength of the linear relationship between two variables X and Y is the *coefficient of correlation*. The sample coefficient of correlation (r) is defined as the square root of the coefficient of determination. The population correlation coefficient measures the strength of the relationship between two variables in the population. Thus, the sample and population correlation coefficients are as follows:

$$r = \sqrt{r^2} \qquad (9.18)$$

$$\rho = \sqrt{\rho^2} \qquad (9.19)$$

The correlation coefficient can be any value between –1 and +1, both values inclusive. When r equals –1, there is a perfect inverse linear correlation between the variables of interest. When r equals 1, there is a perfect direct linear correlation between X and Y. When r equals 0, the variables X and Y are not linearly correlated. Figure 9.3.2 shows different scatter diagrams representing various sample correlation coefficients.

The algebraic sign of r is the same as that of b in the regression equation. Thus, if the regression coefficient b is positive, then r will also have a positive value. Similarly, if the regression coefficient is negative, we will expect r to have a negative value.

The sample coefficient of correlation for Example 9.2.2 is:

$$r = \sqrt{0.798}$$

$$= 0.89$$

Note that the sign of this sample correlation coefficient is positive, as was that of the b coefficient.

A word of caution is needed with respect to the interpretation of the correlation coefficient. In the case of the coefficient of determination r^2, we could interpret that parameter as a proportion or as a percentage. However, when the square root of a percentage is taken, as is the case with the correlation coefficient, the specific meaning attached to it becomes obscure. Therefore, we should merely conclude that the closer the r value is to 1, the better the correlation is between X and Y. Because r^2 is a decimal value, its square root r is a larger number. This may give a false impression that a high degree of correlation exists between X and Y. For example, consider a situation in

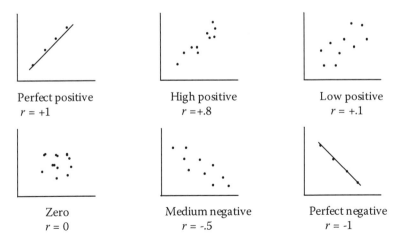

FIGURE 9.3.2
Various scatter diagrams with different correlation coefficients.

which $r = 0.70$, which indicates a relatively high degree of association. However, because $r^2 = 0.49$, the reduction in total variation is only 49%, or less than half. Despite the common use of the correlation coefficient, it is best to use the coefficient of determination to explain the degree of association between X and Y when regression analysis is employed.

The correlation coefficient that is computed without performing regression analysis does provide more meaningful results. In the following section, you will be shown how to compute the correlation coefficient without performing a regression analysis.

There are many instances in which a researcher is not interested in making a prediction but is interested in whether there is a relationship between X and Y. For example, an animal scientist may be interested in finding whether increasing a drug's dosage reduces the symptoms. In such cases, the following formula to compute the value referred to as the *Pearson's sample correlation coefficient* can be used:

$$r = \frac{\sum XY - n\bar{X}\bar{Y}}{\sqrt{\left(\sum X^2 - n\bar{X}^2\right)\left(\sum Y^2 - n\bar{Y}^2\right)}} \qquad (9.20)$$

We may use the intermediate calculations found earlier in Example 9.2.2 to compute the sample correlation coefficient as shown:

$$n = 10 \qquad\qquad \bar{X} = 26.9 \qquad\qquad \bar{Y} = 63$$

$$\sum XY = 17,702 \qquad \sum X^2 = 9,343 \qquad \sum Y^2 = 40,028$$

$$r = \frac{17,702 - 10(26.9)(63)}{\sqrt{\left[(9,343) - 10(26.9)^2\right]\left[40,028 - 10(63)^2\right]}}$$

$$r = \frac{17,702 - 16,947}{\sqrt{712,132.2}}$$

$$= \frac{755}{843.881}$$

$$= 0.89$$

The correlation coefficient 0.89 computed with Equation 9.20 is the same as the value we found earlier by taking the positive square root of r^2.

Inferences regarding regression and correlation coefficients: So far, our discussions have centered around the computation and interpretation of the regression and correlation coefficients. The topic of this section is our degree of confidence that these coefficients do not contain sampling error and that they correspond to the population parameters. Hypothesis testing, or confidence interval estimation, is often used to determine whether the sample data provide sufficient evidence to indicate that the estimated regression coefficient differs from 0. If we can reject the null hypothesis that b is equal to 0 , we can conclude that X and Y are linearly related.

To illustrate the hypothesis-testing procedure for a regression coefficient, let us use the data from Example 9.2.2. The null and alternative hypotheses regarding the regression coefficient may be stated as follows:

$$H_0 = \beta = 0$$

$$H_1 = \beta \neq 0$$

To test this hypothesis, the agronomist wishes to use a 0.05 level of significance. The procedure involves a 2-tailed test in which the test statistic is:

$$t = \frac{b - \beta}{S_b} \tag{9.21}$$

where S_b is the estimated standard error of the regression coefficient and is computed as follows:

$$S_b = \frac{S_{y.x}}{\sqrt{\sum X^2 - n\bar{X}^2}} \tag{9.22}$$

We substitute the appropriate values into Equation 9.22 and Equation 9.21 to perform the test.

$$S_b = \frac{2.90}{\sqrt{9,343 - 10(26.9)^2}}$$

$$= 0.06$$

The calculated test statistic is

$$t = \frac{0.358}{0.06}$$

$$= 5.966$$

Given an = .05 and eight degrees of freedom ($10 - 2 = 8$), the critical value of t from Appendix I is 2.306. Because the computed t exceeds the critical value of 2.306, the null hypothesis is rejected at the .05 level of significance, and we conclude that the slope of the regression line is not 0.

Similar to the test of significance of b, the slope of the regression equation, we can perform a test for the significance of a linear relationship between X and Y. In this test, we are basically interested in knowing whether there is correlation in the population from which the sample was selected. We may state the null and alternative hypotheses as follows:

$$H_0 = \rho = 0$$

$$H_1 = \rho \neq 0$$

The test statistic for samples of small size is

$$t = \frac{r\sqrt{n-2}}{\sqrt{1-r^2}} \tag{9.23}$$

Again, using a .05 level of significance and $n - 2$ degrees of freedom, the critical t from Appendix I is 2.306. The decision rule states that if the computed t falls within ± 2.306, we should accept the null hypothesis; otherwise, we reject it.

$$t = \frac{0.89\sqrt{10-2}}{\sqrt{1-(0.89)^2}}$$

$$= \frac{0.89(2.828)}{\sqrt{1-0.792}}$$

$$= 5.52$$

Because 5.52 exceeds the critical t value of 2.306, the null hypothesis is rejected, and we conclude that X and Y are linearly related.

For large samples, the test of significance of correlation can be performed using the following equation:

$$z = \frac{r}{\frac{1}{\sqrt{n-1}}} \tag{9.24}$$

F-test — an illustration: Instead of using the t distribution to test a coefficient of correlation for significance, we may use an analysis of variance or the F ratio. In computing r^2 and the $S_{y.x}$, we partitioned the total sum of squares into explained and unexplained sums of squares. In order to perform the F-test, we first set up the variance table, determine the degrees of freedom, and then compute F as a test of our hypothesis. The null and alternative hypotheses are

$$H_0 = \rho = 0$$

$$H_1 = \rho \neq 0$$

We will test the hypothesis at the .05 level of significance. Table 9.3.2 shows the computation of the F ratio. The numerical values of the explained and total sums of squares used in Table 9.3.2 were presented earlier in Table 9.3.1. The degree of freedom associated with the explained variation is always equal to the number of independent variables used to explain variations in the dependent variable ($n - k$, where k refers to the number of independent variables). Thus, for the present problem we have only one degree of freedom for the explained variation. The degree of freedom associated with the unexplained variation is found simply by subtracting the degrees of freedom of the explained sum of squares from the degrees of freedom of the total sum of squares. Hence, the unexplained sum of squares is equal to 8. Computing the F ratio, as has been shown before, requires the variance estimate, which is found by dividing the explained and unexplained sums of squares by their

TABLE 9.3.2

Variance Table for Testing Significance of Correlation by F Ratio

(1) Source of Variation	(2) Sum of Squares	(3) Degree of Freedom	(4) Variance Estimate	(5) F
Explained	269.97	1	269.97	31.76
Unexplained	68.03	8	8.50	
Total	338.00	9		

respective degrees of freedom. The variance estimates are shown in column 4 of Table 9.3.2.

The *F*-test is the ratio between the variance explained by the regression and the variance that is not explained by the regression. The *F* ratio is shown in Table 9.3.2.

The computed *F* ratio is compared with the critical value of *F* given in Appendix E. The critical *F* for the .05 level of significance and one and eight degrees of freedom is 5.32. Our tested *F* = 31.76 is much greater than 5.32. Therefore, we reject the null hypothesis of no correlation and conclude that the relationship between the amount of fertilizer applied and the yield is significant.

9.4 Curvilinear Regression Analysis

In our discussions so far, we have only considered the simplest form of a relationship between two variables, namely, the linear relationship. Although the simple linear regression may be considered a powerful tool in analyzing the relationship that may exist between two variables, the assumption of linearity has serious limitations. There are many instances when the straight line does not provide an adequate explanation of the relationship between two variables. When theory suggests that the underlying bivariate relationship is nonlinear, or a visual check of the scatter diagram indicates a curvilinear relationship, it is best to perform a nonlinear regression analysis. For example, the crop yield may increase with increases in application of fertilizer up to a point, beyond which it will decrease. This type of a relationship suggests a curvilinear (in this case, a parabolic) model.

How we decide on which curve to use depends on the nature of the relationship between the variables and our own knowledge and experience of the relationship. When dealing with agricultural data, there may be instances when the relationship between two variables is so complex that we are unable to use a simple equation for it. Under such conditions it is best to find an equation that provides a good fit to the data, without making any claims that the equation expresses any natural relationship.

There are many nonlinear models that can be used with agricultural data. The following are some examples of the nonlinear equations that can easily be transformed to their linear counterparts and analyzed as linear equations.

$$Y = \alpha e^{\beta X} \tag{9.25}$$

$$Y = \alpha \beta^{X} \tag{9.26}$$

$$\frac{1}{Y} = \alpha + \beta X \tag{9.27}$$

$$Y = \alpha + \frac{\beta}{X} \tag{9.28}$$

$$Y = \left(\alpha + \frac{\beta}{X} \right)^{-1} \tag{9.29}$$

For illustration we have selected two of the nonlinear models that are extensively used with agricultural and economic data to perform the analysis.

The exponential growth model: In its simplest form, the model is used for the decay or growth of some variable with time. The general equation for this curve is:

$$Y = \alpha e^{\beta X} \tag{9.25}$$

In this equation, X (time) appears as an exponent, the coefficient describes the rate of growth or decay, and $e \cong 2.718$ is the Euler's constant, which appears in the formula for the normal curve and is also the base for natural logs. The advantage of using this base is that $\beta \cong$ the growth rate.

The assumptions of the model are that the rate of decay is proportional to the current value of Y and that the error term is multiplicative rather than additive, as it is reasonable to assume that large errors are associated with large values of the dependent variable Y. Thus, the statistical model is

$$Y = \alpha e^{\beta X} . u \tag{9.30}$$

The nonlinear models can be transformed (as was discussed in Chapter 2, Section 2.4) to a linear form by taking the logarithm of the equation. For example, transforming Equation 9.30 into its logarithm we get:

$$\log Y = \log \alpha + \beta X + \log u \tag{9.31}$$

where

$$Y' \equiv \log Y$$

$$\alpha \equiv \log \alpha$$

$$e \equiv \log u$$

Then, we can rewrite Equation 9.31 in the standard linear form as:

$$Y' = \alpha' + \beta X + e \qquad (9.32)$$

Exponential models have been extensively used in laboratory experiments in which studies of insects treated with chemicals are conducted. In agricultural field experiments that study the decay of chemicals in the soil or in animals, the exponential model is used. Other situations in which these models have been useful are in growth studies, in which the response variable either increases with time t or as a result of an increasing level of a stimulus variable X. Figure 9.4.1 shows the simple exponential model with growth and decay curves.

When considering different forms of writing a particular relationship, it must be kept in mind that in fitting the model to the data, the choice of the correct error structure assumption is critically important.

Example 9.4.1 illustrates the use of an exponential model and how it can be fitted.

Example 9.4.1

From the early 1900s, there have been major resettlements of the population in many parts of California. Population data for the period 1900 to 1980 show the growth in a southern California community. Figure 9.4.2 shows the growth in population with a continual, though varying, percentage increase. This type of growth is similar to that of compound interest or unrestrained

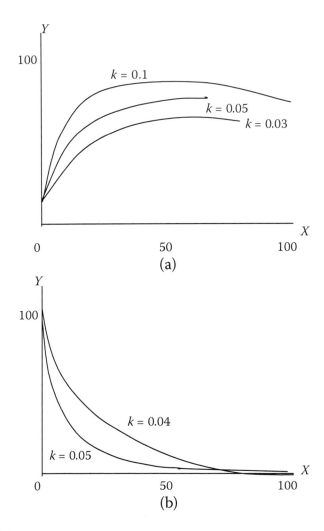

FIGURE 9.4.1
Simple exponential models: (a) gradual approach of yield to an upper limit, (b) decay curve
with time.

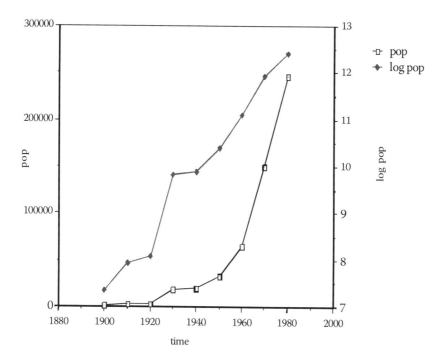

FIGURE 9.4.2
Population growth (1900–1980), and its exponential fit.

biological growth. Therefore, the appropriate model to use in analyzing the data is the exponential model.

Year	Population
1900	1,520
1910	2,800
1920	3,200
1930	18,600
1940	19,800
1950	32,560
1960	65,490
1970	150,450
1980	245,900

SOLUTION

Step 1: Transform the population data into the natural logs and compute the totals as shown in Table 9.4.1.

Step 2: Compute the regression coefficients as shown:

TABLE 9.4.1

Population of a Southern California Community,
1900–1980

Year	Time (X)	Population (Y)	Log Y
1900	0	1,520	7.326
1910	1	2,806	7.940
1920	2	3,210	8.074
1930	3	18,630	9.833
1940	4	19,800	9.893
1950	5	32,560	10.391
1960	6	65,495	11.090
1970	7	150,457	11.921
1980	8	245,904	12.413
Total	36		88.881
Sum of Squares	204		903.627
Sum of X log Y			394.405

$$\sum x^2 = \sum X^2 - \frac{\left(\sum X\right)^2}{n} \qquad (9.33)$$

$$= 204 - \frac{(36)^2}{9} = 60$$

$$\sum y'^2 = \sum Y'^2 - \frac{\left(\sum Y'\right)^2}{n} \qquad (9.34)$$

$$= 903.627 - \frac{(88.881)^2}{9} = 25.867$$

$$\sum xy' = \sum XY' - \frac{\left(\sum X \sum Y'\right)}{n} \qquad (9.35)$$

$$= 394.405 - \frac{36(88.881)}{9} = 38.881$$

$$b' = \frac{\sum xy'}{\sum x^2} \qquad (9.36)$$

$$= \frac{38.881}{60} = 0.648$$

$$a' = \frac{\sum Y'^2}{n} - b' \frac{\sum X}{n} \qquad (9.37)$$

$$= \frac{88.881}{9} - 0.648 \left(\frac{36}{9} \right) = 7.284$$

The resultant regression equation is:

$$\hat{Y}' = 7.284 + 0.6480x$$

Step 3: Transform the above equation back into the exponential model as follows:

$$\log Y = 7.284 + 0.6480x$$

Taking antilogs (exponential):

$$\hat{Y} = e^{7.284} e^{0.648}$$

That is:

$$\hat{Y} = 1,457 e^{0.648}$$

For convenience we have left time (x) in deviation form. Thus, we can interpret the coefficient $\hat{\alpha} = 1,457$ as the estimate of the population in 1900 (when $x = 0$). The coefficient $\hat{\beta} = 0.648 = 64.80\%$ is the appropriate population growth rate every 10 yr.

Step 4: Compute the coefficient of determination as follows:

$$r^2 = \frac{\left(\sum xy'\right)^2}{\sum x^2 \sum y'^2}$$

$$= \frac{(38.881)^2}{(60)(25.867)} \tag{9.38}$$

$$= 0.97$$

From the scatter diagram in Figure 9.4.2 and the analysis performed, it appears that the exponential curve fits the data of past population growth better than any straight line. However, it is important to keep in mind that using it for any short-term prediction of the population is unwarranted. The concern mostly stems from the fact that in this simple growth model, the error u is likely to be serially correlated and, thus, has to be accounted for in any prediction.

Another example of the exponential model is the Cobb–Douglas production function, which is given as:

$$Q = \alpha K^\beta L^\gamma u \tag{9.39}$$

where

$$Q = \text{quantity produced}$$
$$K = \text{capital}$$
$$L = \text{labor}$$
$$u = \text{multiplicative error term}$$
$$\alpha, \beta, \text{ and } \delta = \text{parameters to be estimated}$$

As before, by taking the logs of Equation 9.39, we obtain

$$\log Q = (\log \alpha) + \beta (\log K) + (\delta \log L) + (\log u) \tag{9.40}$$

which is of the standard form:

$$Y = \alpha' + \beta X + \delta Z + e \tag{9.41}$$

The demand function is also an exponential model, i.e.:

$$Q = \alpha P^\beta u \tag{9.42}$$

where

$$Q = \text{quantity demanded}$$
$$P = \text{price}$$

Equation 9.42 can be linearized by taking its logarithms as shown:

$$\log Q = (\log \alpha) + \beta \log P + \log u \tag{9.43}$$

which in standard form is:

$$Y = \alpha' + \beta X + e \tag{9.44}$$

The polynomial curve: The polynomial is by far the most widely used equation to describe the relation between two variables. Snedecor and Cochran (1980) have pointed out that when faced by nonlinear regression, and when one has no knowledge of the theoretical equation to use, the second-degree polynomial, in many instances, provides a satisfactory fit.

The general form of a polynomial equation is:

$$Y = \alpha + \beta_1 X + \beta_2 X^2 + ... + \beta_k X^k \tag{9.45}$$

In its simplest form, the equation will have the first two terms on the right-hand side of the equation and is then known as the equation for a straight line. Adding another term ($_2X^2$) to the straight-line equation gives us the second-degree or a quadratic equation, the graph of which is a parabola. As more terms are added to the equation, the degree or power of X increases. A polynomial equation with a third degree (X^3) is called a cubic and those with fourth and fifth degrees are referred to as quartic and quintic, respectively. Figure 9.4.3 shows examples of the polynomial curves with different degrees.

To illustrate the method and some of its applications, the following example is used.

Example 9.4.2

In order to determine the relationship between yield and the protein content of soybean, an agronomist gathered the following data. The agronomist is particularly interested in estimating the protein content for the different yields. Furthermore, she wishes to test if there is any departure from linearity in the data. Thus, the regression equation is assumed to have the following functional form:

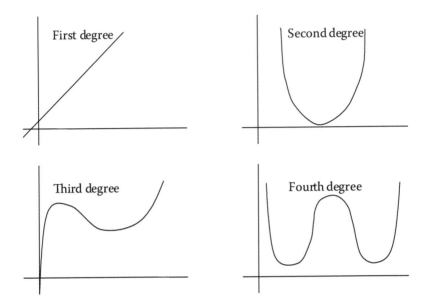

FIGURE 9.4.3
Typical shapes of polynomial curves.

$$Y = \alpha + \beta X + \gamma X^2$$

Soybean Yield (bu/ac)	Protein (%)	Soybean Yield (bu/ac)	Protein (%)
29.50	40.80	31.20	37.60
28.60	42.10	34.50	35.10
28.10	41.30	38.10	35.80
31.20	40.30	33.20	36.90
32.40	39.60	36.10	35.80
30.20	43.40	33.60	36.80
38.30	35.30	33.00	35.20
30.40	41.50	32.30	36.60
30.00	39.70	37.30	34.40
31.30	39.20	35.50	34.00
29.80	41.90	34.30	37.20
34.50	36.05	33.90	37.60
30.20	39.00	39.20	37.80
27.00	43.10	39.15	35.60
32.00	37.30	38.40	36.25

SOLUTION

   Step 1: Because it is assumed by the agronomist that the functional form of the regression equation is quadratic and the data gathered are only on two variables (yield and protein content), we need to add another variable to the regression equation. The added variable is the square of the yield of

TABLE 9.4.2

Soybean Yield (X), and Protein Content (Y) from 30 Plots

Soybean Yield (bu/ac)		Protein (%)	Soybean Yield (bu/ac)		Protein (%)
X	X^2	Y	X	X^2	Y
29.50	870.25	40.80	31.20	973.44	37.60
28.60	817.96	42.10	34.50	1190.25	35.10
28.10	789.61	41.30	38.10	1451.61	35.80
31.20	973.44	40.30	33.20	1102.24	36.90
32.40	1049.76	39.60	36.10	1303.21	35.80
30.20	912.04	43.40	33.60	1128.96	36.80
38.30	1466.89	35.30	33.00	1089.00	35.20
30.40	924.16	41.50	32.30	1043.29	36.60
30.00	900.00	39.70	37.30	1391.29	34.40
31.30	979.69	39.20	35.50	1260.25	34.00
29.80	888.04	41.90	34.30	1176.49	37.20
34.50	1190.25	36.05	33.90	1149.21	37.60
30.20	912.04	39.00	39.20	1536.64	37.80
27.00	729.00	43.10	39.15	1532.72	35.60
32.00	1024.00	37.30	38.40	1474.56	36.25

soybean (column 2 of Table 9.4.2) and is treated as a third variable in the equation. Because of this added variable, we now have two independent variables and, thus, a multiple regression equation. Because the step-by-step analysis of multiple regression is given in Section 9.5, we will not elaborate on the calculation procedures now. For this particular example, we will simply compare the estimated regression equations and their fit to the data.

Step 2: Calculate a simple regression, using the yield and the protein data to determine the fit. The estimated regression equation and the coefficient of determination are as follows:

$$\hat{Y} = 58.923 - 0.629X$$

$$R^2 = 0.64$$

The coefficient of determination shows that the straight line accounts for 64% of the variability in protein content. Figure 9.4.4 shows how the estimated simple linear equation fits the data.

Step 3: To determine whether the relationship between the yield and the protein content is quadratic, as hypothesized by the agronomist, we will estimate a quadratic equation with the square of the yield data serving as the third variable. The estimated regression equation and the multiple coefficient of determination are as follows:

$$\hat{Y} = 148.832 - 6.039X + 0.081X^2$$

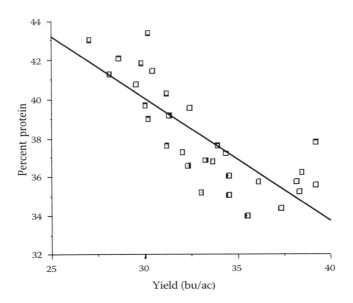

FIGURE 9.4.4
Graph of the soybean data fitted to a linear equation.

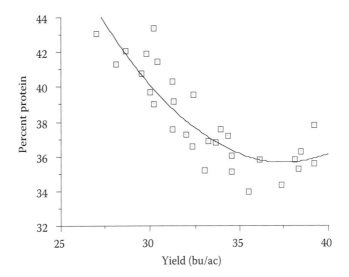

FIGURE 9.4.5
Graph of the soybean data fitted to a quadratic equation.

$$R^2 = 0.77$$

Figure 9.4.5 shows that the quadratic equation accounts for 77% of the variability in the protein content and is a much better fit than the straight line.

TABLE 9.4.3

Test of Significance of Departure from Linear Regression

Source of Variation	Degree of Freedom	Sum of Squares	Mean Square	F
Deviations from linear regression	28	75.357		
Deviations from quadratic regression	27	49.217	1.823	14.34**
Reduction in sum of squares	1	26.140	26.140	

Note: ** = highly significant.

Step 4: Compare the results from the simple and quadratic equations as summarized in Table 9.4.3.

From Table 9.4.3 we observe that the reduction in the sum of squares tested against the mean square remaining after curvilinear regression is highly significant, thus confirming the agronomist's hypothesis that the relationship between yield and protein content is curvilinear.

9.5 Multiple Regression and Correlation

In the previous sections, it was shown how regression and correlation analysis are used in agricultural experiments. The techniques and concepts presented were used as a tool in analyzing the relationship that may exist between two variables. A single independent variable was used to estimate the value of the dependent variable. A brief introduction to multiple regression was made in connection with fitting a curvilinear relationship between variables. In this section, we will discuss the concepts of regression and correlation, in which two or more independent variables are used to estimate the dependent variable in more detail.

Because multiple regression and correlation are simply an extension of simple regression and correlation, we will first show how to derive the multiple regression equation using two or more independent variables. Second, attention will be given to the method of calculating $S_{y.x}$ and related measures. Finally, the computation of the multiple coefficient of determination and correlation will be explained.

The advantage of multiple regression over simple regression analysis is in its enhancing our ability to use more available information in estimating the dependent variable. To describe the relationship between a single variable Y and several variables X, we may write the multiple regression equation as follows:

$$Y = \alpha + \beta_1 X_1 + \beta_2 X_2 + \ldots + \beta_k X_k + \varepsilon \qquad (9.46)$$

where

Y = the dependent variable

$X_1 ... X_k$ = the independent variables

ε = the error term, which is a random variable with a mean of 0 and a standard deviation of ρ.

The numerical constants, α and β_1 to β_k must be determined from the data and are referred to as the partial regression coefficients. The underlying assumptions of the multiple regression model are as follows:

1. The explanatory variables ($X_1 ... X_k$) may be either random or non-random (fixed) variables.

2. The value of Y selected for one value of X is probabilistically independent.

3. The random error has a normal distribution with mean equal to 0 and variance equal to σ^2.

These assumptions imply that the mean, or expected value $E(Y)$, for a given set of values of $X_1 ... X_k$ is equal to

$$E(Y) = \alpha + \beta_1 X_1 + \beta_2 X_2 + ... + \beta_k X_k \tag{9.47}$$

The coefficient α is the Y intercept when the expected value of all independent variables is zero. Equation 9.47 is called a *linear statistical model*. The scatter diagram for a two-independent-variables case is a regression plane, as shown in Figure 9.5.1.

Estimating the multiple regression equation — the least-squares method: In Section 9.2 we mentioned that if a straight line is fitted to a set of data using the least-squares method, that line is the best fit in the sense that the sum of the squared deviations is less than it would be for any other possible line. The least-squares formula provides a best-fitting *plane* to the data. The normal equations for k variables are as follows:

$$\sum x_1 y = b_1 \sum x_1^2 + b_2 \sum x_1 x_2 + ... + b_k \sum x_1 x_k \tag{9.48}$$

$$\sum x_2 y = b_1 \sum x_1 x_2 + b_2 \sum x_2^2 + ... + b_k \sum x_2 x_k$$

$$\cdots \quad \cdots\cdots\cdots\cdots\cdots\cdots\cdots\cdots\cdots\cdots\cdots\cdots$$

$$\cdots \quad \cdots\cdots\cdots\cdots\cdots\cdots\cdots\cdots\cdots\cdots\cdots\cdots$$

$$\sum x_k y = b_1 \sum x_1 x_k + b_2 \sum x_2 x_k + ... + b_k \sum x_2^2$$

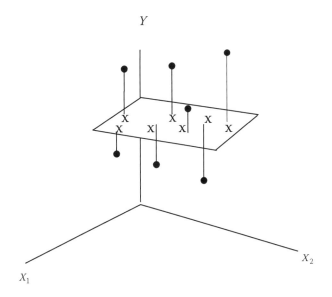

FIGURE 9.5.1
Scatter diagram for multiple regression analysis involving two independent variables.

In Equation 9.48, $b_1, b_2, \dots, b_k$ are estimates of $\beta_1, \beta_2, \dots, \beta_k$, and the number of equations is equal to the number of parameters to be estimated. Given a sample of n observations on (Y, X_1, X_2), the sample regression or prediction equation is:

$$\hat{Y} = a + b_1 X_1 + b_2 X_2 \tag{9.49}$$

With the least-squares model, the resulting estimates a, b_1, b_2, and $\hat{Y}$ are unbiased and have the smallest standard errors of any unbiased estimates that are linear expressions in the Y's. To estimate the value of a, we use the following equation

$$a = \bar{Y} - b_1 \bar{X}_1 - b_2 \bar{X}_2 \tag{9.50}$$

You should note that computing the coefficients, although not theoretically complex, is difficult and tedious even for k as small as 4 or 5. We generally use a computer program such as Excel or MINITAB to solve for the coefficients.

To illustrate the concept and to show how the multiple regression coefficients are computed, let us take an example with two independent variables.

Example 9.5.1

In Section 9.2, an agronomist had taken ten observations on the yield of maize as different amounts of fertilizer were applied. In this example, he wishes to consider an additional explanatory variable — the amount of rainfall. He believes that the amount of fertilizer and rainfall, together, are better predictors of yield. Using the information given in the following table, determine for the agronomist the least-squares multiple regression equation.

Yield of Maize (bu/ac) Y	Fertilizer (lb N/ac) X_1	Rainfall (in.) X_2
50	5	5
57	10	10
60	12	15
62	18	20
63	25	25
65	30	25
68	36	30
70	40	30
69	45	25
66	48	30

SOLUTION

Step 1: Compute the intermediate values needed in the normal equations from the data given in Table 9.5.1.

TABLE 9.5.1

Calulation of Coefficients for Normal Equations

Yield of Maize (bu/ac) Y	Fertilizer (lb N/ac) X_1	Rainfall (in.) X_2
50	5	5
57	10	10
60	12	15
62	18	20
63	25	25
65	30	25
68	36	30
70	40	30
69	45	25
66	48	30
$\bar{Y} = 63$	$\bar{X}_1 = 26.9$	$\bar{X}_2 = 21.5$
$\sum y^2 = 338$	$\sum x_1^2 = 2,106.9$	$\sum x_2^2 = 702.5$
$\sum x_1 y = 755$	$\sum x_2 y = 460$	$\sum x_1 x_2 = 1,110.5$

The normal equations for the two independent variables, as is the case with our example, are written as follows:

$$\sum x_1 y = b_1 \sum x_1^2 + b_2 \sum x_1 x_2 \qquad (9.51)$$

$$\sum x_2 y = b_1 \sum x_1 x_2 + b_2 \sum x_2^2 \qquad (9.52)$$

Step 2: Solve for b_1 and b_2, using the following formulas:

$$b_1 = \frac{\left(\sum x_2^2\right)\left(\sum x_1 y\right) - \left(\sum x_1 x_2\right)\left(\sum x_2 y\right)}{\left(\sum x_1^2\right)\left(\sum x_2^2\right) - \left(\sum x_1 x_2\right)^2} \qquad (9.53)$$

and

$$b_2 = \frac{\left(\sum x_1^2\right)\left(\sum x_2 y\right) - \left(\sum x_1 x_2\right)\left(\sum x_1 y\right)}{\left(\sum x_1^2\right)\left(\sum x_2^2\right) - \left(\sum x_1 x_2\right)^2} \qquad (9.54)$$

Substituting the appropriate values into Equation 9.53 and Equation 9.54, we get:

$$b_1 = \frac{(702.5)(755) - (1,101.5)(460)}{(2,106.9)(702.5) - (1,101.5)^2}$$

$$= 0.088$$

and

$$b_2 = \frac{(2,106.9)(460) - (1,101.5)(755)}{(2,106.9)(702.5) - (1,101.5)^2}$$

$$= 0.516$$

Step 3: Solve for a, using the following formula:

$$a = \bar{Y} - b_1\bar{X}_1 - b_2\bar{X}_2$$

$$= 63 - (0.088)(26.9) - (0.516)(21.5)$$

$$= 63 - 2.367 - 11.094$$

$$= 49.539$$

The estimated multiple linear regression is:

$$\hat{Y} = 49.539 + 0.088X_1 + 0.516X_2$$

The interpretation of the coefficients a, b_1, and b_2 is analogous to simple linear regression. The constant a is the intercept of the regression line. However, we interpret it as the value of $\hat{Y}$ when both X_1 and X_2 are 0. The values b_1 and b_2 are called the *partial regression coefficients*. Coefficient b_1 simply measures the change in $\hat{Y}$ per unit change in X_1 when X_2 is held constant. Likewise, coefficient b_2 measures the change in $\hat{Y}$ per unit change in X_2 when X_1 is held constant. Thus, we may say that the b coefficients measure the net influence of each independent variable on the estimate of the dependent variable.

In the present example, the b_1 value of 0.088 indicates that for each increase of 1 lb of fertilizer, the yield increases by 0.088 bu, regardless of the rainfall, (i.e., the amount of rainfall is held constant). The b_2 coefficient indicates that for each increase of 1 in. of rainfall, the yield increases by 0.516 bu regardless of the amount of fertilizer used.

Before we can compute $S_{y.x}$ and the coefficient of multiple determination, we need to partition the sum of squares for the dependent variable.

The total sum of squares (SST) has already been computed before as:

$$SST = \sum y^2 = 338$$

The explained or regression sum of squares (SSR) is computed as follows:

$$SSR = b_1 \sum x_1 y + b_2 \sum x_2 y \qquad (9.55)$$

Substituting the appropriate values into Equation 9.55 we get:

$$SSR = 0.088\left(755\right) + 0.516\left(460\right)$$

$$= 303.80$$

The unexplained or error sum of squares (*SSE*) is the difference between the *SST* and *SSR*:

$$SSE = SST - SSR \tag{9.56}$$

Therefore, the *SSE* is

$$SSE = 338 - 303.80 = 34.20$$

Step 4: Compute $S_{y.x}$, which measures the standard deviation of the residuals about the regression plane and thus specifies the amount of error incurred when the least-squares regression equation is used to predict values of the dependent variable. The smaller $S_{y.x}$ is, the closer the fit of the regression equation is to the scatter of observations.

$S_{y.x}$ is computed by using the following equation:

$$S_{y.12} = \sqrt{\frac{SSE}{n-k}} \tag{9.57}$$

where

$$SSE = \text{the error sum of squares}$$
$$n = \text{the number of observations}$$
$$k = \text{the number of parameters}$$

Hence, in a multiple regression analysis involving two independent variables and a dependent variable, the divisor will be $n - 3$. Having computed the *SSE*, we substitute its value into Equation 9.57 to compute $S_{y.x}$ as follows:

$$S_{y.12} = \sqrt{\frac{34.20}{7}}$$

$$= 2.21 \text{ bu}$$

$S_{y.x}$ about the regression plane may be compared with $S_{y.x}$ of the simple regression.

In Section 9.2, when only fertilizer was used to explain the variation in yield, we computed an $S_{y.x}$ of 2.92. Including an additional variable (rainfall) to explain the variation in yield has given us an $S_{y.x}$ value of 2.21. As was mentioned earlier, the standard error expresses the amount of variation in the dependent variable that is left unexplained by regression analysis. Because the standard error of the regression plane is smaller than the standard error of the regression line, inclusion of this additional variable will enable better prediction.

Step 5: Compute the multiple coefficient of determination, using the following equation:

$$R^2 = \frac{SSR}{SST} \tag{9.58}$$

where

SSR = the regression sum of squares

$SST = \sum y^2$ = the total sum of squares

Substituting the values of SSR and SST into Equation 9.58, we get:

$$R^2 = \frac{303.80}{338} = 0.90$$

The coefficient of determination measures the contribution of the k independent variables to the variation in Y. This means that 90% of the variation in the yield of maize is explained by the amount of fertilizer applied and the amount of rainfall in a locality. The preceding R^2 value is not adjusted for degrees of freedom. Hence, we may overestimate the impact of adding another independent variable in explaining the amount of variability in the dependent variable. Thus, it is recommended that an adjusted R^2 be used in interpreting the results.

The adjusted coefficient of multiple determination is computed as follows:

$$R_a^2 = 1 - \left(1 - R^2\right)\frac{n-1}{n-k} \tag{9.59}$$

where

R_a^2 = adjusted coefficient of multiple determination
n = number of observations
k = total number of parameters

For the present example, the adjusted R^2 is

$$R_a^2 = 1 - \left(1 - 0.90\right)\frac{10-1}{10-3}$$

$$= 0.87$$

We may wish to compare the coefficient of determination of the simple regression model in which one independent variable, namely, the impact of application of fertilizer, was analyzed with the coefficient of multiple determination in which, in addition to fertilizer use, the impact of rainfall on yield was observed. The adjusted coefficient of determination for the simple regression was $r^2 = 0.76$, whereas the adjusted coefficient of multiple determination was 0.87. The difference of 0.11 indicates that an additional 11% of the variation in the yield of maize (beyond that already explained by the amount of fertilizer applied) is explained by the amount of rainfall.

Step 6: Test the significance of R^2 by computing the F value as:

$$F = \frac{SSR / k}{SSE / (n - k - 1)} \tag{9.60}$$

Substituting the appropriate values into Equation 9.60, we get:

$$F = \frac{303.80 / 2}{34.20 / (10 - 2 - 1)}$$

$$= 31.09$$

Step 7: Compare the computed F with the tabular F value given in Appendix E. For this example, the tabular F (for two and seven degrees of freedom) is 8.65 at the 1% level of significance. Because the computed F value is greater than the tabular F value, the estimated multiple linear regression is highly significant.

Assumptions and problems in multiple linear regression: As with simple regression, a number of assumptions apply to the case of the multiple regression. These assumptions are

1. The regression model is linear and of the form

$$E(Y) = a + b_1 X_1 + b_2 X_2 + \dots + \beta_k X_k \tag{9.61}$$

2. The values of Y are independent of each other.
3. The values of Y are normally distributed.
4. The variance of Y values is the same for all values of $X_1, X_2, \dots, X_k$.

Violation of the preceding assumptions leads to a number of problems such as serial or autocorrelation, heteroscedasticity, and multicollinearity, which are explained in the text that follows.

Serial or autocorrelation: This problem arises when the assumption of the independence of Y values is not met. That is, there is dependence between

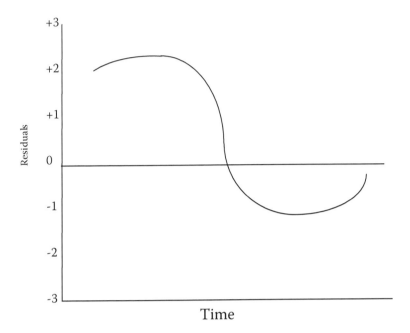

FIGURE 9.5.2
Positive and negative autocorrelation in the residuals.

successive values. This problem is often observed when time-series data are employed in the analysis. To be sure that there is no autocorrelation, plotting the residuals against time is helpful. Figure 9.5.2 suggests the presence of autocorrelated terms. Methods for measuring serial correlation, such as the Durbin–Watson test, should be employed.

Multicollinearity: The problem of multicollinearity arises when two or more independent variables are highly correlated with each other. This implies that the regression model specified is unable to separate out the effect of each individual variable on the dependent variable. When multicollinearity exists between the independent variables, estimates of the parameters have larger standard errors and the regression coefficients tend to be unreliable.

How do we know whether we have a problem of multicollinearity? When a researcher observes a large coefficient of determination (R^2) accompanied by statistically insignificant estimates of the regression coefficients, the chances are that there is *imperfect multicollinearity*. When one (or more) independent variable(s) is an exact linear combination of the others, we have *perfect multicollinearity*.

Remedying the serial correlation problems requires proper transformation of the dependent variable. Such a transformation is discussed elsewhere, and the reader should consult the references given at the end of this chapter.

Equal variances: One of the assumptions of the regression model was that the error terms all have equal variances. This condition of equal variance is known as *homoscedasticity*. When this assumption is violated, the problem of heteroscedasticity arises. For example, in studying the yield of milk with different high-energy rations, we might find that yield rises with different rations. In this case, the yield function is probably heteroscedastic. When heteroscedasticity exists, it is difficult to make statistical inferences regarding the regression equation. Kelejian and Oats (1989) suggested a test for heteroscedasticity and defined a procedure to overcome problems arising from this error.

Once it is determined that multicollinearity exists between the independent variables, a number of possible steps can be taken to remedy this problem:

1. Drop the correlated variable from the equation: Which independent variable to drop from the equation depends on the test of significance of the regression coefficient and the judgment of the researcher. If the *t*-test indicates that the regression coefficient of an independent variable is statistically insignificant, that variable may be dropped from the equation. Dropping a highly correlated independent variable from the equation will not affect the value of R^2 very much.

2. Change the form of one or more independent variables: For example, an agricultural economist, in a demand equation for beef (Y), finds that income (X_1) and another independent variable (X_2) are highly correlated. In such a situation, dividing the income by the variable of population yields the per capita income, which may result in less correlated independent variables. Other approaches are suggested in Kelejian and Oats (1989) and Hamburg (1983).

References and Suggested Readings

Claus, M.P.L. and Veerkamp, R.E. 2003. Estimation of environmental sensitivity of genetic merit for milk production traits using a random regression model. *J. Dairy Sci.* 86: 3756–3764.

Galton, F. 1908. *Memories of My Life*. New York: E.P. Dutton.

Hamburg, M. 1983. *Statistical Analysis for Decision Making*. 3rd ed. New York: Harcourt Brace Jovanovich. Chap. 12.

Hoshmand, A.R. 1997. *Statistical Methods for Environmental and Agricultural Sciences.*2nd ed. Boca Raton: CRC Press. Chap. 11 and Chap. 12.

Kelejian, H.H. and Oats, W.E. 1989. *Introduction to Econometrics*. 3rd ed. New York: Harper & Row.

Kolmodin, R., Strandberg, E., Danell, B., and Jorjani, H. 2004. Reaction norms for protein yield and days open in Swedish red and white dairy cattle in relation to various environmental variables. *Acta Agriculturae Scandinavica. Sect. A, Anim. Sci.* 54: 139–151.

Mead, R. and Curnow, R.N. 1983. *Statistical Methods in Agriculture and Experimental Biology*. New York: Chapman & Hall.

Netter, J. and Kutner, M.H. 1983. *Applied Linear Regression Models*. Homewood, IL: Irwin.

Ping, J.L., Bronson, K.F., Zartman, R.E., and Dobermann, A. 2004. Identification of relationships between cotton yield, quality, and soil properties. *Agron. J.* 96: 1588–1597.

Snedecor, G.W. and Cochran, W.G. 1980. *Statistical Methods*. Ames, IA: Iowa State University Press.

Exercises

1. A researcher interested in the relationship between the rate of germination of warm-season forage grasses and temperature has postulated a linear regression model in which the number of seeds germinated per day is dependent on the average daily temperature. The following data were collected by the researcher.

Germinated Seed (number/day)	Temperature (°C)	Germinated Seed (number/day)	Temperature (°C)
5	10	28	22
7	11	31	23
9	13	35	24
10	15	38	26
14	16	49	27
20	18	55	29
24	20	61	30
25	21	73	32

(a) Use the least-squares technique in estimating the equation.

(b) Compute $S_{y.x}$. What is the interpretation of $S_{y.x}$?

(c) Is there a significant correlation between the two variables? Use a 5% level of significance.

2. In a two-factor experiment, an environmental horticulturist wishes to determine the impact of nitrogen fertilization and hourly exposure to sunlight on leaf thickness in the rubber plant. After 2 months of experimentation, the following average observations of leaf thickness, amount of light in the greenhouse, and fertilization were recorded from an experiment with four replications. Plants of the same age were grown in the same environment, varying only the two factors of interest.

Treatment Number	Leaf Thickness (mm)	Hours of Sunlight	Nitrogen Fertilization (mg/pot)
1	2.15	5.0	0.20
2	2.28	5.0	0.30
3	4.56	5.0	0.40
4	7.68	5.0	0.50
5	8.76	5.0	0.60
6	12.80	5.0	0.70
7	15.62	5.0	0.80
8	16.54	5.0	0.90
9	3.24	6.0	0.20
10	6.32	6.0	0.30
11	8.76	6.0	0.40
12	7.48	6.0	0.50
13	12.69	6.0	0.60
14	16.87	6.0	0.70
15	20.92	6.0	0.80
16	18.66	6.0	0.90
17	5.33	7.0	0.20
18	6.38	7.0	0.30
19	6.87	7.0	0.40
20	8.88	7.0	0.50
21	9.89	7.0	0.60
22	20.87	7.0	0.70
23	21.92	7.0	0.80
24	22.96	7.0	0.90

(a) Estimate the regression equation.

(b) Compute the coefficient of multiple determination.

(c) Test the significance of R^2.

(d) Compute the various coefficients of correlation.

3. An animal researcher is interested in the relationship between urea nitrogen and the energy balance and the phosphorous balance. The researcher has randomly selected 15 cows that have been fed rations that contain high and low levels of energy and phosphorous. The data for the experiment are given below.

Treatment Number	Y Urea N (mg/100 ml)	X_1 Energy Balance (kcal/d)	X_2 P Balance (g/d)
1	9.5	8.1	21.3
2	10.2	8.4	25.9
3	11.0	8.5	24.8
4	11.2	10.9	28.4
5	12.3	7.5	−2.7
6	10.8	7.8	22.2
7	10.8	11.3	19.0
8	11.1	−5.4	8.5
9	11.7	7.6	17.9
10	11.8	8.3	−9.9
11	10.9	9.6	−8.4
12	12.4	9.5	−10.1
13	12.6	10.5	12.4
14	10.9	−5.4	7.8
15	11.3	6.4	8.9
16	10.8	5.7	9.4

(a) Estimate the least-squares regression equation.

(b) Compute $S_{y.x}$.

(c) Test the significance of the regression coefficients.

(d) Compute the correlation coefficients.

4. An agricultural engineer wishes to determine the relationship between the monthly electricity usage in a greenhouse and the size of the greenhouse. He believes that the relationship is best exemplified by the following model:

$$Y = a + b_1 X_1 + b_2 X_2^2 + \varepsilon$$

Given the following data:

Monthly Electricity Usage (kW/h)	Greenhouse Size (ft²)
2,000	2,800
2,225	2,900
2,540	3,100
2,678	3,200
2,700	3,250
2,890	3,300
2,980	3,400
3,000	3,500
3,220	3,550
3,454	3,600
3,590	3,650
3,760	3,850
4,000	3,900
4,325	4,000
4,550	5,050

(a) Estimate the multiple regression equation. (Use a MINITAB or Excel to solve this problem.)

(b) Is the overall model useful in this investigation?

(c) Compute the correlation coefficients.

(d) Test the overall significance of the computed coefficients at = .01.

5. Suppose that in the previous example, the agricultural engineer hypothesized that there is a strong relationship between the monthly electricity usage in the greenhouse and the size of the greenhouse, as well as the average monthly temperature. Data gathered are shown:

Monthly Electricity Usage (kW/h)	Greenhouse Size (ft²)	Average Monthly Temperature (°F)
2,000	2,800	68
2,225	2,900	70
2,540	3,100	78
2,678	3,200	80
2,700	3,250	82
2,890	3,300	84
2,980	3,400	86
3,000	3,500	75
3,220	3,550	75
3,454	3,600	78
3,590	3,650	72
3,760	3,850	70
4,000	3,900	69
4,325	4,000	65
4,550	5,050	62

(a) Compute the estimated regression equation. (Use a computer to solve this problem.)

(b) Test the overall significance of the regression model when = .05.

6. An equine researcher has postulated that there is a nonlinear relationship between the lifespan of Arabian horses and the gestation period. The researcher believes that the following model best exemplifies the relationship:

$$Y = a + b_1 X_1 + b_2 X_2^2 + \varepsilon$$

Lifespan (yr)	Gestation Period (d)
15.2	210
17.8	230
18.2	240
20.0	265
22.1	285
23.4	370
20.5	345
22.2	340
21.5	385
19.6	295
18.8	279
20.4	305
22.3	315
20.0	395
18.4	290
15.5	300

(a) Estimate the multiple regression equation. (Use a computer to solve this problem.)

(b) Interpret the meaning of the coefficients of multiple determination.

(c) Are the regression coefficients significant at = .01?

(d) Test the overall significance of the model at = .01.

7. To determine if there is a relationship between the rate of nitrogen fertilization and rice yield, an agronomist has conducted research and collected the following data:

Nitrogen Rate (kg/ha)	Grain Yield (kg/ha)
0	4,890
50	5,230
100	6,250
150	7,030
200	7,350
0	4,650
50	5,100
100	5,950
150	6,950
200	7,200
0	4,450
50	5,250
150	6,890
200	7,150
0	4,450
50	5,025
100	5,675
150	5,980
200	7,150

(a) Estimate the regression equation.

(b) Compute the coefficient of determination.

(c) Test the significance of R^2.

(d) Compute the coefficient of correlation and interpret its meaning.

8. A rice researcher has postulated that there is a relationship between grain yield, plant height, and tiller. Data gathered for this research are as follows:

Grain Yield (kg/ha) Y	Plant Height (cm) X_1	Tiller (number/hill) X_2
5,550	95.2	12.8
5,620	110.0	14.9
5,432	90.2	12.8
4,325	87.6	13.6
5,210	118.2	14.0
6,320	119.7	18.2
6,100	105.6	16.8
5,890	111.3	17.9
4,890	90.0	14.5
4,765	78.6	12.7
7,100	84.5	19.2
6,995	88.6	18.2
5,234	91.2	16.9
5,320	89.5	17.0
7,030	92.4	19.8

(a) Compute the estimated regression equation. (Use a computer to solve this problem.)

(b) Test the overall significance of the regression model when = .05.

10

Covariance Analysis

10.1 Introduction

In our discussion in Chapter 3, we mentioned that to minimize error, the design of an experiment is a critical factor. So far, we have discussed the various designs and how they minimize experimental error. You may recall that grouping of experimental units into homogenous blocks was used to increase the uniformity of the experiment and thus reduce the amount of error directly in an experiment. In addition to using properly designed experiments, researchers also have statistical tools to help reduce the amount of error in an experiment. In this chapter, we will introduce you to the analysis of covariance as an indirect measure to control error.

The purpose of the covariance analysis is to allow the researcher to use supplementary data to reduce experimental errors by eliminating the effects of variations within a block. We refer to this approach as an indirect or statistical control to increase the precision of an experiment.

To perform covariance analysis, we need a measurement of the character of primary interest along with the measurement of one or more variables known as the *covariates*. Furthermore, we should know the functional relationship that exists between the covariates and the character of primary interest. Often, characteristics such as the biophysical features (soil heterogeneity in a field or unpredictable insect infestation) of an experimental plot are functionally related to each other. To analyze the effect of such characteristics separately, the researcher can examine the variances and covariances among the selected variables of interest simultaneously.

For example, field experiments may be conducted on a *Striga hermonthica*-infested field to determine the effect of *Striga* on growth characteristics of sorghum intercropped with groundnut varieties. The infestation by *Striga* is used as a covariate. *Striga* weakens the host, wounding its outer root tissues and absorbing its supplies of moisture, sugar, and minerals (Bebawi and Mitwali, 1991; Carsky, Singh, and Ndikawa, 1994). With the known functional relationship between *Striga* and grain yield, which is the character of primary interest, the researcher can use the analysis of covariance to adjust the grain yield in each plot to a common level of *Striga* incidence and effectively separate variation in yield due to infestation and varietal differences.

To illustrate how the adjustment is made, suppose that the means of the variate in a single-factor experiment are denoted as $\bar{Y}_1, \bar{Y}_2, ..., \bar{Y}_k$ and the means of the covariate are denoted as $\bar{X}_1, \bar{X}_2, ..., \bar{X}_k$. Our primary interest lies in the differences among the $\bar{Y}_j$. Suppose that differences in $\bar{X}_j$ are due to sources of variation related to $\bar{Y}_j$ but not directly related to the treatment effects. In such a situation, more precise information on the treatment effects may be obtained by adjusting $\bar{Y}_j$ for the association with the $\bar{X}_j$. Furthermore, suppose that the adjusted variate means are denoted as $\bar{Y}_1', \bar{Y}_2', ..., \bar{Y}_k'$. Given this information, we are now able to make the adjustment in several different ways. In some cases, the adjustment may take the form of a simple difference between variate and covariate as shown:

$$\bar{Y}_j' = \bar{Y}_j - \bar{X}_j \tag{10.1}$$

In other cases, the adjusted mean may take the form:

$$\bar{Y}_j' = \bar{Y}_j / \bar{X}_j \tag{10.2}$$

Whether to use Equation 10.1 or Equation 10.2 is usually decided by the prior knowledge about the interrelationship between the variate and the covariate. By adjusting the data, two important improvements are made in our analysis of experimental results. First, the treatment mean is adjusted to a value that it would have had, had there been no differences in the value of the covariate. Second, we have reduced the experimental error and, hence, increased the precision for comparing the treatment means. Keep in mind that when analysis of covariance is used for adjustment, it essentially combines regression analysis procedures with analysis of variance procedures.

A word of caution is in order when using covariance analysis for error control and adjustment of treatment means. The covariate must not be affected by the treatments being tested. A good example of covariates that are free of treatment effects are those that are measured before the treatments are applied; for example, when measuring the effects of different diets on the weights of animals, their weights are measured prior to the introduction of the new diets. Similarly, an agronomist may want to perform soil analysis to determine the level of various nutrients in the soil prior to amending it with fertilizers.

Analysis of covariance was introduced by Fisher (1932). Readers may wish to consult an excellent and quite readable summary of the wide variety of uses of the analysis of covariance in a special issue of *Biometrics* (1957), which is devoted entirely to this topic. In the same issue, H.F. Smith provides a lengthy discussion on the interpretation of the adjusted treatment means as it affects the covariate. Cox and McCullough (1982) provide a more recent presentation of the analysis of covariance and its applications.

The use of covariance analysis is especially helpful in agricultural research when we face soil heterogeneity, stand irregularities, nonuniformity in pest incidence (see Park and Cho, 2004), nonuniformity in environmental stress, and competition effects in greenhouse trials. Furthermore, covariance analysis is helpful in estimation of missing data. We will explore these next.

Soil heterogeneity: In agricultural research we often face soil conditions that are not homogeneous from plot to plot. In our earlier discussions in Chapter 2, we suggested that blocking should be used as a way to reduce error in an experiment. However, spotty soil heterogeneity makes blocking ineffective. In such circumstances, the researcher needs to measure, from individual experimental plots, a covariate that can identify the differences in the native soil fertility between plots. The covariate must be linearly related to the character of primary interest, such as yield of grains. To distinguish such differences in soil fertility, two types of covariates are commonly used to reduce error in an experiment: crop-performance data collected prior to treatment implementation and uniformity-trial data.

Crop-performance data gathered just prior to the application of the treatment often provide researchers with a good measure of a common characteristic. This is often done when there is a time lag between crop establishment and treatment application. The researcher measures some crop characteristics that are related to crop growth, such as the tiller number or plant height prior to the application of the treatment. If there are differences between plots in yield, it is assumed to be due to differences in soil characteristics. The other covariate analysis, uniformity-trial data, is ideal for soil heterogeneity conditions, as they allow researchers to measure differences in crop performance from one plot to another before treatments are applied. Even though uniformity-trial data used as covariates allow more precision in the analysis of treatments, its use is limited. Often-cited reasons for lack of its use have been the cost of conducting such trials and the requirement of more complex data collection. In comparison to the uniformity-trial data crop-performance data is clearly easier and cheaper to gather.

Stand irregularities: When conducting field experiments, variation in the number of plants per plot often becomes an important source of variation. Such variation is due to loss of plants from mechanical errors in planting, damage during cultivation, or random causes such as grazing by cattle or other animals and possible rat infestation in the field. Often researchers use such approaches as mathematical corrections, where the corrected yield (Y_c) is a function of the actual yield (Y_a) and a correction factor (C), that is, $Y_c = C(Y_a)$. The correction factor is derived from the estimate of yield increase (over normal plants) of those plants adjacent to one or more missing hills. So, if the total number of plants in the plot is m and the number of missing hills n, then the actual grain yield would have to be adjusted by ($m-n$) plants in the plot. Researchers may also exclude from harvest those plants surrounding the missing hill to accommodate variability in yield.

The covariance technique is a better technique that provides researchers with the tools to remedy stand irregularities. Using stand count as the covariate allows the researcher to properly account for the treatment effects.

Nonuniformity in environmental stress: Research scientists are constantly experimenting with crop-breeding programs to find the right combination of traits for broadly adapted crop varieties that will yield well, even when there are environmental stresses such as drought conditions, high salinity, iron toxicity, or prevalence of disease. Substantial research into rust resistance was conducted when the high-yielding varieties of wheat were introduced to the world. In all of these conditions, the researcher is interested in knowing what would be the performance of a variety to such stresses. Covariance analysis allows for gauging such differences and adjusting for the variability that comes from environmental stresses. To measure variations in stress conditions over an experimental area, one could plant a check variety whose reaction to the stress of interest is well established. Often, we use a susceptible variety as a check. The resultant yield from the check is then used as the covariate to adjust for the variability in the stress level of different test plots.

Competition effects in greenhouse experiments: In order to minimize the environmental effects on yield data, researchers often use greenhouses as experimental plots in their research. Although this strategy is well worth the effort, there are instances in which the variability in such trials is still as significant as those in field trials. What gives rise to such variability has to do with the competition effect between adjacent units in the greenhouse. The size of greenhouses creates a limitation. Experimental-unit size is small and one cannot adjust yield data appropriately, as is done in field experiments. In field experiments, researchers are able to remove border plants from the measurement so that a more accurate assessment of treatment effects is made. Given this limitation, the covariance analysis provides a good alternative tool to control experimental error due to competition effects. Researchers often use the average height of plants in pots within the greenhouse as a covariate to adjust for the competition effect.

Handling of missing data: We have pointed out in earlier chapters that when a valid observation is not available to researchers, they have to contend with the problems associated with missing data in their analysis. At least two difficulties are encountered when dealing with missing data. First, loss of information and, second, the nonapplicability of the standard analysis of variance. In both of these situations, the covariance analysis provides an alternative to the missing-data-formula technique as will be seen in the subsequent sections of this chapter. Procedural approaches to covariance analysis are discussed next.

10.2 Covariance Analysis Procedures

The computational procedure for the covariance analysis is similar to that of the analysis of variance. Hence, the assumptions applied to the analysis of variance hold true for covariance analysis as well. Furthermore, the covariance analysis assumes that the relationship between the primary character of interest and the covariate is linear and that the linear relationship remains constant over other known sources of variation, such as treatments and blocks.

As was mentioned in the introduction section, in applying the covariance analysis an important task is to identify the covariate. This means gathering data on the primary characteristic of interest. Once a covariate is identified, the computational procedures for error control or estimation of missing data are very similar.

The procedure for controlling error in the various designs (completely randomized, randomized complete block, and split-plot) are discussed first, and then we will address the procedure for the missing data. To control error, the covariance analysis requires data that are paired observation of X and Y, where X is the covariate and Y is the primary character of interest. Keep in mind that the covariate must satisfy the requirement that it is not affected by treatments and that it represents the source of variation that the analysis aims to control. We illustrate the procedure for the different designs below.

10.2.1 Completely Randomized Design (CRD)

To illustrate the computational procedure for error control in a completely randomized design let us take a look at the following example.

Example 10.2.1.1

An animal scientist, in performing a study of vitamin supplementation, randomly assigned five heifers to each of the following eight treatment groups. The main character of interest was the weight gain (Y) after the treatment. Weight gain before the treatment is used as a covariate (X). The weight gains (in kg) were recorded from four replicates and are shown in Table 10.2.1.1.

The computational procedure for the covariance analysis is as follows:

Step 1: Compute the correction factor (C). Keep in mind that a correction factor for each of the variables, namely, X, Y, and XY has to be computed. You will recall from Chapter 4 that the generalized form for computing the correction factor is:

$$C = \frac{G^2}{N} \tag{4.1}$$

TABLE 10.2.1.1

Weight Gain from Vitamin Supplementation

Treatment	Rep. I		Rep. II		Rep. III		Rep. IV		Treatment Total	
	X	Y	X	Y	X	Y	X	Y	X	Y
A	4.5	4.6	5.2	5.2	6.2	6.5	3.9	4.2	19.8	20.5
B	5.6	5.6	4.7	4.7	4.3	4.3	4.4	4.7	19.0	19.3
C	6.4	6.5	6.7	6.9	6.8	6.3	6.1	7.0	26.0	26.7
D	5.2	5.2	5.0	4.7	6.8	6.5	3.6	4.0	20.6	20.4
E	4.0	4.7	4.9	5.2	4.3	4.9	4.8	5.2	18.0	20.0
F	7.1	6.9	6.5	6.9	6.2	6.5	6.8	6.5	26.6	26.8
G	6.1	6.5	4.9	5.3	4.2	4.9	3.9	3.8	19.1	20.5
H (Control)	4.6	4.4	4.0	4.0	4.9	5.8	3.8	3.9	17.3	18.1
Total	43.5	44.4	41.9	42.9	43.7	45.7	37.3	39.3		
Grand Total									166.4	172.3
Grand Mean									5.20	5.38

where

G = grand total

N = the total number of experimental plots $[(r)\,(t)]$

For the cross product the correction factor is:

$$C = \frac{G_x G_y}{rt} \tag{10.1}$$

$$C = \frac{(166.4)(172.3)}{32}$$

$$C = 895.96$$

Using Equation 4.1 the correction factor for variables X and Y are 865.28 and 927.73, respectively.

Step 2: Calculate the various sums of squares (SS) for each of the two variables using the procedures outlined as follows. For expository purposes we have only shown the procedure for the total sum of squares for the XY cross product. The total sum of squares for the variables XX, YY, and XY are shown in Table 10.2.1.2.

$$\text{Total sum of cross product} = \sum_{i=1}^{t}\sum_{j=1}^{r}(X_{ij})(Y_{ij}) - C \tag{10.2}$$

where

TABLE 10.2.1.2

Analysis of Covariance of the Completely Randomized Data

Source of Variation	Degree of Freedom	XX	XY	YY
		Sum of Cross Product		
Treatment	7	21.98	20.19	19.24
Error	24	14.08	12.55	13.98
Total	31	36.06	32.74	33.22

X_{ij} = value of the X variable for the ith treatment and jth replication

Y_{ij} = value of the Y variable for the ith treatment and jth replication

C = correction factor

$$\text{Total } S_{XY} = [(4.5)(4.6) + (5.2)(5.2) + \dots + (3.8)(3.9)] - 895.96$$
$$= 928.7 - 895.96$$
$$= 32.74$$

Step 3: Compute the sum of the cross products for treatment as:

$$\text{Treatment } S_{XY} = \frac{\sum (T_X)(T_Y)}{r} - C \tag{10.3}$$

$$= \frac{(19.8)(20.5) + (19.0)(19.3) + \dots + (17.3)(18.1)}{4} - 895.96$$

$$= 916.15 - 895.96$$

$$= 20.19$$

Step 4: Compute the sum of the cross products for error as:

$$\text{Error } S_{XY} = \text{Total } SS_{XY} - \text{Treatment } SS_{XY} \tag{10.4}$$

$$\text{Error } S_{XY} = 32.74 - 20.19$$

$$= 12.55$$

Step 5: For each source of variation, compute the adjusted sum of squares of the Y variable as:

$$\text{Total adjusted } SS \text{ of } Y = \text{Total } SS \text{ of } Y - \frac{\left(\text{Total } S_{XY}\right)^2}{\text{Total } SS \text{ of } X} \qquad (10.5)$$

$$= 33.22 - \frac{\left(32.74\right)^2}{36.06}$$

$$= 33.22 - 29.73$$

$$= 3.49$$

$$\text{Error adjusted } SS \text{ of } Y = \text{Error } SS \text{ of } Y - \frac{\left(\text{Error } S_{XY}\right)^2}{\text{Error } SS \text{ of } X} \qquad (10.6)$$

$$= 13.98 - \frac{\left(12.55\right)^2}{14.08}$$

$$= 13.98 - 11.19$$

$$= 2.79$$

Treatment adjusted SS of Y = Total adjusted SS of Y − Error adjusted SS of Y

$$= 3.49 - 2.79 = 0.7 \qquad (10.7)$$

Step 6: Now we compute the degree of freedom (*d.f.*) for each of the adjusted sum of squares computed in step 5.

$$\text{Adjusted error } d.f. = \text{Error } d.f. - 1 = 24 - 1 = 23$$

$$\text{Adjusted total } d.f. = \text{Total } d.f. - 1 = 31 - 1 = 30$$

$$\text{Adjusted treatment } d.f. = \text{Treatment } d.f. = 7$$

Step 7: Compute the adjusted mean squares of Y for treatment and adjusted error as:

$$\text{Treatment adjusted } MS \text{ of } Y = \frac{\text{Treatment adjusted } SS \text{ of } Y}{\text{Adjusted treatment } d.f.} \qquad (10.8)$$

$$= \frac{0.7}{7} = 0.1$$

TABLE 10.2.1.3

Analysis of Covariance

Source of Variation	Degree of Freedom	Y Adjusted for X		
		SS	MS	F
Treatment-adjusted	7	0.70	0.1	0.83*
Error	23	2.79	0.12	
Total	30	3.49		

Note: * = Not significant at 5% level; cv = 6.44%.

$$\text{Error adjusted } MS \text{ of } Y = \frac{\text{Error adjusted } SS \text{ of } Y}{\text{Adjusted error } d.f.}$$

$$= \frac{2.79}{23} = 0.12$$

Step 8: Compute the F value as shown:

$$F = \frac{\text{Treatment adjusted } MS \text{ of } Y}{\text{Error adjusted } MS \text{ of } Y} \tag{10.9}$$

$$= \frac{0.1}{0.12} = 0.83$$

Table 10.2.1.3 summarizes the findings of the analysis of covariance of the CRD data.

Step 9: To determine if the computed F value is statistically significant we compare it to the tabular F values in Appendix E. Note that the tabular F value for our example with f_1 = adjusted treatment $d.f.$ of 7 and f_2 = adjusted error $d.f.$ of 23 is 2.45 at the 5% level of significance. Because the computed F value is smaller than the tabular F, we conclude that there is no significant difference between the adjusted treatment means at the 5% level of significance.

Step 10: Compute the relative efficiency (*R.E.*) of covariance analysis compared to the standard analysis of variance as:

$$R.E. = \frac{\left(\text{Error } MS \text{ of } Y\right)\left(100\right)}{\left(\text{Error adjusted } MS \text{ of } Y\right)\left[1 + \dfrac{\text{Treatment } MS \text{ of } X}{\text{Error } SS \text{ of } X}\right]} \quad (10.10)$$

$$= \frac{\left(13.98 / 24\right)\left(100\right)}{\left(0.12\right)\left[1 + \dfrac{21.98 / 7}{14.08}\right]}$$

$$= 8.5\%$$

The *R.E.* indicates that using the initial weight as the covariate has increased the precision of the analysis by 8.5%.

Step 11: Calculate the coefficient of variation (*cv*) as:

$$cv = \frac{\sqrt{\text{Error-adjusted } MS \text{ of } Y}}{\text{Grand mean of } Y} \times 100 \quad (10.11)$$

$$= \frac{\sqrt{0.12}}{5.38} \times 100$$

$$= 6.44\%$$

As has been discussed before, the *cv* indicates the degree of precision with which the experiment was conducted. It expresses the experimental error as a percentage of the mean. In this example, the reduction in error that occurs from the use of the analysis of covariance is 6.44%.

10.2.2   Randomized Complete Block (RCB) Design

To illustrate the covariance analysis procedure for an RCB design, we consider the following experiment that we encountered earlier in Chapter 4. The agronomist designed the experiment to evaluate five different fertilizer placement methods on the yield of corn.

Example 10.2.2.1

Seed yield of early-maturing high-protein soybean lines adapted to the mid-Atlantic area of the U.S. were evaluated in field tests in an RCB with three replications. Each plot contained four 20-ft rows, 30 in. apart. Each plot was evaluated for seed yield, and the covariate used was the seed-yield data for the region as a whole. The data gathered are recorded in Table 10.2.2.1.

TABLE 10.2.2.1

Seed Yield (bu/ac) of Early-Maturing Soybean Varieties with Variety and Replication Totals

Variety	Rep. I		Rep. II		Rep. III		Variety Total	
	X	Y	X	Y	X	Y	X	Y
CX797-115	32.2	30.9	33.5	32.8	33.3	29.5	99.0	93.2
CX797-21	35.8	33.4	36.2	34.1	36.8	35.6	108.8	103.1
CX804-3	34.5	32.9	35.4	33.8	36.6	34.3	106.5	101.0
K1085	37.2	35.8	36.4	35.9	38.3	37.2	111.9	108.9
K1091	39.8	38.5	36.2	37.1	40.2	41.2	116.2	116.8
Williams	37.8	38.0	38.2	39.1	41.1	40.9	117.1	118.0
Douglas	36.4	36.2	38.6	37.9	40.2	41.0	115.2	115.1
Total	253.7	245.7	254.5	250.7	266.5	259.7	774.7	756.1
Mean							36.89	36.00

SOLUTION

Step 1: Compute the correction factor. Keep in mind that a correction factor for each of the variables, namely, X, Y, and XY has to be computed as was done in Example 10.2.1.1.

For the cross product the correction factor is:

$$C = \frac{G_x G_y}{rt} \tag{10.1}$$

$$C = \frac{(774.7)(756.1)}{21}$$

$$C = 27,892.89$$

Using Equation 10.1 the correction factor for variables X and Y are 28,579.05 and 27,223.2, respectively.

Step 2: Calculate the various sums of squares for each of the two variables, using the procedures outlined as follows. For expository purposes we have only shown the procedure for the total sum of squares for the XY cross product. The total sum of squares for the variables XX, YY, and XY are shown in Table 10.2.2.2.

$$\text{Total sum of cross product} = \sum_{i=1}^{t}\sum_{j=1}^{r}(X_{ij})(Y_{ij}) - C \tag{10.2}$$

$$\begin{aligned}\text{Total } S_{XY} &= [(32.2)(30.9) + (35.8)(33.4) + \ldots + (40.2)(41.0)] - 27,892.89 \\ &= 28,037.97 - 27,892.89 \\ &= 145.08\end{aligned}$$

TABLE 10.2.2.2

Analysis of Covariance of the Completely Randomized Block Data

Source of Variation	Degree of Freedom	Sum of Cross Product			Y Adjusted for X			
		XX	XY	YY	d.f.	SS	MS	F
Treatment	6	83.68	119.10	173.57				
Replication	2	14.69	13.87	14.38				
Error	12	14.31	12.11	20.68	11	10.43	0.95	
Treatment + error	18	97.99	131.21	194.25	17	18.55		
Treatment-adjusted					6	8.12	1.35	1.42**
Total	20	112.68	145.08	208.63				

Note: ** = Not significant at 1% level.

Step 3: Compute the sum of cross products for treatment as:

$$\text{Treatment } S_{XY} = \frac{\sum (T_X)(T_Y)}{r} - C \tag{10.3}$$

$$= \frac{(99.0)(93.2) + (108.8)(103.1) + \ldots + (115.2)(115.1)}{3} - 27{,}892.89$$

$$= 28{,}011.99 - 27{,}892.89$$

$$= 119.10$$

Step 4: Compute the sum of cross products for replication as:

$$\text{Replication } S_{XY} = \frac{\sum (Rep._X)(Rep._Y)}{t} - C \tag{10.12}$$

$$= \frac{\left[(253.7)(245.7) + (254.5)(250.7) + (266.5)(259.7)\right]}{7} - 27{,}892.89$$

$$= 27{,}906.76 - 27{,}892.89$$

$$= 13.87$$

Step 5: Compute the error sum of cross product as:

$$\text{Error } S_{XY} = \text{Total } S_{XY} - \text{Treatment } S_{XY} - Rep.S_{XY} \qquad (10.13)$$

$$= 145.08 - 119.10 - 13.87$$

$$= 12.11$$

Step 6: Compute the error-adjusted sum of squares of the Y variable as:

$$\text{Error adjusted } SS \text{ of } Y = \text{Error } SS \text{ of } Y - \frac{\left(\text{Error } S_{XY}\right)^2}{\text{Error } SS \text{ of } X} \qquad (10.6)$$

$$= 20.68 - \frac{\left(12.11\right)^2}{14.31}$$

$$= 20.68 - 10.25$$

$$= 10.43$$

Step 7: Compute the (treatment + error) adjusted sum of squares of the variable Y as:

$$\left(\text{Treatment} + \text{error}\right) \text{ adjusted } SS \text{ of } Y = A - \frac{C^2}{B} \qquad (10.14)$$

where:

$$A = (\text{treatment} + \text{error}) \, SS \text{ of } Y$$
$$= \text{Treatment } SS \text{ of } Y + \text{Error } SS \text{ of } Y$$
$$B = (\text{treatment} + \text{error}) \, SS \text{ of } X$$
$$= \text{Treatment } SS \text{ of } X + \text{Error } SS \text{ of } X$$
$$C = (\text{treatment} + \text{error}) \, SS \text{ of } XY$$
$$= \text{Treatment } SS \text{ of } XY + \text{Error } SS \text{ of } XY$$

Using Equation 10.14 and computing A, B, and C, we have:

$$A = 173.57 + 20.68 = 194.25$$
$$B = 83.68 + 14.31 = 97.99$$
$$C = 119.10 + 12.11 = 131.21$$

So the (treatment + error) adjusted sum of squares of Y is:

$$\left(\text{Treatment} + \text{error}\right) \text{ adjusted } SS \text{ of } Y = 194.25 - \frac{\left(131.21\right)^2}{97.99}$$

$$= 194.25 - 175.69$$

$$= 18.55$$

Step 8: Compute the treatment adjusted sum of squares of Y as:

$$\text{Treatment adjusted } SS \text{ of } Y = (\text{Treatment} + \text{error}) \text{ adjusted } SS \text{ of } Y - \text{Error}$$
$$\text{adjusted } SS \text{ of } Y$$
$$= 18.55 - 10.43 = 8.12 \tag{10.15}$$

Step 9: Compute the degree of freedom for each of the adjusted sum of squares as:

$$\text{Adjusted error } d.f. = \text{Error } d.f. - 1$$

$$= 12 - 1 = 11$$

$$\text{Adjusted (treatment} + \text{error) } d.f. = 6 + 12 - 1 = 17$$

Step 10: Compute the treatment adjusted mean square of Y and the error adjusted means square of Y as:

$$\text{Treatment adjusted } MS \text{ of } Y = \frac{\text{Treatment adjusted } SS \text{ of } Y}{\text{Adjusted treatment } d.f.} \tag{10.8}$$

$$= \frac{8.12}{6} = 1.35$$

$$\text{Error adjusted } MS \text{ of } Y = \frac{\text{Error adjusted } SS \text{ of } Y}{\text{Adjusted error } d.f.}$$

$$= \frac{10.43}{11} = 0.95$$

Step 11: Compute the F value as shown below:

$$F = \frac{\text{Treatment adjusted } MS \text{ of } Y}{\text{Error adjusted } MS \text{ of } Y} \tag{10.9}$$

$$= \frac{1.35}{0.95} = 1.42$$

Table 10.2.2.2 summarizes the findings of the analysis of covariance of the RCB data.

Step 12: To determine if the computed F value is statistically significant we compare it to the tabular F values in Appendix E. Note that the tabular F value for our example with f_1 = adjusted treatment *d.f.* of 6 and f_2 = adjusted error *d.f.* of 11 is 3.09 at the 5% level of significance. Because the computed F value is smaller than the tabular F, we conclude that there is no significant difference between the adjusted treatment means at the 5% level of significance.

Step 13: Compute the *R.E.* of covariance analysis compared to the standard analysis of variance as:

$$R.E. = \frac{\left(\text{Error } MS \text{ of } Y\right)\left(100\right)}{\left(\text{Error adjusted } MS \text{ of } Y\right)\left(1 + \dfrac{\text{Treatment } MS \text{ of } X}{\text{Error } SS \text{ of } X}\right)} \quad (10.10)$$

$$= \frac{\left(20.68 / 12\right)\left(100\right)}{\left(0.95\right)\left(1 + \dfrac{83.68 / 6}{14.31}\right)}$$

$$= 92.15\%$$

The *R.E.* indicates that using the regional yield as the covariate has increased the precision of the analysis by 92.15%.

Step 14: Calculate the *cv* as:

$$cv = \frac{\sqrt{\text{Error adjusted } MS \text{ of } Y}}{\text{Grand mean of } Y} \times 100 \quad (10.11)$$

$$= \frac{\sqrt{0.95}}{36.0} \times 100$$

$$= 2.7\%$$

As has been discussed before, the *cv* indicates the degree of precision with which the experiment was conducted. It expresses the experimental error as a percentage of the mean. In this example, the reduction in error that occurs from the use of the analysis of covariance is 2.7%.

10.2.3 Split-Plot Design

To illustrate the use of covariance analysis using the split-plot design, we use the following example and the steps to perform the analysis.

Example 10.2.3.1

An agronomist is interested in determining if major differences in yield response to nitrogen (N) fertilization exist among widely grown hybrids in the northern "Corn Belt." The subplot treatments were 5 hybrids, and the main plot treatments were N rates of 70, 140, and 210 lb/acre broadcast applied before planting. The study was replicated two times. The covariate X is the yield/acre before the application, and the Y variable represents the yield after the application of the fertilizer. The data gathered by the agronomist are shown in Table 10.2.3.1.

TABLE 10.2.3.1

Grain-Yield (bu/ac) Data of Five Corn Hybrids Grown with Four Levels of Nitrogen (N) in a Split-Plot Experiment with Two Replications

Fertilizer N rate, lb/ac	Hybrid	Replication I		Replication II	
		X	Y	X	Y
70	P3747	110	150	112	17
	P3732	109	150	113	160
	Mo17 × A634	103	140	118	155
	A632 × LH38	115	140	113	150
	LH74 × LH51	111	170	109	180
140	P3747	103	170	109	190
	P3732	106	160	112	180
	Mo17 × A634	112	155	110	165
	A632 × LH38	110	160	109	175
	LH74 × LH51	105	160	107	195
210	P3747	110	165	108	185
	P3732	102	165	111	200
	Mo17 × A634	105	150	113	175
	A632 × LH38	103	140	106	170
	LH74 × LH51	107	170	105	200

SOLUTION

Step 1: Compute the correction factor. Keep in mind that a correction factor for each of the variables, namely, X, Y, and XY has to be computed as was done in Example 10.2.1.1.

For the cross product the correction factor is:

$$C = \frac{(3266)(4995)}{30}$$

$$C = 543{,}789.0$$

Using Equation 10.1 the correction factor for variables X and Y are 355,559.0 and 831,667.5, respectively.

TABLE 10.2.3.2

The Replication × Fertilizer Table of Totals (RA) Computed from Data in Table 10.2.3.1

| | N Applied lb/ac | | | | | | Rep. Total (R) | |
| | 70 | | 140 | | 210 | | | |
Replication	X	Y	X	Y	X	Y	X	Y
I	548	750	536	805	527	790	1611	2345
II	565	815	547	905	543	930	1655	2650
Fertilizer Total	1113	1565	1083	1710	1070	1720		
Grand Total							3266	4995

TABLE 10.2.3.3

The Fertilizer × Hybrid Table of Totals (AB) Computed from Data in Table 10.2.3.1

| | N Applied (lb/ac) | | | | | | Fertilizer Total (B) | |
| | 70 | | 140 | | 210 | | | |
Hybrid	X	Y	X	Y	X	Y	X	Y
P3747	222	320	212	360	218	350	652	1,030
P3732	222	310	218	340	213	365	653	1,015
Mo17 × A634	221	295	222	320	218	325	661	940
A632 × LH38	228	290	219	335	209	310	656	935
LH74 × LH51	220	350	212	355	212	370	644	1,075

Step 2: Prior to computing the various sums of squares for each of the two variables, we need to construct the two-way tables of the totals for X and Y as shown in Table 10.2.3.2 and Table 10.2.3.3.

For expository purposes we have only shown the procedure for the total sum of squares for the XY cross product. The total sum of squares for the variables XX, YY, and XY are shown in Table 10.2.3.4.

Step 3: Compute the sum of squares as:

$$\text{Total sum of cross product} = \sum_{i=1}^{t}\sum_{j=1}^{r}(X_{ij})(Y_{ij}) - C \qquad (10.2)$$

$$\text{Total } S_{XY} = [(110)(150) + (109)(150) + \ldots + (105)(200)] - 543,789.0$$

$$= 543,725.00 - 543,789.0$$

$$= -64.0$$

TABLE 10.2.3.4

Summary Table of Data for Example 10.2.3.1

Source of Variation	Degree of Freedom	Sum of Cross Products		
		XX	XY	YY
Replication	1	64.53	447.33	3,100.83
Main Plot Factor (A)	2	97.27	−371.50	1,505.00
Error (a)	2	2.07	−2.83	281.67
A + error (a)	4	99.33	−374.33	1,786.67
Subplot Factor (B)	4	25.80	−229.83	2,428.33
A × B	8	64.40	14.83	561.67
Error (b)	12	613.40	78.00	20,116.70
B + error (b)	16	639.20	−151.83	22,545.00
A × B + error (b)	20	677.80	92.83	20,678.33
Total	29	867.47	−64.00	27,994.17

$$\text{Replication } S_{XY} = \frac{\sum (R_X)(R_Y)}{ab} - C \tag{10.16}$$

$$= \frac{(1611)(2345)+(1655)(2650)}{15} - 543,789.0$$

$$= 544,236.33 - 543,789.0$$

$$= 447.33$$

$$\text{Fertilizer rates (Factor A) } S_{XY} = \frac{\sum (A_{\cdot X})(A_{\cdot Y})}{rb} - C \tag{10.17}$$

$$= \frac{\left[(1113)(1565)+(1083)(1710)+(1070)(1720)\right]}{10} - 543,789.0$$

$$= 543,417.5 - 543,789.0$$

$$= -371.5$$

$$\text{Error (a) } S_{XY} = \frac{\sum (RA_X)(RA_Y)}{b} - C - \text{Rep. } S_{XY} - \text{Factor}(A)S_{XY} \tag{10.18}$$

$$= \frac{(548)(750)+(536)(805)+\ldots+(543)(930)}{5} - 543,789.0 - 447.33 - (-371.5) = -2.83$$

Step 4: Compute the various sums of cross products of X and Y for the subplot analysis as:

$$\text{Hybrid (Factor B) } S_{XY} = \frac{\sum (B._X)(B._Y)}{ra} - C \qquad (10.19)$$

$$= \frac{(652)(1030)+(653)(1015)+...+(644)(1075)}{6} - 543,789.0$$

$$= 543,559.2 - 543,789.0$$

$$= -299.8$$

$$\text{Fertilizer} \times \text{Hybrid (or } A \times B) \, S_{XY} = \frac{\sum (AB._X)(AB._Y)}{r} \qquad (10.20)$$

$$-C - \textit{Factor (B)} S_{XY} - \textit{Factor (A)} S_{XY}$$

$$= \frac{(222)(320)+(222)(310)+...+(212)(370)}{2} - 543,789.0 - (-299.8) - (-371.5)$$

$$= 543,202.5 - 543,789.0$$

$$= 14.83$$

$$\text{Error } (b) \, S_{XY} = \text{Total } S_{XY} - [\text{Rep. } S_{XY} + \text{Factor (A) } S_{XY} + \text{Error } (a) \, S_{XY} + \qquad (10.21)$$
$$\text{Factor (B) } S_{XY} + (A \times B) \, S_{XY}$$
$$= -64 - [447.33 + (-371.5) + (-2.83) + (-299.8) + (-5.83) + 14.83]$$
$$= 78.0$$

Table 10.2.3.5 shows the computed values of the various sums of squares.
Step 4: Compute the adjusted sum of squares of the variable Y that includes an error term as:

$$\text{Adjusted SS of } Y = YY - \frac{(XY)^2}{XX} \qquad (10.22)$$

Because the adjusted sum of squares for Y includes two error terms, that is, Error (a) and Error (b), we compute them using Equation 10.22 as follows:

TABLE 10.2.3.5

Analysis of Covariance of Split-Plot Data in Table 10.2.3.2

Source of Variation	Degree of Freedom	Y Adjusted for X		
		SS	MS	F
Error (a)	1	277.80	277.80	
A + error (a)	3	375.99		
A-adjusted	2	98.19	49.10	0.18[ns]
Error (b)	11	20,106.78	1,827.89	0.33[ns]
B + error (b)	15	22,508.94	1,500.60	
B-adjusted	4	2,402.18	600.55	
A × B + error (b)	19	20,665.62		
(A × B)-adjusted	8	558.86	69.86	0.04[ns]

Note: ns = Nonsignificant.

$$\text{Error (a) adjusted } SS \text{ of } Y = \text{Error (a) } SS \text{ of } Y - \frac{\left[\text{Error (a) } S_{XY}\right]^2}{\text{Error (a) } SS_{XX}} \quad (10.23)$$

$$= 281.67 - \frac{\left[-2.83\right]^2}{2.07}$$

$$= 277.80$$

$$\left[A + \text{Error (a)}\right] \text{ adjusted } SS \text{ of } Y = \left[A + \text{Error (a)}\right] \text{ Error (a) } SS \text{ of } Y \quad (10.24)$$

$$- \frac{\left[A + \text{Error (a) } S_{XY}\right]^2}{\left[A + \text{Error (a)}\right] SS_{XX}}$$

$$= 1786.67 - \frac{\left(-374.33\right)^2}{99.33}$$

$$= 375.99$$

$$\text{Error (b) adjusted } SS \text{ of } Y = \text{Error (b) } SS \text{ of } Y - \frac{\left[\text{Error (b) } S_{XY}\right]^2}{\text{Error (b) } SS_{XX}} \quad (10.25)$$

$$= 20,116.7 - \frac{\left(78\right)^2}{613.40}$$

$$= 20,106.78$$

$$\left[B + \text{ Error (b)} \right] \text{ adjusted } SS \text{ of } Y = \left\{ B + \text{ Error (b)} \right\} \text{ Error (b) } SS \text{ of } Y \quad (10.26)$$

$$- \frac{\left\{ \left[B + \text{Error (b)} \right] SS_{XY} \right\}^2}{\left[B + \text{Error (b)} \right] SS_{XX}}$$

$$= 22{,}545 - \frac{(-151.83)^2}{639.2}$$

$$= 22{,}508.94$$

$$\left[A \times B + \text{ Error (b)} \right] \text{ adjusted } SS \text{ of } Y = \left[A \times B + \text{ Error (b)} \right] SS \text{ of } Y \quad (10.27)$$

$$- \frac{\left\{ \left[A \times B + \text{Error (b)} \right] S_{XY} \right\}^2}{\left[A \times B + \text{Error (b)} \right] SS_{XX}}$$

$$= 20{,}678.33 - \frac{(92.83)^2}{677.8}$$

$$= 20{,}665.62$$

Step 5: Calculate the adjusted sum of squares of Y for the main effect associated with each factor and their interactions as shown in the following text:

$$A \text{ adjusted } SS \text{ of } Y = [A + \text{error (a)}] \text{ adjusted } SS \text{ of } Y \quad (10.28)$$

$$- \text{Error (a) adjusted } SS \text{ of } Y$$

$$= 375.99 - 277.80 = 98.19$$

$$B \text{ adjusted } SS \text{ of } Y = [B + \text{error (b)}] \text{ adjusted } SS \text{ of } Y \quad (10.29)$$

$$- \text{Error (b) adjusted } SS \text{ of } Y$$

$$= 22{,}508.94 - 20{,}106.78 = 2{,}402.16$$

$$(A \times B) \text{ adjusted } SS \text{ of } Y = [A \times B + \text{error (b)}] \text{ adjusted } SS \text{ of } Y \quad (10.30)$$

$$- \text{Error (b) adjusted } SS \text{ of } Y$$

$$= 20{,}665.62 - 20{,}106.78 = 558.84$$

Step 6: Compute the degree of freedom for each of the adjusted sum of squares as:

$$\text{Adjusted error (a) } d.f. = \text{Error (a) } d.f. - 1 = 2 - 1 = 1$$

$$\text{Adjusted } [A + \text{Error (a)}] \ d.f. = A \ d.f. + \text{Error (a) } d.f. - 1 = 2 + 2 - 1 = 3$$

$$\text{Adjusted } A \text{ d.f.} = A \ d.f. = 2$$

$$\text{Adjusted error (b) } d.f. = \text{Error (b) } d.f. - 1 = 12 - 1 = 11$$

$$\text{Adjusted } B \text{ d.f.} = B \ d.f. = 4$$

$$\text{Adjusted } [B + \text{Error (b)}] \ d.f. = B \ d.f. + \text{Error (b) } d.f. - 1 = 4 + 12 - 1 = 15$$

$$\text{Adjusted } [A \times B + \text{Error (b)}] \ d.f. = A \times B \ d.f. + \text{Error (b) } d.f. - 1 = 8 + 12 - 1 = 19$$

Step 7: Compute the adjusted mean square of Y as:

$$\text{Error (a) adjusted } MS = \frac{\text{Error (a) adjusted } SS}{\text{Adjusted error (a) } d.f.} \qquad (10.31)$$

$$= \frac{277.80}{1} = 277.80$$

$$A \text{ adjusted } MS = \frac{A \text{ adjusted } SS}{\text{Adjusted } A \text{ d.f.}} \qquad (10.32)$$

$$= \frac{98.19}{2} = 49.10$$

$$\text{Error (b) adjusted } MS = \frac{\text{Error (b) adjusted } SS}{\text{Adjusted error (b) } d.f.} \qquad (10.33)$$

$$= \frac{20,106.78}{11} = 1,827.89$$

$$B \text{ adjusted } MS = \frac{B \text{ adjusted } SS}{\text{Adjusted } B \text{ d.f.}} \tag{10.34}$$

$$= \frac{2,402.18}{4} = 600.55$$

$$(A \times B) \text{ adjusted } MS = \frac{[A \times B] \text{ adjusted } SS}{\text{Adjusted } A \times B \text{ d.f.}} \tag{10.35}$$

$$= \frac{558.86}{8} = 69.86$$

Step 8: Compute the F value as shown below:

$$F(A) = \frac{A \text{ adjusted } MS}{\text{Error (a) adjusted } MS} \tag{10.36}$$

$$= \frac{49.10}{277.80} = 0.18$$

$$F(B) = \frac{B \text{ adjusted } MS}{\text{Error (b) adjusted } MS} \tag{10.37}$$

$$= \frac{600.55}{1,827.89} = 0.33$$

$$F(A \times B) = \frac{(A \times B) \text{ adjusted } MS}{\text{Error (b) adjusted } MS} \tag{10.38}$$

$$= \frac{69.86}{1,827.89} = 0.04$$

Table 10.2.3.5 summarizes the findings of the analysis of covariance of the adjusted data.

Step 9: To determine if the computed F value is statistically significant we compare it to the tabular F values in Appendix E. Note that all three computed F values are less than the tabular F. Therefore, we concluded that the main effect for factor A, the main effect for factor B, and the interaction effect are not significant at the 5% level.

Step 10: Calculate the *cv* for the main and subplot as:

$$cv\ (a) = \frac{\sqrt{\text{Error }(a)\text{ adjusted }MS\text{ of }Y}}{\text{Grand mean of }Y} \times 100$$

$$= \frac{\sqrt{277.80}}{166.5} \times 100$$

$$= 10.01\%$$

$$cv\ (b) = \frac{\sqrt{\text{Error }(b)\text{ adjusted }MS\text{ of }Y}}{\text{Grand mean of }Y} \times 100$$

$$= \frac{\sqrt{1,827.89}}{166.5} \times 100$$

$$= 25.66\%$$

As discussed previously, the *cv* indicates the degree of precision with which the experiment was conducted. It expresses the experimental error as a percentage of the mean. In this example, the reduction in error that occurs from the use of the analysis of covariance is 10.01% and 25.66% for each factor, respectively.

10.3 Estimating Missing Data

We mentioned at the beginning of this chapter that we use covariance analysis to reduce error in an experiment. You have observed in the previous sections that when we used covariance analysis to control error and to adjust treatment means, we measured the covariate X along with the Y variable for each experimental unit. In the case of missing data, the covariate is not measured but assigned, one each to a missing observation. In some ways this is similar to the inclusion of a qualitative variable in a regression model. You may recall that in such a situation we assigned a value of 0 and 1 for the absence or presence of a characteristic in a variable of interest. For example, to include a variable such as a home with a swimming pool in the regression model we assigned a value of 1 for those homes that had a pool and 0 for those that did not. We do the same here in the covariance analysis. For simplicity, we are going to illustrate the case of one missing observation in a covariance analysis. We suggest that the reader consult Steel and Torrie

TABLE 10.3.1

Assignment of Values for the Missing Data of Corn Yield (Y) and the Covariates (X)

Treatment	Grain Yield (bu/acre)						Treatment Total (T)	
	Rep. I		Rep. II		Rep. III			
	X	Y	X	Y	X	Y	X	Y
Control	0	147.0	0	130.1	0	142.2	0	419.3
2 × 2-in. band	0	159.4	0	167.3	0	150.5	0	477.2
Broadcast	0	158.9	0	166.2	0	159.1	0	484.2
Deep band	0	173.6	0	170.8	1^a	0	1	344.4
Disk applied	0	158.4	0	169.3	0	160.2	0	487.9
Sidedress	0	157.1	0	148.8	0	139.0	0	444.9
Rep. total (R)	0	954.4	0	952.5	0	751.0		
Grand Total (G)							1	2,657.9

Note: ^{a} = Missing data.

(1980) and Snedecor and Cochran (1980) for the analysis procedure in cases with more than one observation missing.

The step-by-step procedure for the computation of missing data in covariance analysis is outlined in the text that follows, using the data from Example 4.2.2.1 as shown in Table 10.3.1.

Step 1: Assign the missing plot value to the covariate as shown in Table 10.3.1. You will note that the missing value is for deep-band treatment in replication 3. Hence, we have assigned a value of 1 for the X variable; Y is assigned 0 as its value is missing.

Step 2: Compute the various sums of squares for Y using the analysis of variance procedure for a CRB design as was shown in Chapter 4. The results are shown in Table 10.3.2 under the column YY.

TABLE 10.3.2

Sum of Cross Products for Missing Observation in a Covariance Analysis

Source of Variation	Degree of Freedom	Sum of Cross Products		
		XX	XY	YY
Replication	2	0.11	−22.49	4,554.30
Treatment	5	0.27	−32.86	5,056.78
Error	10	0.56	−92.31	15,778.20
Treatment + error	15	0.83	−125.17	20,834.98
Total	17	0.94	−147.66	25,389.30

Step 3: Calculate the various sums of squares for the X variable using the following procedure:

$$\text{Total } SS_{XX} = 1 - \frac{1}{rt} \qquad\qquad (10.39)$$

$$= 1 - \frac{1}{18} = 0.94$$

$$\text{Replication } SS_{XX} = \frac{1}{t} - \frac{1}{rt} \qquad\qquad (10.40)$$

$$= \frac{1}{6} - \frac{1}{18} = 0.11$$

$$\text{Treatement } SS_{XX} = \frac{1}{r} - \frac{1}{rt} \qquad\qquad (10.41)$$

$$= \frac{1}{3} - \frac{1}{18} = 0.27$$

$$\text{Error } SS_{XX} = \text{Total } SS_{XX} - \text{Replication } SS_{XX} - \text{Treatment } SS_{XX} \qquad (10.42)$$
$$= 0.94 - 0.11 - 0.27 = 0.56$$

Step 4: Compute the various sums of squares of the cross products XY as:

$$C = \frac{G_Y}{(r)(t)} \qquad\qquad (10.43)$$

$$= \frac{2,657.90}{(3)(6)} = 147.66$$

$$\text{Total } SS_{XY} = - (C) \qquad\qquad (10.44)$$

$$= - 147.66$$

$$\text{Replication } S_{XY} = \frac{B_Y}{t} - C \qquad\qquad (10.45)$$

$$= \frac{751}{6} - 147.66 = -22.49$$

Note that B_Y in Equation 10.45 is the total for the Y variable in the replication where we have a missing observation. In our case, it is the third replication.

$$\text{Treatment } S_{XY} = \frac{T_Y}{r} - C \tag{10.46}$$

$$= \frac{344.4}{3} - 147.66 = -32.86$$

You will also observe that in Equation 10.46, T_Y is the treatment total for the Y variable that corresponds to the treatment with the missing data. In this example, it is the fourth treatment (deep band).

$$\text{Error } S_{XY} = \text{Total } S_{XY} - \text{Replication } S_{XY} - \text{Treatment } S_{XY} \tag{10.47}$$
$$= -147.66 - (-22.49) \, (-32.86)$$
$$= -92.31$$

Step 5: Compute the various sums of squares for YY as:

$$C = \frac{(2,657.9)(2,657.9)}{18} = 392,468.47$$

$$\text{Total } SS_{YY} = [(147)(147) + (159.4)(159.4) + \ldots + (139)(139)] - 543,789.0$$
$$= 417,857.75 - 392,468.47$$
$$= 25,389.28$$

$$\text{Treatment } SS_{YY} = \frac{\sum (T_Y)(T_Y)}{r} - C$$

$$= \frac{(419.3)(419.3) + (477.2)(477.2) + \ldots + (444.9)(444.9)}{3} - 392,468.47$$

$$= 397,525.25 - 392,468.47$$

$$= 5,056.78$$

$$\text{Replication } SS_{YY} = \frac{\sum (Rep._Y)(Rep._Y)}{t} - C$$

$$= \frac{\left[(954.4)(954.4)+(952.5)(952.5)+(751.0)(751.0)\right]}{6} - 392,468.47$$

$$= 397,022.77 - 392,468.47$$

$$= 4,554.30$$

$$\text{Error } SS_{YY} = \text{Total } SS_{YY} - \text{Treatment } SS_{YY} - \text{Re } p.S_{YY}$$

$$= 25,389.28 - 5,056.78 - 4,554.30$$

$$= 15,778.20$$

Step 6: Compute the error adjusted *SS* of *Y* as:

$$\text{Error adjusted } SS \text{ of } Y = 15,778.20 - \frac{(-92.31)^2}{0.56} = 561.89$$

To compute the (treatment + error) adjusted sum of squares of *Y*, we use Equation 10.14:

$$\text{(Treatment + error) adjusted } SS \text{ of } Y = A - \frac{C^2}{B} \qquad (10.14)$$

where:

A = (treatment + error) *SS* of *Y* = Treatment *SS* of *Y* + Error *SS* of *Y*

B = (treatment + error) *SS* of *X* = Treatment *SS* of *X* + Error *SS* of *X*

C = (treatment + error) *SS* of *XY* = Treatment *SS* of *XY* + Error *SS* of *XY*

Using Equation 10.14 we compute *A*, *B*, and *C* to be:

$$A = 5,056.78 + 15,778.20 = 20,834.98$$
$$B = 0.27 + 0.56 = 0.83$$
$$C = \text{-}92.31 + (-32.86) = -125.17$$

So the (treatment + error) adjusted sum of squares of Y is:

$$\left(\text{Treatment} + \text{error}\right) \text{ adjusted } SS \text{ of } Y = 20,834.98 - \frac{\left(-125.17\right)^2}{0.83}$$

$$=20,834.98 - 18,876.54$$

$$= 1,958.44$$

Step 7: Compute the treatment adjusted sum of squares of Y as:

Treatment adjusted SS of Y = (treatment + error) adjusted SS of Y

$$- \text{Error adjusted } SS \text{ of } Y \qquad (10.15)$$

$$= 1,958.44 - 561.89$$

$$= 1,396.55$$

Step 8: Compute the degree of freedom for each of the adjusted sum of squares as:

Adjusted error $d.f.$ = Error $d.f.$ $- 1 = 10 - 1 = 9$

Adjusted (treatment + error) $d.f.$ = $5 + 10 - 1 = 14$

Step 9: Compute the treatment adjusted mean square of Y and the error adjusted mean square of Y as:

$$\text{Treatment adjusted } MS \text{ of } Y = \frac{\text{Treatment adjusted } SS \text{ of } Y}{\text{Adjusted treatment } d.f.} \qquad (10.8)$$

$$= \frac{1,396.55}{5} = 279.31$$

$$\text{Error adjusted } MS \text{ of } Y = \frac{\text{Error adjusted } SS \text{ of } Y}{\text{Adjusted error } d.f.}$$

$$= \frac{561.89}{9} = 62.43$$

TABLE 10.3.3

Analysis of Covariance with Missing Data When Y Has Been Adjusted for X

Source of Variation	Degree of Freedom	Y Adjusted for X		
		SS	MS	F
Treatment-adjusted	5	1,396.55	279.31	
Error	9	561.89	62.43	4.47*
Treatment + error	14	1,958.44	139.89	

Note: * = Significant at 5% level.

Step 10: Compute the F value as shown below:

$$F = \frac{\text{Treatment adjusted } MS \text{ of } Y}{\text{Error adjusted } MS \text{ of } Y}$$

(10.9)

$$= \frac{279.31}{62.43} = 4.47$$

Table 10.3.3 summarizes the findings of the analysis of covariance with missing data.

Step 11: To determine if the computed F value is statistically significant we compare it to the tabular F values in Appendix E. Note that the tabular F value for our example with f_1 = adjusted treatment *d.f.* of 5 and f_2 = adjusted error *d.f.* of 9 is 3.38 at the 5% level of significance. Because the computed F value is greater than the tabular F, we conclude that there is significant difference between the adjusted treatment means at the 5% level of significance.

Step 12: Compute the estimate of the missing data as:

$$\text{Estimate of missing data} = -b_{y.x} = \text{Error } S_{XY}/\text{Error } SS \text{ of } X \quad (10.48)$$

$$= \frac{-(-92.31)}{0.56} = 164.84 \text{ bu/ac}$$

The covariance analysis, as well as other techniques, can be used to estimate the value of a missing observation.

References and Suggested Reading

Bubawi, F.F. and Mitwali, E.M. 1991. Witch-weed management by sorghum-sudan grass seed size and stage of harvest. *Agron. J.* 83: 781–785.

Carsky, R.J., Singh, L., and Ndikawa, R. 1994. Suppression of *Striga hermonthica* on sorghum using a cowpea intercrop. *Exp. Agric.* 30: 349–358.

Cox, D.R. and McCullough, P. 1982. Some aspects of analysis of covariance. *Biometrics.* 38: 541–561.

Fisher, R.A. 1932. *Statistical Methods for Research Workers.* Edinburgh: Oliver & Boyd.

Park, H. and Cho, K. 2004. Use of covariates in Taylor's power law for sequential sampling in pest management. *J. Agric. Biol. Environ. Stat.* 9(4): 462–478 .

Smith, H.F. 1957. Interpretation of adjusted treatment means in regression and analysis of covariance. *Biometrics* 13(3): 282–297.

Snedecor, G.W. and Cochran, W.G. 1980. *Statistical Methods.* Ames, IA: Iowa State University Press. pp. 388–391.

Steel, R.G.D. and Torrie, J.A. 1980. *Principles and Procedures of Statistics.* 2nd ed. New York: McGraw-Hill. pp. 428–434.

Veerkamp. R.F. and Goddard, M.E. 1998. Covariance functions across herd production levels for test day records on milk, fat, and protein yield. *J. Dairy Sci.* 81: 1690–1701.

Exercises

1. A researcher obtained the following data on ten animals under each of three treatment conditions:

Treatment					
A		B		C	
X	Y	X	Y	X	Y
8	17	5	16	3	14
4	7	3	18	4	25
4	8	4	16	4	13
5	18	6	22	5	24
5	16	5	17	3	22
7	17	4	14	4	25
4	15	4	12	3	23
5	18	3	15	4	19
6	22	4	18	3	17
6	20	5	19	4	22

(a) Complete an analysis of variance for Y.

(b) Complete an analysis of covariance for (Y fixed).

(c) Compare the outcomes of the two analyses.

2. To study the effect of expectation upon reactions to a TV commercial on a new agricultural product, a researcher asked ten respondents about their reactions to the commercial. Each subject evaluated the product's desirability on a 100-point scale after watching the commercial. A covariate — attitude toward the company — was measured prior to treatment. The data gathered were as follows:

Expectation					
Excellent		Good		Poor	
X	Y	X	Y	X	Y
65	80	55	75	53	55
73	82	65	77	60	65
71	85	62	78	52	55
67	83	60	62	50	52
77	90	68	72	61	67
52	77	53	60	60	65
67	89	56	67	66	86
77	85	64	68	54	68
62	81	63	78	52	65
66	86	53	73	54	55

(a) Complete an analysis of covariance.

(b) What is the estimate of variation due to error, given the model.

3. An environmental horticulturist is interested in finding out whether
   (1) stress-adapted landscapes save water, (2) whether irrigation
   equal to 15% (or less) reference evapotranspiration (ET_0) can be
   applied to established shrubs and ground cover without any
   drought-related injury. This is a 3-factor experiment designed to test
   the effect of 3 irrigation regimes (no irrigation, 12 in., and 24 in. of
   water) and 2 different irrigation methods (drip and furrow) on the
   growth of shrubs and ground covers such as Xylosma, Oleander,
   Cotoneaster, Juniper, Ice plant, and Hedera. The experiment is strip-
   split-plot design replicated three times. The data collected at the end
   of a 2-yr period are as follows:

Growth of Shrubs and Ground Cover as a Function of Irrigation Water Received
from April to August

Plantings	Irrigation Method	Inches of Water Applied	Rep. I		Rep. II		Rep. III	
			X	Y	X	Y	X	Y
Xylosma	Drip	0.0	4.0	8.0	3.5	8.4	3.0	9.5
		12.0	10.0	19.5	12.0	20.1	10.0	20.2
		24.0	12.0	30.6	14.0	31.0	12.2	31.4
	Furrow	0.0	3.5	6.0	4.0	5.4	3.8	5.8
		12.0	8.5	12.8	12.0	16.9	11.5	17.4
		24.0	14.0	28.2	15.0	27.6	13.2	29.4
Oleander	Drip	0.0	6.5	18.0	8.2	19.4	9.5	19.5
		12.0	8.5	39.5	12.2	40.1	16.0	40.3
		24.0	14.3	60.6	18.5	59.0	24.0	61.4
	Furrow	0.0	3.2	16.0	2.5	15.4	3.3	15.7
		12.0	5.4	22.8	5.8	36.9	10.0	37.4
		24.0	16.0	48.2	22.8	47.6	21.9	49.4
Cotoneaster	Drip	0.0	2.1	6.0	2.3	6.4	3.2	6.5
		12.0	14.0	35.5	16.8	31.1	22.0	30.6
		24.0	28.0	40.6	26.2	41.0	23.8	41.3
	Furrow	0.0	1.2	4.0	1.0	4.4	2.3	4.8
		12.0	3.8	19.8	4.2	16.9	4.6	18.4
		24.0	6.2	25.2	8.6	27.6	10.2	29.5
Juniper	Drip	0.0	3.0	12.0	4.2	12.4	5.2	12.7
		12.0	4.4	10.5	4.8	10.1	4.3	10.2
		24.0	6.8	20.6	8.2	18.0	10.1	19.3
	Furrow	0.0	2.1	10.0	2.3	11.1	2.2	10.8
		12.0	3.5	9.8	2.2	6.9	3.1	7.4
		24.0	4.5	13.2	4.5	14.8	4.6	15.4
		12.0	6.6	22.5	8.2	26.9	8.1	25.4
		24.0	7.2	28.8	8.8	29.6	10.0	29.9

(a) Perform the analysis of covariance.

(b) What conclusions can you draw from the analysis?

4. Field measurements were made to study the response of field-grown cassava (*Manihot esculenta* Crantz) to changes in the application of a fertilizer. The researcher conducted a completely randomized design to find out if there were any differences in the amount of the dry matter produced under five different fertilizer regimes. Yield data prior to fertilization are used as the covariate variable. The following data were collected from the experiment with four replications:

Dry Matter Production (t/ha) of Cassava as a Result of Five Different Fertilizer Applications

	Rep. I		Rep. II		Rep. III		Rep. IV	
Treatment	X	Y	X	Y	X	Y	X	Y
Control	1.8	2.20	1.5	2.10	1.8	2.25	1.6	2.01
50 kg/ha	1.7	2.40	1.9	2.56	1.1	2.66	1.1	2.52
100 kg/ha	1.8	2.60	1.7	2.68	1.3	2.79	1.2	2.66
150 kg/ha	1.5	3.00	1.8	3.56	1.9	4.00	2.1	4.66
200 kg/ha	1.8	3.50	2.1	4.98	2.1	5.00	2.3	4.20

(a) Perform an analysis of covariance.

(b) What conclusions can you draw from this experiment?

Appendix A

Chi-Square Distribution

Degrees of Freedom	Upper-Tail Area						
	.99	.98	.95	.90	.80	.70	.50
1	.0³157	.0³628	.00393	.0158	.0642	.148	.455
2	.0201	.0404	.103	.211	.446	.713	1.386
3	.115	.185	.352	.584	1.005	1.424	2.366
4	.297	.429	.711	1.064	1.649	2.195	3.357
5	.554	.752	1.145	1.610	2.343	3.000	4.351
6	.872	1.134	1.635	2.204	3.070	3.828	5.348
7	1.239	1.564	2.167	2.833	3.822	4.671	6.346
8	1.646	2.032	2.733	3.490	4.594	5.527	7.344
9	2.088	2.532	3.325	4.168	5.380	6.393	8.343
10	2.558	3.059	3.940	4.865	6.179	7.267	9.342
11	3.053	3.609	4.575	5.578	6.989	8.148	10.341
12	3.571	4.178	5.226	6.304	7.807	9.034	11.340
13	4.107	4.765	5.892	7.042	8.634	9.926	12.340
14	4.660	5.368	6.571	7.790	9.467	10.821	13.339
15	5.229	5.985	7.261	8.547	10.307	11.721	14.339
16	5.812	6.614	7.962	9.312	11.152	12.624	15.338
17	6.408	7.255	8.672	10.085	12.002	13.531	16.338
18	7.015	7.906	9.390	10.865	12.857	14.440	17.338
19	7.633	8.567	10.117	11.651	13.716	15.352	18.338
20	8.260	9.237	10.851	12.443	14.578	16.266	19.337
21	8.897	9.915	11.591	13.240	15.445	17.182	20.337
22	9.542	10.600	12.338	14.041	16.314	18.101	21.337
23	10.196	11.293	13.091	14.848	17.187	19.021	22.337
24	10.856	11.992	13.848	15.659	18.062	19.943	23.337
25	11.524	12.697	14.611	16.473	18.940	20.867	24.337
26	12.198	13.409	15.379	17.292	19.820	21.792	25.336
27	12.879	14.125	16.151	18.114	20.703	22.719	26.336
28	13.565	14.847	16.928	18.939	21.588	23.647	27.336
29	14.256	15.574	17.708	19.768	22.475	24.577	28.336
30	14.953	16.306	18.493	20.599	23.364	25.508	29.336

Degrees of Freedom	Upper-Tail Area						
	.30	.20	.10	.05	.02	.01	.001
1	1.074	1.642	2.706	3.841	5.412	6.635	10.827
2	2.408	3.219	4.605	5.991	7.824	9.210	13.815
3	3.665	4.642	6.251	7.815	9.837	11.345	16.268
4	4.878	5.989	7.779	9.488	11.668	13.277	18.465
5	6.064	7.289	9.236	11.070	13.388	15.086	20.517
6	7.231	8.558	10.645	12.592	15.033	16.812	22.457
7	8.383	9.803	12.017	14.067	16.622	18.475	24.322
8	9.524	11.030	13.362	15.507	18.168	20.090	26.125
9	10.656	12.242	14.684	16.919	19.679	21.666	27.877
10	11.781	13.442	15.987	18.307	21.16)	23.209	29.588
11	12.899	14.631	17.275	19.675	22.618	24.725	31.264
12	14.011	15.812	18.549	21.026	24.054	26.217	32.909
13	15.119	16.985	19.812	22.362	25.472	27.688	34.528
14	16.222	18.151	21.064	23.685	26.873	29.141	36.123
15	17.322	19.311	22.307	24.996	28.259	30.578	37.697
16	18.418	20.465	23.542	26.296	29.633	32.000	39.252
17	19.511	21.615	24.769	27.587	30.995	33.409	40.790
18	20.601	22.760	25.989	28.869	32.346	34.805	42.312
19	21.689	23.900	27.204	30.144	33.687	36.191	43.820
20	22.775	25.038	28.412	31.410	35.020	37.566	45.315
21	23.858	26.171	29.615	32.671	36.343	38.932	46.797
22	24.939	27.301	30.813	33.924	37.659	40.289	48.268
23	26.018	28.429	32.007	35.172	38.968	41.638	49.728
24	27.096	29.553	33.196	36.415	40.270	42.980	51.179
25	28.172	30.675	34.382	37.652	41.566	44.314	52.620
26	29.246	31.795	35.563	38.885	42.856	45.642	54.052
27	30.319	32.912	36.741	40.113	44.140	46.963	55.476
28	31.391	34.027	37.916	41.337	45.419	48.278	56.893
29	32.461	35.139	39.087	42.557	46.693	49.588	58.302
30	33.530	36.250	40.256	43.773	47.962	50.892	59.703

Source: From Table IV of Ronals A. Fisher and Frank Yates, *Statistical Tables for Biological, Agricultural and Medical Research*, published by Longman Group, London (previously published by Oliver & Boyd, Edinburgh, 1963). Reproduced with permission of the authors and publishers.

Appendix B

The Arc Sine $\sqrt{\text{Percentage}}$ Transformation

%	0	1	2	3	4	5	6	7	8	9
0.0	0	0.57	0.81	0.99	1.15−	1.28	1.40	1.52	1.62	1.72
0.1	1.81	1.90	1.99	2.07	2.14	2.22	2.29	2.36	2.43	2.50
0.2	2.56	2.63	2.69	2.75−	2.81	2.87	2.32	2.38	3.03	3.09
0.3	3.14	3.19	3.24	3.29	3.34	3.39	3.44	3.49	3.53	3.58
0.4	3.63	3.67	3.72	3.76	3.80	3.85−	3.89	3.93	3.97	4.01
0.5	4.05+	4.09	4.13	4.17	4.21	4.25+	4.29	4.33	4.37	4.40
0.6	4.44	4.48	4.52	4.55+	4.59	4.62	4.66	4.69	4.73	4.76
0.7	4.80	4.83	4.87	4.90	4.93	4.97	5.00	5.03	5.07	5.10
0.8	5.13	5.16	5.20	5.23	5.26	5.29	5.32	5.35+	5.38	5.41
0.9	5.44	5.47	5.50	5.53	5.56	5.59	5.62	5.65+	5.68	5.71
1	5.74	6.02	6.29	6.55−	6.80	7.04	7.27	7.49	7.71	7.92
2	8.13	8.33	8.53	8.72	8.91	9.10	9.28	9.46	9.63	9.81
3	9.98	10.14	10.31	10.47	10.63	10.78	10.94	11.09	11.24	11.39
4	11.54	11.68	11.83	11.97	12.11	12.25−	12.39	12.52	12.66	12.79
5	12.92	13.05+	13.18	13.31	13.44	13.56	13.69	13.81	13.94	14.06
6	14.18	14.30	14.42	14.54	14.65+	14.77	14.89	15.00	15.12	1523
7	15.34	15.45+	15.56	15.68	15.79	15.89	16.00	16.11	1622	16.32
6	16.43	16.54	16.64	16.74	16.85−	16.95+	17.05+	17.16	17.26	17.36
9	17.46	17.56	17.66	17.76	17.85+	17.95+	18.05−	18.15−	18.24	18.34
10	18.44	18.53	18.63	18.72	18.81	18.91	19.00	19.09	19.19	19.28
11	19.37	19.46	19.55+	19.64	19.73	19.82	19.91	20.00	20.09	20.18
12	20.27	20.35	20.44	20.53	20.62	20.70	20.79	20.88	20.96	21.05−
13	21.13	21.22	21.30	21.39	21.47	21.56	21.64	21.72	21.81	21.89
14	21.97	22.06	22.14	22.22	22.30	22.38	22.46	22.55−	22.63	22.71
15	22.79	22.87	22.95−	23.03	23.11	23.19	23.26	23.34	23.42	23.50
16	23.58	23.66	23.73	23.81	23.89	23.97	24.04	24.12	2420	2427
17	24.35+	24.43	24.50	24.58	24.65+	24.73	24.80	24.88	24.95	25.03
16	25.10	25.18	25.25+	25.33	25.40	25.48	25.55−	25.62	25.70	25.77
19	25.84	25.92	25.99	26.06	26.13	26.21	2628	26.35−	26.42	26.49
20	26.56	26.64	26.71	26.78	26.85+	26.92	26.99	27.06	27.13	2720
21	27.28	27.35π	27.42	27.49	27.56	27.63	27.69	27.76	27.83	27.90
22	27.97	28.04	28.11	28.18	28.25−	28.32	28.38	28.45+	28.52	28.59
23	28.66	26.73	28.79	28.86	28.93	29.00	29.06	29.13	2920	2927
24	29.33	29.40	29.47	29.53	2960	29.67	29.73	29.80	29.87	29.93
25	30.00	30.07	30.13	30.20	30.26	30.33	30.40	30 46	30.53	30.59
26	30.66	30.72	30.79	30.85+	30.92	30.98	31.05−	31.11	31.18	31.24
27	31.31	31.37	31.44	31.50	31.56	31.63	31.63	3176	31.82	31.88

%	0	1	2	3	4	5	6	7	8	9
28	31.95–	32.01	32.08	32.14	32.20	32.27	32.33	32.39	32.46	32.52
29	32.58	32.65–	32.71	32.77	32.83	32.90	3296	33.02	33.09	33.15–
30	33.21	33.27	33.34	33.40	33.46	33.52	33.58	33.65–	33.71	33.77
31	33.83	33.89	33.96	34.02	34.08	34.14	34.20	34.27	34.33	34.39
32	34.45–	34.51	34.57	34.63	34.70	34.76	34.82	34.88	34.94	35.00
33	35.06	35.12	35.18	35.24	35.30	35.37	35.43	35.49	35.55–	35.61
34	35.67	35.73	35.79	35.85–	35.91	35.97	36.03	36.09	36.15≠	36.21
35	36.27	36.33	36.39	36.45+	36.51	36.57	36.63	36.69	36.75+	36.81
36	36.87	36.93	36.99	37.05–	37.11	37.17	37.23	3729	37.35–	37.41
37	37.47	37.52	37.58	37.64	37.70	37.76	37.82	37.88	37.94	38.00
38	38.06	38.12	38.17	38.23	38.29	38.35+	38.41	38.47	38.53	38.59
39	38.65–	38.70	38.76	38.82	38.88	38.94	39.00	39.06	39.11	39.17
40	39.23	39.29	39.35–	39.41	39.47	39.52	39.58	39.64	39.70	39.76
41	39.82	39.87	39.93	39.99	40.05–	40.11	40.16	4022	4028	40.34
42	40.40	40.46	40.51	40.57	40.63	40.69	40.74	40.80	40.86	40.92
43	40.98	41.03	41.09	41.15–	41.21	41.27	41.32	41.38	41.44	41.50
44	41.55+	41.61	41.67	41.73	41.78	41.84	41.90	41.96	42.02	42.07
45	42.13	42.19	42.25–	42.30	42.36	42.42	42.48	42.53	42.59	42.65–
46	42.71	42.76	42.82	42.88	42.94	42.99	43.05–	43.11	43.17	43.22
47	43.28	43.34	43.39	43.45+	43.51	43.57	43.62	43.68	43.74	43.80
48	43.85+	43.91	43.97	44.03	44.08	44.14	44.20	44.25+	44.51	44.37
49	44.43	44.46	44.54	44.60	44.66	44.71	44.77	44.83	44.89	44.94
50	45.00	45.06	45.11	45.17	45.23	45.29	45.34	45.40	45.46	45.52
51	45.57	45.63	45.89	45.75–	45.80	45.86	45.92	45.97	46.03	46.09
52	46.15–	46.20	46.26	46.32	46.38	46.43	46.49	46.55–	46.61	46.66
53	46.72	46.78	46.83	46.89	46.95+	47.01	47.06	47.12	47.18	47.24
54	47.29	47.35+	47.41	47.47	47.52	47.58	47.64	47.70	47.75+	47.81
55	47.87	47.93	47.98	48.04	48.10	48.16	48.22	48.27	48.33	48.39
56	48.45–	48.50	48.56	48.62	48.68	48.73	48.79	48.85+	48.91	48.97
57	49.02	49.08	49.14	49.20	49.26	49.31	49.37	49.43	49.49	49.54
58	49.60	49.66	49.72	49.78	49.84	49.89	49.95+	50.01	50.07	50.13
59	50.18	50.24	50.30	50.36	50.42	50.48	50.53	50.59	50.65+	50.71
60	50.77	50.83	50.89	50.94	51.00	51.06	51.12	51.18	51.24	51.30
61	51.35+	51.41	51.47	51.53	51.59	51.65–	51.71	51.77	51.83	51.88
62	51.94	52.00	52.06	52.12	52.18	52.24	52.30	52.36	52.42	52.48
63	52.53	52.59	52.65+	52.71	52.77	52.83	52.89	52.95+	53.01	53.07
64	53.13	53.19	53.25–	53.31	53.37	53.43	53.49	53.55–	5361	53.67
65	53.73	53.79	53.85–	53.91	53.97	54.03	54.09	54.15+	54.21	54.27
66	54.33	54.39	54.45+	54.51	54.57	54.63	54.70	54.76	54.82	54.88
67	54.94	55.00	55.06	55.12	55.18	55.24	55.30	55.37	55.43	55.49
68	55.55+	55.61	55.67	58.73	55.80	55.86	55.92	55.98	56.04	56.11
69	56.17	56.23	56.29	56.35+	56.42	56.48	56.54	56.60	56.66	56.73
70	56.79	56.85 +	66.91	56.98	57.04	57.10	57.17	57.23	57.29	57.35 +
71	57.42	57.48	57.54	57.61	57.67	57.73	57.80	57.86	57.92	57.99

72	58.05+	56.12	58.18	56.24	58.31	58.37	58.44	58.60	5856	58.63
%	0	1	2	3	4	5	6	7	8	9
73	58.69	58.76	58.82	58.89	58.95+	59.02	59.08	59.15−	59.21	59.28
74	59.34	59.41	59.47	59.54	59.60	59.67	59.74	59.80	59.87	59.93
75	60.00	60.07	60.13	60.20	60.27	60.33	60.40	60.47	60.53	60.60
76	60.67	60.73	60.80	60.87	60.94	61.00	61.07	61.14	61.21	61.27
77	61.34	61.41	61.48	61.55−	61.62	61.68	61.75+	61.82	61.89	61.96
78	62.03	62.10	62.17	62.24	62.31	62.37	62.44	62.51	62.58	62.65−
79	62.72	62.80	62.87	62.94	63.01	63.08	63.15	63.22	63.29	63.36
80	63.44	63.51	63.58	63.65+	63.72	63.79	63.87	63.94	64.01	64.08
81	64.16	64.23	64.30	64.38	64.45+	64.52	64.60	64.67	64.75−	64.82
82	64.90	64.97	65.05−	65.12	65.20	65.27	65.35−	65.42	65.50	65.57
83	65.65−	65.73	65.80	65.88	65.96	66.03	66.11	66.19	66.27	66.34
84	66.42	66.50	66.58	66.66	66.74	66.81	66.89	66.97	67.05 +	67.13
85	67.21	67.29	67.37	67.45+	67.54	67.62	67.70	67.78	67.86	67.94
86	68.03	68.11	68.19	68.28	68.36	68.44	68.53	68.61	68.70	68.78
87	68.87	68.95+	69.04	69.12	69.21	69.30	69.38	69.47	69.56	69.64
88	69.73	69.82	69.91	70.00	70.09	70.18	70.27	70.36	70.45−	70.54
89	70.63	70.72	70.81	70.91	71.00	71.09	71.19	71.23	71.37	71.47
90	71.56	71.66	71.76	71.85+	71.95+	72.05−	72.15−	72.24	72.34	72.44
91	72.54	72.64	72.74	72.84	72.95−	73.05−	73.15+	73.26	73.36	73.46
92	73.57	73.68	73.78	73.89	74.00	74.11	74.21	74.32	74.44	74.55−
93	74.66	74.77	74.88	75.00	75.11	75.23	75.35−	75.46	75.58	75.70
94	75.82	75.94	76.06	76.19	76.31	76.44	76.56	76.69	76.82	76.95−
95	77.08	77.21	77.34	77.48	77.61	77.75 +	77.89	78.03	78.17	78.32
96	78.46	78.61	78.76	78.91	79.06	79.22	79.37	79.53	79.69	79.86
97	80.02	80.19	80.37	80.54	80.72	80.90	81.09	81.28	81.47	81.67
98	81.87	82.08	82.29	82.51	82.73	82.96	83.20	83.45+	83.71	83.98
99.0	84.26	84.29	84.32	84.35−	84.38	84.41	84.44	84.47	84.50	84.53
99.1	84.56	84.59	84.62	84.65−	84.68	84.71	84.74	84.77	84.80	84.84
99.2	84.87	84.90	84.93	84.97	85.00	65.03	85.07	65.10	85.13	85.17
99.3	85.20	86.24	85.27	85.31	85.34	85.36	85.41	85.45−	85.48	85.52
99.4	85.56	85.60	85.63	85.67	85.71	85.75−	85.79	85.83	85.87	85.91
99.5	85.96−	85.99	86.03	88.07	86.11	86.15−	86.20	86.24	86.28	86.33
99.6	86.37	86.42	86.47	86.51	66.56	86.61	86.66	86.71	86.76	86.81
99.7	86.86	86.91	86.97	07.02	87.08	87.13	87.19	87.25+	87.31	87.37
99.8	87.44	87.50	87.57	87.64	87.71	87.78	87.86	87.93	88.01	88.10
99.9	88.19	88.28	88.38	88.48	88.60	88.72	88.85+	89.01	89.19	89.43
100.0	90.00									

Note: Transformation of binomial percentages in the margins to angles of equal information in degrees. The + or − signs following angles ending in 5 are for guidance in rounding to one decimal.

Source: Printed with the permission of C.I. Bliss of Biometric Society, pp. 448–449.

Appendix C

Selected Latin Squares

3×3	4×4			
	(1)	(2)	(3)	(4)

3×3			(1)				(2)				(3)				(4)			
A	B	C	A	B	C	D	A	B	C	D	A	B	C	D	A	B	C	D
B	C	A	B	A	D	C	B	C	D	A	B	D	A	C	B	A	D	C
C	A	B	C	D	B	A	C	D	A	B	C	A	D	B	D	C	B	A
			D	C	A	B	D	A	B	C	D	C	B	A	D	C	B	A

5×5					6×6						7×7						
A	B	C	D	E	A	B	C	D	E	F	A	B	C	D	E	F	G
B	A	E	C	D	B	F	D	C	A	E	B	C	D	E	F	G	A
C	D	A	E	B	C	D	E	F	B	A	C	D	E	F	G	A	B
D	E	B	A	C	D	A	F	E	C	B	D	E	F	G	A	B	C
E	C	D	B	A	E	C	A	B	F	D	E	F	G	A	B	C	D
					F	E	B	A	D	C	F	G	A	B	C	D	E
											G	A	B	C	D	E	F

8×8								9×9								
A	B	C	D	E	F	G	H	A	B	C	D	E	F	G	H	I
B	C	D	E	F	G	H	A	B	C	D	E	F	G	H	I	A
C	D	E	F	G	H	A	B	C	D	E	F	G	H	I	A	B
D	E	F	G	H	A	B	C	D	E	F	G	H	I	A	B	C
E	F	G	H	A	B	C	D	E	F	G	H	I	A	B	C	D
F	G	H	A	B	C	D	E	F	G	H	I	A	B	C	D	E
G	H	A	B	C	D	E	F	G	H	I	A	B	C	D	E	F
H	A	B	C	D	E	F	G	H	I	A	B	C	D	E	F	G
								I	A	B	C	D	E	F	G	H

10×10										11×11										
A	B	C	D	E	F	G	H	I	J	A	B	C	D	E	F	G	H	I	J	K
B	C	D	E	F	G	H	I	J	A	B	C	D	E	F	G	H	I	J	K	A
C	D	E	F	G	H	I	J	A	B	C	D	E	F	G	H	I	J	K	A	B
D	E	F	G	H	I	J	A	B	C	D	E	F	G	H	I	J	K	A	B	C
E	F	G	H	I	J	A	B	C	D	E	F	G	H	I	J	K	A	B	C	D
F	G	H	I	J	A	B	C	D	E	F	G	H	I	J	K	A	B	C	D	E
G	H	I	J	A	B	C	D	E	F	G	H	I	J	K	A	B	C	D	E	F
H	I	J	A	B	C	D	E	F	G	H	I	J	K	A	B	C	D	E	F	G
I	J	A	B	C	D	E	F	G	H	I	J	K	A	B	C	D	E	F	G	H
J	A	B	C	D	E	F	G	H	I	J	K	A	B	C	D	E	F	G	H	I
										K	A	B	C	D	E	F	G	H	I	J

$$12 \times 12$$

A	B	C	D	E	F	G	H	I	J	K	L
B	C	D	E	F	G	H	I	J	K	L	A
C	D	E	F	G	H	I	J	K	L	A	B
D	E	F	G	H	I	J	K	L	A	B	C
E	F	G	H	I	J	K	L	A	B	C	D
F	G	H	I	J	K	L	A	B	C	D	E
G	H	I	J	K	L	A	B	C	D	E	F
H	I	J	K	L	A	B	C	D	E	F	G
I	J	K	L	A	B	C	D	E	F	G	H
J	K	L	A	B	C	D	E	F	G	H	I
K	L	A	B	C	D	E	F	G	H	I	J
L	A	B	C	D	E	F	G	H	I	J	K

Greco-Latin Squares

$$3 \times 3 \qquad\qquad 4 \times 4 \qquad\qquad 5 \times 5$$

3×3

A_1	B_3	C_2
B_2	C_1	A_3
C_3	A_2	B_1

4×4

A_1	B_3	C_4	D_2
B_2	A_4	D_3	C_1
C_3	D_1	A_2	B_4
D_4	C_2	B_1	A_3

5×5

A_1	B_3	C_5	D_2	E_4
B_2	C_4	D_1	E_3	A_5
C_3	D_5	E_2	A_4	B_1
D_4	E_1	A_3	B_5	C_2
E_5	A_2	B_4	C_1	D_3

$$7 \times 7 \qquad\qquad\qquad 8 \times 8$$

7×7

A_1	B_5	C_2	D_6	E_3	F_7	G_4
B_2	C_6	D_3	E_7	F_4	G_1	A_5
C_3	D_7	E_4	F_1	G_5	A_2	B_6
D_4	E_1	F_5	G_2	A_6	B_3	C_7
E_5	F_2	G_6	A_3	B_7	C_4	D_1
F_6	G_3	A_7	B_4	C_1	D_5	E_2
G_7	A_4	B_1	C_5	D_2	E_6	F_3

8×8

A_1	B_5	C_2	D_3	E_7	F_4	G_8	H_6
B_2	A_8	G_1	F_7	H_3	D_6	C_5	E_4
C_3	G_4	A_7	E_1	D_2	H_5	B_6	F_8
D_4	F_3	E_6	A_5	C_8	B_1	H_7	G_2
E_5	H_1	D_8	C_4	A_6	G_3	F_2	B_7
F_6	D_7	H_4	B_8	G_5	A_2	E_3	C_1
G_7	C_6	B_3	H_2	F_1	E_8	A_4	D_5
H_8	E_2	F_5	G_6	B_4	C_7	D_1	A_3

$$9 \times 9$$

A_1	B_3	C_2	D_7	E_9	F_8	G_4	H_6	I_5
B_2	C_1	A_3	E_8	F_7	D_9	H_5	I_4	G_6
C_3	A_2	B_1	F_9	D_8	7_4	I_6	G_5	H_4
D_4	E_6	F_5	G_1	H_3	I_2	A_7	B_9	C_8
E_5	F_4	D_6	H_2	I_1	G_3	B_8	C_7	A_9
F_6	D_5	E_4	I_3	G_2	H_1	C_9	A_8	B_7
G_7	H_9	I_8	A_4	B_6	C_5	D_1	E_3	F_2
H_8	I_7	G_9	B_5	C_4	A_6	E_2	F_1	D_3
I_9	G_8	H_7	C_6	A_5	B_4	F_3	D_2	E_1

11×11

A_1	B_7	C_2	D_8	E_3	F_9	G_4	H_{10}	I_5	J_{11}	K_6
B_2	C_8	D_3	E_9	F_4	G_{10}	H_5	I_{11}	J_6	K_1	A_7
C_3	D_9	E_4	F_{10}	G_5	H_{11}	I_6	J_1	K_7	A_2	B_8
D_4	E_{10}	F_5	G_{11}	H_6	I_1	J_7	K_2	A_8	B_3	C_9
E_5	F_{11}	G_6	H_1	I_7	J_2	K_8	A_3	B_9	C_4	D_{10}
F_6	G_1	H_7	I_2	J_8	K_3	A_9	B_4	C_{10}	D_5	E_{11}
G_7	H_2	I_8	J_3	K_9	A_4	B_{10}	C_5	D_{11}	E_6	F_1
H_8	I_3	J_9	K_4	A_{10}	B_5	C_{11}	D_6	E_1	F_7	G_2
I_9	J_4	K_{10}	A_6	B_{11}	C_6	D_1	E_7	F_2	G_8	H_3
J_{10}	K_5	A_{11}	B_6	C_1	D_7	E_2	F_8	G_3	H_9	I_4
K_{11}	A_6	B_1	C_7	D_2	E_8	F_3	G_9	H_4	I_{10}	J_5

12×12

A_1	B_{12}	C_6	D_7	I_5	J_4	K_{10}	L_{11}	E_9	F_8	G_2	H_3
B_2	A_{11}	D_5	C_8	J_6	I_3	L_9	K_{12}	F_{10}	E_7	H_1	G_4
C_3	D_{10}	A_8	B_5	K_7	L_2	I_{12}	J_9	G_{11}	H_6	E_4	F_1
D_4	C_9	B_7	A_6	L_8	K_1	J_{11}	I_{10}	H_{12}	G_5	F_3	E_2
E_5	F_4	G_{10}	H_{11}	A_9	B_8	C_2	D_3	I_1	J_{12}	K_6	L_7
F_6	E_3	H_9	G_{12}	B_{10}	A_7	D_1	C_4	J_2	I_{11}	L_5	K_8
G_7	H_2	E_{12}	F_9	C_{11}	D_6	A_4	B_1	K_3	L_{10}	I_8	J_5
H_8	G_1	F_{11}	E_{10}	D_{12}	C_5	B_3	A_2	L_4	K_9	J_7	I_6
I_9	J_8	K_2	L_3	E_1	F_{12}	G_6	H_7	A_5	B_4	C_{10}	D_{11}
J_{10}	I_7	L_1	K_4	F_2	E_{11}	H_5	G_8	B_6	A_3	D_9	C_{12}
K_{11}	L_6	I_4	J_1	G_3	H_{10}	E_8	F_5	C_7	D_2	A_{12}	B_9
L_{12}	K_5	J_2	I_2	H_4	G_9	F_7	E_6	D_8	C_1	B_{11}	A_{10}

Source: From Cochran, William G. and Cox, Gertrude M., *Experimental Designs*, 2nd ed., John Wiley and Sons, 1957. With permission.

Appendix D

Random Digits

85967	73152	14511	85285	36009	95892	36962	67835	63314	50162
07483	51453	11649	86348	76431	81594	95848	36738	25014	15460
96283	01898	61414	83525	04231	13604	75339	11730	85423	60698
49174	12074	98551	37895	93547	24769	09404	76548	05393	96770
97366	39941	21225	93629	19574	71565	33413	56087	40875	13351
90474	41469	16812	81542	81652	45554	27931	93994	22375	00953
28599	64109	09497	76235	41383	31555	12639	00619	22909	29563
25254	16210	89717	65997	82667	74624	36348	44018	64732	93689
28785	02760	24359	99410	77319	73408	58993	61098	04393	48245
84725	86576	86944	93296	10081	82454	76810	52975	10324	15457
41059	66456	47679	66810	15941	84602	14493	65515	19251	41642
67434	41045	82830	47617	36932	46728	71183	36345	41404	81110
72766	68816	37643	19959	57550	49620	98480	25640	67257	18671
92079	46784	66125	94932	64451	29275	57669	66658	30818	58353
29187	40350	62533	73603	34075	16451	42885	03448	37390	96328
74220	17612	65522	80607	19184	64164	66962	82310	18163	63495
03786	02407	06098	92917	40434	60602	82175	04470	78754	90775
75085	55558	15520	27038	25471	76107	90832	10819	56797	33751
09161	33015	19155	11715	00551	24909	31894	37774	37953	78837
75707	43992	64998	87080	39333	00767	45637	12538	67439	94914
21333	48660	31288	00086	79889	75532	28704	62844	92337	99695
65626	50061	42539	14812	48895	11196	34335	60492	70650	51108
84380	07389	87891	76255	89604	41372	10837	66992	93183	56920
46479	32072	80083	63868	70930	89654	05359	47196	12452	38234
59847	97197	55147	76639	76971	55928	36441	95141	42333	67483
31416	11231	27904	57383	31852	69137	96667	14315	01007	31929
82066	83436	67914	21465	99605	83114	97885	74440	99622	87912
01850	42782	39202	18582	46214	99223	79541	78298	75404	63648
32315	89276	89582	87138	16165	15984	21466	63830	30475	74729

59388	42703	55198	80380	67067	97155	34160	85019	03527	78140
58089	27632	50987	91373	07736	20436	96130	73483	85332	24384
61705	57285	30392	23660	75841	21931	04295	00875	09114	32101
18914	98982	60199	99275	41967	35208	30357	76772	92656	62318
11965	94089	34803	48941	69709	16784	44642	89761	66864	62803
85251	48111	80936	81781	93248	67877	16498	31924	51315	79921
66121	96986	84844	93873	46352	92183	51152	85878	30490	15974
53972	96642	24199	58080	35450	03482	66953	49521	63719	57515
14509	16594	78883	43222	23093	58645	60257	89250	63266	90858
37700	07688	65533	72126	23611	93993	01848	03910	38552	17472
85466	59392	72722	15473	73295	49759	56157	60477	83284	56367
52969	55863	42312	67842	05673	91878	82738	36563	79540	61935
42744	68315	17514	02878	97291	74851	42725	57894	81434	62041
26140	13336	67726	61876	29971	99294	96664	52817	90039	53211
95589	56319	14563	24071	06916	59555	18195	32280	79357	04224
39113	13217	59999	49952	83021	47709	53105	19295	88318	41626
41332	17622	18994	98283	07249	52289	24209	91139	30715	06604
54684	53645	79246	70183	87731	19185	08541	33519	07223	97413
89442	61001	36658	67444	95388	36682	38052	46719	09428	94012
36751	16778	54888	15357	68003	43564	90976	58904	40512	07725
98159	02564	21416	74944	53049	88749	02865	25772	89853	88714

Source: Computer generated by the author.

Appendix E

Points for the Distribution of F

n_2	1	2	3	4	5	6	7	8	9	10	11	12	14	16	20	24	30	40	50	75	100	200	500	∞
1	161	200	216	225	230	234	237	239	241	242	243	244	245	246	248	249	250	251	252	253	253	254	254	254
	4,052	4,999	5,403	5,635	5,764	5,859	5,928	5,981	6,022	6,056	6,082	6,106	6,142	6,169	6,208	6,234	6,261	6,286	6,302	6,323	6,334	6,352	6,361	6,366
2	18.51	19.00	19.16	19.25	19.30	19.33	19.36	19.37	19.38	19.39	19.40	19.41	19.42	19.43	19.44	19.45	19.46	19.47	19.47	19.48	19.49	19.49	19.50	19.50
	98.49	99.00	99.17	99.25	99.30	99.33	99.36	99.37	99.39	99.40	99.41	99.42	99.43	99.44	99.45	99.46	99.47	99.48	99.48	99.49	99.49	99.49	99.50	99.50
3	10.13	9.55	9.28	9.12	9.01	8.94	8.88	8.84	8.81	8.78	8.76	8.74	8.71	8.69	8.66	8.64	8.62	8.60	8.58	8.57	8.56	8.54	8.54	8.53
	34.12	30.82	29.46	28.71	28.24	27.91	27.67	27.49	27.34	27.23	27.13	27.05	26.92	26.83	26.69	26.60	26.50	26.41	26.35	26.27	26.23	26.18	26.14	26.12
4	7.71	6.94	6.59	6.39	6.26	6.16	6.09	6.04	6.00	5.96	5.93	5.91	5.87	5.84	5.80	5.77	5.74	5.71	5.70	5.68	5.66	5.65	5.64	5.63
	21.20	18.00	16.69	15.98	15.52	15.21	14.98	14.80	14.66	14.54	14.45	14.37	14.24	14.15	14.02	13.93	13.83	13.74	13.69	13.61	13.57	13.52	13.48	13.46
5	6.61	5.79	5.41	5.19	5.05	4.95	4.88	4.82	4.78	4.74	4.70	4.68	4.64	4.60	4.56	4.53	4.50	4.46	4.44	4.42	4.40	4.38	4.37	4.36
	16.26	13.29	12.06	11.39	10.97	10.67	10.45	10.29	10.15	10.05	9.96	9.89	9.77	9.68	9.55	9.47	9.38	9.29	9.24	9.24	9.13	9.07	9.04	9.02
6	5.99	5.14	4.76	4.53	4.39	4.28	4.21	4.15	4.10	4.06	4.03	4.00	3.96	3.92	3.87	3.84	3.81	3.77	3.75	3.72	3.71	3.69	3.68	3.67
	13.74	10.92	9.78	9.15	8.75	8.47	8.26	8.10	7.98	7.87	7.79	7.72	7.60	7.52	7.39	7.31	7.23	7.14	7.09	7.02	6.99	6.94	6.90	6.88
7	5.59	4.74	4.35	4.12	3.97	3.87	3.79	3.73	3.68	3.63	3.60	3.57	3.52	3.49	3.44	3.41	3.38	3.34	3.32	3.29	3.28	3.25	3.24	3.22
	12.25	9.55	8.45	7.85	7.46	7.19	7.00	6.84	6.71	6.62	6.54	6.47	6.36	6.27	6.15	6.07	5.98	5.90	5.85	5.78	5.75	5.70	5.67	5.65
8	5.32	4.46	4.07	3.84	3.69	3.58	3.50	3.44	3.39	3.34	3.31	3.28	3.23	3.20	3.15	3.12	3.08	3.05	3.03	3.00	2.98	2.96	2.94	2.93
	11.26	8.65	7.59	7.01	6.63	6.37	6.19	6.03	5.91	5.82	5.74	5.67	5.56	5.48	5.36	5.28	5.20	5.11	5.06	5.00	4.96	4.91	4.88	4.86
9	5.12	4.26	3.86	3.63	3.48	3.37	3.29	3.23	3.18	3.13	3.10	3.07	3.02	2.98	2.93	2.90	2.86	2.82	2.80	2.77	2.76	2.73	2.72	2.71
	10.56	8.02	6.99	6.42	6.06	5.80	5.62	5.47	5.35	5.26	5.18	5.11	5.00	4.92	4.80	4.73	4.56	4.56	4.51	4.45	4.41	4.36	4.33	4.31

10	4.96	4.10	3.71	3.48	3.33	3.22	3.14	3.07	3.02	2.97	2.94	2.91	2.86	2.82	2.77	2.74	2.70	2.67	2.64	2.61	2.59	2.56	2.55	2.54	10
	10.04	**7.56**	**6.55**	**5.99**	**5.64**	**5.39**	**5.21**	**5.06**	**4.95**	**4.85**	**4.78**	**4.71**	**4.60**	**4.52**	**4.41**	**4.33**	**4.25**	**4.17**	**4.12**	**4.05**	**4.01**	**3.96**	**3.93**	**3.91**	
11	4.84	3.98	3.59	3.36	3.20	3.09	3.01	2.95	2.90	2.86	2.82	2.79	2.74	2.70	2.65	2.61	2.57	2.53	2.50	2.47	2.45	2.42	2.41	2.40	11
	9.65	**7.20**	**6.22**	**5.67**	**5.32**	**5.07**	**4.88**	**4.74**	**4.63**	**4.54**	**4.46**	**4.40**	**4.29**	**4.21**	**4.10**	**4.02**	**3.94**	**3.86**	**3.80**	**3.74**	**3.70**	**3.66**	**3.62**	**3.60**	
12	4.75	3.88	3.49	3.26	3.11	3.00	2.92	2.85	2.80	2.76	2.72	2.69	2.64	2.60	2.54	2.50	2.46	2.42	2.40	2.36	2.35	2.32	2.31	2.30	12
	9.33	**6.93**	**5.95**	**5.41**	**5.06**	**4.82**	**4.65**	**4.50**	**4.39**	**4.30**	**4.22**	**4.16**	**4.05**	**3.98**	**3.86**	**3.78**	**3.70**	**3.61**	**3.56**	**3.49**	**3.46**	**3.41**	**3.38**	**3.36**	
13	4.67	3.80	3.41	3.18	3.02	2.92	2.84	2.77	2.72	2.67	2.63	2.60	2.55	2.51	2.46	2.42	2.38	2.34	2.32	2.28	2.26	2.24	2.22	2.21	13
	9.07	**6.70**	**5.74**	**5.20**	**4.86**	**4.62**	**4.44**	**4.30**	**4.19**	**4.10**	**4.02**	**3.96**	**3.85**	**3.78**	**3.67**	**3.59**	**3.51**	**3.42**	**3.37**	**3.30**	**3.27**	**3.21**	**3.18**	**3.16**	
14	4.60	3.74	3.34	3.11	2.96	2.85	2.77	2.70	2.65	2.60	2.56	2.53	2.48	2.44	2.39	2.35	2.31	2.27	2.24	2.21	2.19	2.16	2.14	2.13	14
	8.86	**6.51**	**5.56**	**5.03**	**4.69**	**4.46**	**4.28**	**4.14**	**4.03**	**3.94**	**3.86**	**3.80**	**3.70**	**3.62**	**3.51**	**3.43**	**3.34**	**3.26**	**3.21**	**3.14**	**3.11**	**3.06**	**3.02**	**3.00**	
15	4.54	3.68	3.29	3.06	2.90	2.79	2.70	2.64	2.59	2.55	2.51	2.48	2.43	2.39	2.33	2.29	2.25	2.21	2.18	2.15	2.12	2.10	2.08	2.07	15
	8.68	**6.36**	**5.42**	**4.89**	**4.56**	**4.32**	**4.14**	**4.00**	**3.89**	**3.80**	**3.73**	**3.67**	**3.56**	**3.48**	**3.36**	**3.29**	**3.20**	**3.12**	**3.07**	**3.00**	**2.97**	**2.92**	**2.89**	**2.87**	
16	4.49	3.63	3.24	3.01	2.85	2.74	2.66	2.59	2.54	2.49	2.45	2.42	2.37	2.33	2.28	2.24	2.20	2.16	2.13	2.09	2.07	2.04	2.02	2.01	16
	8.53	**6.23**	**5.29**	**4.77**	**4.44**	**4.20**	**4.03**	3.89	**3.78**	**3.69**	**3.61**	**3.55**	**3.45**	**3.37**	**3.25**	**3.18**	**3.10**	**3.01**	**2.96**	**2.89**	**2.86**	**2.80**	**2.77**	**2.75**	
17	4.45	3.59	3.20	2.96	2.81	2.70	2.62	2.55	2.50	2.45	2.41	2.38	2.33	2.29	2.23	2.19	2.15	2.11	2.08	2.04	2.02	1.99	1.97	1.96	17
	8.40	**6.11**	**5.18**	**4.67**	**4.34**	**4.10**	**3.93**	**3.79**	**3.68**	**3.59**	**3.52**	**3.45**	**3.35**	**3.27**	**3.16**	**3.08**	**3.00**	**2.92**	**2.86**	**2.79**	**2.76**	**2.70**	**2.67**	**2.65**	
18	4.41	3.55	3.16	2.93	2.77	2.66	2.58	2.51	2.46	2.41	2.37	2.34	2.29	2.25	2.19	2.15	2.11	2.07	2.04	2.00	1.98	1.95	1.93	1.92	18
	8.28	**6.01**	**5.09**	**4.58**	**4.25**	**4.01**	**3.85**	**3.71**	**3.60**	**3.51**	**3.44**	**3.37**	**3.27**	**3.19**	**3.07**	**3.00**	**2.91**	**2.83**	**2.78**	**2.71**	**2.68**	**2.62**	**2.59**	**2.57**	
19	4.38	3.52	3.13	2.90	2.74	2.63	2.55	2.48	2.43	2.38	2.34	2.31	2.26	2.21	2.15	2.11	2.07	2.02	2.00	1.96	1.94	1.91	1.90	1.88	19
	8.18	**5.93**	**5.01**	**4.50**	**4.17**	**3.94**	**3.77**	**3.63**	**3.52**	**3.43**	**3.36**	**3.30**	**3.19**	**3.12**	**3.00**	**2.92**	**2.84**	**2.76**	**2.70**	**2.63**	**2.60**	**2.54**	**2.51**	**2.49**	

Points for the Distribution of F (continued)

n_1

n_2	1	2	3	4	5	6	7	8	9	10	11	12	14	16	20	24	30	40	50	75	100	200	500	∞
20	4.35	3.49	3.10	2.87	2.71	2.60	2.52	2.45	2.40	2.35	2.31	2.28	2.23	2.18	2.12	2.08	2.04	1.99	1.96	1.92	1.90	1.87	1.85	1.84
	8.10	5.85	4.94	4.43	4.10	3.87	3.71	3.56	3.45	3.37	3.30	3.23	3.13	3.05	2.94	2.86	2.77	2.69	2.63	2.56	.53	2.47	2.44	2.42
21	4.32	3.47	3.07	2.84	2.68	2.57	2.49	2.42	2.37	2.32	2.28	2.25	2.20	2.15	2.09	2.05	2.00	1.96	1.93	1.89	1.87	1.84	1.82	1.81
	8.02	5.78	4.87	4.37	4.04	3.81	3.65	3.51	3.40	3.31	3.24	3.17	3.07	2.99	2.88	2.80	2.72	2.63	2.58	2.51	2.47	2.42	2.38	2.36
22	4.30	3.44	3.05	2.82	2.66	2.55	2.47	2.40	2.35	2.30	2.26	2.23	2.18	2.13	2.07	2.03	1.98	1.93	1.91	1.87	1.84	1.81	1.80	1.78
	7.94	5.72	4.82	4.31	3.99	3.76	3.59	3.45	3.35	3.26	3.18	3.12	3.02	2.94	2.83	2.75	2.67	2.58	2.53	2.46	2.42	2.37	2.33	2.31
23	4.28	3.42	3.03	2.80	2.64	2.53	2.45	2.38	2.32	2.28	2.24	2.20	2.14	2.10	2.04	2.00	1.96	1.91	1.88	1.84	1.82	1.79	1.77	1.76
	7.88	5.66	4.76	4.26	3.94	3.71	3.54	3.41	3.30	3.21	3.14	3.07	2.97	2.89	2.78	2.70	2.62	2.53	2.48	2.41	2.37	2.32	2.28	2.26
24	4.26	3.40	3.01	2.78	2.62	2.51	2.43	2.36	2.30	2.26	2.22	2.18	2.13	2.09	2.02	1.98	1.94	1.89	1.86	1.82	1.80	1.76	1.74	1.73
	7.82	5.61	4.72	4.22	3.90	3.67	3.50	3.36	3.25	3.17	3.09	3.03	2.93	2.85	2.74	2.66	2.58	2.49	2.44	2.36	2.33	2.27	2.23	2.21
25	4.24	3.38	2.99	2.76	2.60	2.49	2.41	2.34	2.28	2.24	2.20	2.16	2.11	2.06	2.00	1.96	1.92	1.87	1.84	1.80	1.77	1.74	1.72	1.71
	7.77	5.57	4.68	4.18	3.86	3.63	3.46	3.32	3.21	3.13	3.05	2.99	2.89	2.81	2.70	2.62	2.54	2.45	2.40	2.32	2.29	2.23	2.19	2.17
26	4.22	3.37	2.98	2.74	2.59	2.47	2.39	2.32	2.27	2.22	2.18	2.15	2.10	2.05	1.99	1.95	1.90	1.85	1.82	1.78	1.76	1.72	1.70	1.69
	7.72	5.53	4.64	4.14	3.82	3.59	3.42	3.29	3.17	3.09	3.02	2.96	2.86	2.77	2.66	2.58	2.50	2.41	2.36	2.28	2.25	2.19	2.15	2.13
27	4.21	3.35	2.96	2.73	2.57	2.46	2.37	2.30	2.25	2.20	2.16	2.13	2.08	2.03	1.97	1.93	1.88	1.84	1.80	1.76	1.74	1.71	1.68	1.67
	7.68	5.49	4.60	4.11	3.79	3.56	3.39	3.26	3.14	3.06	2.98	2.93	2.83	2.74	2.63	2.55	2.47	2.38	2.33	2.25	2.21	2.16	2.12	2.10
28	4.20	3.34	2.95	2.71	2.56	2.44	2.36	2.29	2.24	2.19	2.15	2.12	2.06	2.02	1.96	1.91	1.87	1.81	1.78	1.75	1.72	1.69	1.67	1.65
	7.64	5.45	4.57	4.07	3.76	3.53	3.36	3.23	3.11	3.03	2.95	2.90	2.80	2.71	2.60	2.52	2.44	2.35	2.30	2.22	2.18	2.13	2.09	2.06

Critical values (upper entries = .05 level, **lower bold entries = .01 level**); row labels are denominator degrees of freedom.

df																									
29	1.64	1.65	1.68	1.71	1.73	1.77	1.80	1.85	1.90	1.94	2.00	2.05	2.10	2.14	2.18	2.22	2.28	2.35	2.43	2.54	2.70	2.93	3.33	4.18
29	**2.03**	**2.06**	**2.10**	**2.15**	**2.19**	**2.27**	**2.32**	**2.41**	**2.49**	**2.57**	**2.68**	**2.77**	**2.87**	**2.92**	**3.00**	**3.08**	**3.20**	**3.33**	**3.50**	**3.73**	**4.04**	**4.54**	**5.42**	**7.60**
30	1.62	1.64	1.66	1.69	1.72	1.76	1.79	1.84	1.89	1.93	1.99	2.04	2.09	2.12	2.16	2.21	2.27	2.34	2.42	2.53	2.69	2.92	3.32	4.17
30	**2.01**	**2.03**	**2.07**	**2.13**	**2.16**	**2.24**	**2.29**	**2.38**	**2.47**	**2.55**	**2.66**	**2.74**	**2.84**	**2.90**	**2.98**	**3.06**	**3.17**	**3.30**	**3.47**	**3.70**	**4.02**	**4.51**	**5.39**	**7.56**
32	1.59	1.61	1.64	1.67	1.69	1.74	1.76	1.82	1.86	1.91	1.97	2.02	2.07	2.10	2.14	2.19	2.25	2.32	2.40	2.51	2.67	2.90	3.30	4.15
32	**1.96**	**1.98**	**2.02**	**2.08**	**2.12**	**2.20**	**2.25**	**2.34**	**2.42**	**2.51**	**2.62**	**2.70**	**2.80**	**2.86**	**2.94**	**3.01**	**3.12**	**3.25**	**3.42**	**3.66**	**3.97**	**4.46**	**5.34**	**7.50**
34	1.57	1.59	1.61	1.64	1.67	1.71	1.74	1.80	1.84	1.89	1.95	2.00	2.05	2.08	2.12	2.17	2.23	2.30	2.38	2.49	2.65	2.88	3.28	4.13
34	**1.91**	**1.94**	**1.98**	**2.04**	**2.08**	**2.15**	**2.21**	**2.30**	**2.38**	**2.47**	**2.58**	**2.66**	**2.76**	**2.82**	**2.89**	**2.97**	**3.08**	**3.21**	**3.38**	**3.61**	**3.93**	**4.42**	**5.29**	**7.44**
36	1.55	1.56	1.59	1.62	1.65	1.69	1.72	1.78	1.82	1.87	1.93	1.98	2.03	2.06	2.10	2.15	2.21	2.28	2.36	2.48	2.63	2.86	3.26	4.11
36	**1.87**	**1.90**	**1.94**	**2.00**	**2.04**	**2.12**	**2.17**	**2.26**	**2.35**	**2.43**	**2.54**	**2.62**	**2.72**	**2.78**	**2.86**	**2.94**	**3.04**	**3.18**	**3.35**	**3.58**	**3.89**	**4.38**	**5.25**	**7.39**
38	1.53	1.54	1.57	1.60	1.63	1.67	1.71	1.76	1.80	1.85	1.92	1.96	2.02	2.05	2.09	2.14	2.19	2.26	2.35	2.46	2.62	2.85	3.25	4.10
38	**1.84**	**1.86**	**1.90**	**1.97**	**2.00**	**2.08**	**2.14**	**2.22**	**2.32**	**2.40**	**2.51**	**2.59**	**2.69**	**2.75**	**2.82**	**2.91**	**3.02**	**3.15**	**3.32**	**3.54**	**3.86**	**4.34**	**5.21**	**7.35**
40	1.51	1.53	1.55	1.59	1.61	1.66	1.69	1.74	1.79	1.84	1.90	1.95	2.00	2.04	2.07	2.12	2.18	2.25	2.34	2.45	2.61	2.84	3.23	4.08
40	**1.81**	**1.84**	**1.88**	**1.94**	**1.97**	**2.05**	**2.11**	**2.20**	**2.29**	**2.37**	**2.49**	**2.56**	**2.66**	**2.73**	**2.80**	**2.88**	**2.99**	**3.12**	**3.29**	**3.51**	**3.83**	**4.31**	**5.18**	**7.31**
42	1.49	1.51	1.54	1.57	1.60	1.64	1.68	1.73	1.78	1.82	1.89	1.94	1.99	2.02	2.06	2.11	2.17.	2.24	2.32	2.44	2.59	2.83	3.22	4.07
42	**1.78**	**1.80**	**1.85**	**.91**	**1.94**	**2.02**	**2.08**	**2.17**	**2.26**	**2.35**	**2.46**	**2.54**	**2.64**	**2.70**	**2.77**	**2.86**	**2.96**	**3.10**	**3.26**	**3.49**	**3.80**	**4.29**	**5.15**	**7.27**
44	1.48	1.50	1.52	1.56	1.58	1.63	1.66	1.72	1.76	1.81	1.88	1.92	1.98	2.01	2.05	2.10	2.16	2.23	2.31	2.43	2.58	2.82	3.21	4.06
44	**1.75**	**1.78**	**1.82**	**1.88**	**1.92**	**2.00**	**2.06**	**2.15**	**2.24**	**2.32**	**2.44**	**2.52**	**2.62**	**2.68**	**2.75**	**2.84**	**2.94**	**3.07**	**3.24**	**3.46**	**3.78**	**4.26**	**5.12**	**7.24**
46	1.46	1.48	1.51	1.54	1.57	1.62	1.65	1.71	1.75	1.80	1.87	1.91	1.97	2.00	2.04	2.09	2.14	2.22	2.30	2.42	2.57	2.81	3.20	4.05
46	**1.72**	**1.76**	**1.80**	**1.86**	**1.90**	**1.98**	**2.04**	**2.13**	**2.22**	**2.30**	**2.42**	**2.50**	**2.60**	**2.66**	**2.73**	**2.82**	**2.92**	**3.05**	**3.22**	**3.44**	**3.76**	**4.24**	**5.10**	**7.21**

Points for the Distribution of F (continued)

n_1

n_2	1	2	3	4	5	6	7	8	9	10	11	12	14	16	20	24	30	40	50	75	100	200	500	∞	n_2
48	4.04	3.19	2.80	2.56	2.41	2.30	2.21	2.14	2.08	2.03	1.99	1.96	1.90	1.86	1.79	1.74	1.70	1.64	1.61	1.56	1.53	1.50	1.47	1.45	48
	7.19	**5.08**	**4.22**	**3.74**	**3.42**	**3.20**	**3.04**	**2.90**	**2.80**	**2.71**	**2.64**	**2.58**	**2.48**	**2.40**	**2.28**	**2.20**	**2.11**	**2.02**	**1.96**	**1.88**	**1.84**	**1.78**	**1.73**	**1.70**	
50	4.03	3.18	2.79	2.56	2.40	2.29	2.20	2.13	2.07	2.02	1.98	1.95	1.90	1.85	1.78	1.74	1.69	1.63	1.60	1.55	1.52	1.48	1.46	1.44	50
	7.17	**5.06**	**4.20**	**3.72**	**3.41**	**3.18**	**3.02**	**2.88**	**2.78**	**2.70**	**2.62**	**2.56**	**2.46**	**2.39**	**2.26**	**2.18**	**2.10**	**2.00**	**1.94**	**1.86**	**1.82**	**1.76**	**1.71**	**1.68**	
55	4.02	3.17	2.78	2.54	2.38	2.27	2.18	2.11	2.05	2.00	1.97	1.93	1.88	1.83	1.76	1.72	1.67	1.61	1.58	1.52	1.50	1.46	1.43	1.41	55
	7.12	**5.01**	**4.16**	**3.68**	**3.37**	**3.15**	**2.98**	**2.85**	**2.75**	**2.66**	**2.59**	**2.53**	**2.43**	**2.35**	**2.23**	**2.15**	**2.06**	**1.96**	**1.90**	**1.82**	**1.78**	**1.71**	**1.66**	**1.64**	
60	4.00	3.15	2.76	2.52	2.37	2.25	2.17	2.10	2.04	1.99	1.95	1.92	1.86	1.81	1.75	1.70	1.65	1.59	1.56	1.50	1.48	1.44	1.41	1.39	60
	7.08	**4.98**	**4.13**	**3.65**	**3.34**	**3.12**	**2.95**	**2.82**	**2.72**	**2.63**	**2.56**	**2.50**	**2.40**	**2.32**	**2.20**	**2.12**	**2.03**	**1.93**	**1.87**	**1.79**	**1.74**	**1.68**	**1.63**	**1.60**	
65	3.99	3.14	2.75	2.51	2.36	2.24	2.15	2.08	2.02	1.98	1.94	1.90	1.85	1.80	1.73	1.68	1.63	1.57	1.54	1.49	1.46	1.42	1.39	1.37	65
	7.04	**4.95**	**4.10**	**3.62**	**3.31**	**3.09**	**2.93**	**2.79**	**2.70**	**2.61**	**2.54**	**2.47**	**2.37**	**2.30**	**2.18**	**2.09**	**2.00**	**1.90**	**1.84**	**1.76**	**1.71**	**1.64**	**1.60**	**1.56**	
70	3.98	3.13	2.74	2.50	2.35	2.23	2.14	2.07	2.01	1.97	1.93	1.89	1.84	1.79	1.72	1.67	1.62	1.56	1.53	1.47	1.45	1.40	1.37	1.35	70
	7.01	**4.92**	**4.08**	**3.60**	**3.29**	**3.07**	**2.91**	**2.77**	**2.67**	**2.59**	**2.51**	**2.45**	**2.35**	**2.28**	**2.15**	**2.07**	**1.98**	**1.88**	**1.82**	**1.74**	**1.69**	**1.62**	**1.56**	**1.53**	
80	3.96	3.11	2.72	2.48	2.33	2.21	2.12	2.05	1.99	1.95	1.91	1.88	1.82	1.77	1.70	1.65	1.60	1.54	1.51	1.45	1.42	1.38	1.35	1.32	80
	6.96	**4.88**	**4.04**	**3.56**	**3.25**	**3.04**	**2.87**	**2.74**	**2.64**	**2.55**	**2.48**	**2.41**	**2.32**	**2.24**	**2.11**	**2.03**	**1.94**	**1.84**	**1.78**	**1.70**	**1.65**	**1.57**	**1.52**	**1.49**	
100	3.94	3.09	2.70	2.46	2.30	2.19	2.10	2.03	1.97	1.92	1.88	1.85	1.79	1.75	1.68	1.63	1.57	1.51	1.48	1.42	1.39	1.34	1.30	1.28	100
	6.90	**4.82**	**3.98**	**3.51**	**3.20**	**2.99**	**2.82**	**2.69**	**2.59**	**2.51**	**2.43**	**2.36**	**2.26**	**2.19**	**2.06**	**1.98**	**1.89**	**1.79**	**1.73**	**1.64**	**1.59**	**1.51**	**1.46**	**1.43**	
125	3.92	3.07	2.68	2.44	2.29	2.17	2.08	2.01	1.95	1.90	1.86	1.83	1.77	1.72	1.65	1.60	1.55	1.49	1.45	1.39	1.36	1.31	1.27	1.25	125
	6.84	**4.78**	**3.94**	**3.47**	**3.17**	**2.95**	**2.79**	**2.65**	**2.56**	**2.47**	**2.40**	**2.33**	**2.23**	**2.15**	**2.03**	**1.94**	**1.85**	**1.75**	**1.68**	**1.59**	**1.54**	**1.46**	**1.40**	**1.37**	

n₂																								
150	3.91	3.06	2.67	2.43	2.27	2.16	2.07	2.00	1.94	1.89	1.85	1.82	1.76	1.71	1.64	1.59	1.54	1.47	1.44	1.37	1.34	1.29	1.25	1.22
	6.81	**4.75**	**3.91**	**3.44**	**3.14**	**2.92**	**2.76**	**2.62**	**2.53**	**2.44**	**2.37**	**2.30**	**2.20**	**2.12**	**2.00**	**1.91**	**1.83**	**1.72**	**1.66**	**1.56**	**1.51**	**1.43**	**1.37**	**1.33**
200	3.89	3.04	2.65	2.41	2.26	2.14	2.05	1.98	1.92	1.87	1.83	1.80	1.74	1.69	1.62	1.57	1.52	1.45	1.42	1.35	1.32	1.26	1.22	1.19
	6.76	**4.71**	**3.88**	**3.41**	**3.11**	**2.90**	**2.73**	**2.60**	**2.50**	**2.41**	**2.34**	**2.28**	**2.17**	**2.09**	**1.97**	**1.88**	**1.79**	**1.69**	**1.62**	**1.53**	**1.48**	**1.39**	**1.33**	**1.28**
400	3.86	3.02	2.62	2.39	2.23	2.12	2.03	1.96	1.90	1.85	1.81	1.78	1.72	1.67	1.60	1.54	1.49	1.42	1.38	1.32	1.28	1.22	1.16	1.13
	6.70	**4.66**	**3.83**	**3.36**	**3.06**	**2.85**	**2.69**	**2.55**	**2.46**	**2.37**	**2.29**	**2.23**	**2.12**	**2.04**	**1.92**	**1.84**	**1.74**	**1.64**	**1.57**	**.47**	**1.42**	**1.32**	**1.24**	**1.19**
1000	3.85	3.00	2.61	2.38	2.22	2.10	2.02	1.95	1.89	1.84	1.80	1.76	1.70	1.65	1.58	1.53	1.47	1.4]	1.36	1.30	1.26	1.19	1.13	1.08
	6.66	**4.62**	**3.80**	**3.34**	**3.04**	**2.82**	**2.66**	**2.53**	**2.43**	**2.34**	**2.26**	**2.20**	**2.09**	**2.01**	**1.89**	**1.81**	**1.71**	**1.61**	**1.54**	**1.44**	**1.38**	**1.28**	**1.19**	**1.11**
∞	3.84	2.99	2.60	2.37	2.21	2.09	2.01	1.94	1.88	1.83	1.79	1.75	1.69	1.64	1.57	1.52	1.46	1.40	1.35	1.28	1.24	1.17	1.11	1.00
	6.64	**4.60**	**3.78**	**3.32**	**3.02**	**2.80**	**2.64**	**2.51**	**2.41**	**2.32**	**2.24**	**2.18**	**2.07**	**1.99**	**1.87**	**1.79**	**1.69**	**1.59**	**1.52**	**1.41**	**1.36**	**1.25**	**1.15**	**1.00**

Note: Light type = 5%; bold type = 1%; n_1 = degrees of freedom (for greater mean square).

Source: From Snedecor, George W. and Cochran, William G., *Statistical Methods*, 6th ed., Iowa State University Press, Ames, 1967. With permission.

Appendix F

Basic Plans for Balanced and Partially Balanced Lattice Designs

Plans

Plan 1								3 × 3 Balanced Lattice							

$t = 9, k = 3, r = 4, b = 12, \lambda = 1$[a]

Block	Rep. I				Rep. II				Rep. III				Rep. IV		
(1)	1	2	3	(4)	1	4	7	(7)	1	5	9	(10)	1	8	6
(2)	4	5	6	(5)	2	5	8	(8)	7	2	6	(11)	4	2	9
(3)	7	8	9	(6)	3	6	9	(9)	4	8	3	(12)	7	5	3

Plan 2							4 × 4 Balanced Lattice							

$t = 16, k = 4, r = 5, b = 20, \lambda = 1$

Block	Rep. I					Rep. II					Rep. III			
(1)	1	2	3	4	(5)	1	5	9	13	(7)	1	6	11	16
(2)	5	6	7	8	(6)	2	6	10	14	(8)	5	2	15	12
(3)	9	10	11	12	(7)	3	7	11	15	(9)	9	14	3	8
(4)	13	14	15	16	(8)	4	8	12	16	(12)	13	10	7	4

	Rep. IV					Rep. V			
(13)	1	14	7	12	(17)	1	10	15	8
(14)	13	2	11	8	(18)	9	2	7	16
(15)	5	10	3	16	(19)	13	6	3	12
(16)	9	6	15	4	(20)	5	14	11	4

Plan 3   5 × 5 Balanced Lattice

$$t = 25, k = 5, r = 6, b = 30, \lambda = 1$$

Block	Rep. I						Rep. II						Rep. III				
(1)	1	2	3	4	5	(6)	1	6	11	16	21	(11)	1	7	13	19	25
(2)	6	7	8	9	10	(7)	2	7	12	17	22	(12)	21	2	8	14	20
(3)	11	12	13	14	15	(8)	3	8	13	18	23	(13)	16	22	3	9	15
(4)	16	17	18	19	20	(9)	4	9	14	19	24	(14)	11	17	23	4	10
(5)	21	22	23	24	25	(10)	5	10	15	20	25	(15)	6	12	18	24	5

	Rep. IV						Rep. V						Rep. VI				
(16)	1	12	23	9	20	(21)	1	17	8	24	15	(26)	1	22	18	14	10
(17)	16	2	13	24	10	(22)	11	2	18	9	25	(27)	6	2	23	19	15
(18)	6	17	3	14	25	(23)	21	12	3	19	10	(28)	11	7	3	24	20
(19)	21	7	18	4	15	(24)	6	22	13	4	20	(29)	16	12	8	4	25
(20)	11	22	8	19	5	(25)	16	7	23	14	5	(30)	21	17	13	9	5

Plan 4 7×7 Balanced Lattice

$t = 49, k = 7, r = 8, b = 56, \lambda = 1$

Block	Rep. I								Rep. II						
(1)	1	2	3	4	5	6	7	(8)	1	8	15	22	29	36	43
(2)	8	9	10	11	12	13	14	(9)	2	9	16	23	30	37	44
(3)	15	16	17	18	19	20	21	(10)	3	10	17	24	31	38	45
(4)	22	23	24	25	26	27	28	(11)	4	11	18	25	32	39	46
(5)	29	30	31	32	33	34	35	(12)	5	12	19	26	33	40	47
(6)	36	37	38	39	40	41	42	(13)	6	13	20	27	34	41	48
(7)	43	44	45	46	47	48	49		7	14	21	28	35	42	49
								(14)							

Block	Rep. III								Rep. IV						
(15)	1	9	17	25	33	41	49	(22)	1	37	24	11	47	34	21
(16)	43	2	10	18	26	34	42	(23)	15	2	38	25	12	48	35
(17)	36	44	3	11	19	27	35	(24)	29	16	3	39	26	13	49
(18)	29	37	45	4	12	20	28	(25)	43	30	17	4	40	27	14
(19)	22	30	38	46	5	13	21	(26)	8	44	31	18	5	41	28
(20)	15	23	31	39	47	6	14	(27)	22	9	45	32	19	6	42
(21)	8	16	24	32	40	48	7	(28)	36	23	10	46	33	20	7

Block	Rep. V								Rep. VI						
(29)	1	30	10	39	18	48	28	(36)	1	23	45	18	40	13	35
(30)	22	2	31	11	40	20	49	(37)	29	2	24	46	19	41	14
(31)	43	23	3	32	12	41	21	(38)	8	30	3	25	47	20	42
(32)	15	44	24	4	33	13	42	(39)	36	9	31	4	26	48	21
(33)	36	16	45	25	5	34	14	(40)	15	37	10	32	5	27	28
(34)	8	37	17	46	26	6	35	(41)	43	16	38	11	33	6	28
(35)	29	9	38	18	47	27	7	(42)	22	44	17	39	12	34	7

Block	Rep. VII								Rep. VIII						
(43)	1	16	31	46	12	27	42	(50)	1	44	38	32	26	20	14
(44)	36	2	17	32	47	13	28	(51)	8	2	45	39	33	27	21
(45)	22	37	3	18	33	48	14	(52)	15	9	3	46	40	34	28
(46)	8	23	38	4	19	34	49	(53)	22	16	10	4	47	41	35
(47)	43	9	24	39	5	20	35	(54)	29	23	17	11	5	48	42
(48)	29	44	10	25	40	6	21	(55)	36	30	24	18	12	6	49
(49)	15	30	45	11	26	41	7	(56)	43	37	31	25	19	13	7

Plan 5							8 × 8 Balanced Lattice								

$t = 64, k = 8, r = 9, b = 72, \lambda = 1$

Block				Rep. I					Block				Rep. II				
(1)	1	2	3	4	5	6	7	8	(9)	1	9	17	25	33	41	49	57
(2)	9	10	11	12	13	14	15	16	(10)	2	10	18	26	34	42	50	58
(3)	17	18	19	20	21	22	23	24	(11)	3	11	19	27	35	43	51	59
(4)	25	26	27	28	29	30	31	32	(12)	4	12	20	28	36	44	52	60
(5)	33	34	35	36	37	38	39	40	(13)	5	13	21	29	37	45	53	61
(6)	41	42	43	44	45	46	47	48	(14)	6	14	22	30	38	46	54	62
(7)	49	50	51	52	53	54	55	56	(15)	7	15	23	31	39	47	55	63
(8)	57	58	59	60	61	62	63	64	(16)	8	16	24	32	40	48	56	64

Block				Rep. III					Block				Rep. IV				
(17)	1	10	19	28	37	46	55	64	(25)	1	18	27	44	13	62	39	56
(18)	9	2	51	44	61	30	23	40	(26)	17	2	35	60	53	46	31	16
(19)	17	50	3	36	29	62	15	48	(27)	25	34	3	12	45	54	23	64
(20)	25	42	35	4	21	14	63	56	(28)	41	58	11	4	29	22	55	40
(21)	33	58	27	20	5	54	47	16	(29)	9	50	43	28	5	38	63	24
(22)	41	26	59	12	53	6	39	24	(30)	57	42	51	20	37	6	15	32
(23)	49	18	11	60	45	38	7	32	(31)	33	26	19	52	61	14	7	48
(24)	57	34	43	52	13	22	31	8	(32)	49	10	59	36	21	30	47	8

Block				Rep. V					Block				Rep. VI				
(33)	1	26	43	60	21	54	15	40	(41)	1	34	11	20	53	30	63	48
(34)	25	2	11	52	37	62	47	24	(42)	33	2	59	28	45	22	15	56
(35)	41	10	3	20	61	38	31	56	(43)	9	58	3	52	21	46	39	32
(36)	57	50	19	4	45	30	39	16	(44)	17	26	51	4	13	38	47	64
(37)	17	34	59	44	5	14	55	32	(45)	49	42	19	12	5	62	31	40
(38)	49	58	35	28	13	6	23	48	(46)	25	18	43	36	61	6	55	16
(39)	9	42	27	36	53	22	7	64	(47)	57	10	35	44	29	54	7	24
(40)	33	18	51	12	29	46	63	8	(48)	41	50	27	60	37	14	23	8

Block				Rep. VII					Block				Rep. VIII				
(49)	1	42	59	52	29	38	23	16	(57)	1	50	35	12	61	22	47	32
(50)	41	2	19	36	13	54	63	32	(58)	49	2	43	20	29	14	39	64
(51)	57	18	3	28	53	14	47	40	(59)	33	42	3	60	13	30	55	24
(52)	49	34	27	4	61	46	15	24	(60)	9	18	59	4	37	54	31	48
(53)	25	10	51	60	5	22	39	48	(61)	57	26	11	36	5	46	23	56
(54)	33	50	11	44	21	6	31	64	(62)	17	10	27	52	45	6	63	40
(55)	17	58	43	12	37	30	7	56	(63)	41	34	51	28	21	62	7	16
(56)	9	26	35	20	45	62	55	8	(64)	25	58	19	44	53	38	15	8

Block				Rep. IX				
(65)	1	58	51	36	45	14	31	24
(66)	57	2	27	12	21	38	55	48
(67)	49	26	3	44	37	22	63	16
(68)	33	10	43	4	53	62	23	32
(69)	41	18	35	52	5	30	15	64
(70)	9	34	19	60	29	6	47	56
(71)	25	50	59	20	13	46	7	40
(72)	17	42	11	28	61	54	39	8

Plan 6 | 9 × 9 Balanced Lattice

$t = 81,\ k = 9,\ r = 10,\ b = 90,\ \lambda = 1$

Block				Rep. I											Rep. II				
(1)	1	2	3	4	5	6	7	8	9	(10)	1	10	19	28	37	46	55	64	73
(2)	10	11	12	13	14	15	16	17	18	(11)	2	11	20	29	38	47	56	65	74
(3)	19	20	21	22	23	24	25	26	27	(12)	3	12	21	30	39	48	57	66	75
(4)	28	29	30	31	32	33	34	35	36	(13)	4	13	22	31	40	49	58	67	76
(5)	37	38	39	40	41	42	43	44	45	(14)	5	14	23	32	41	50	59	68	77
(6)	46	47	48	49	50	51	52	53	54	(15)	6	15	24	33	42	51	60	69	78
(7)	55	56	57	58	59	60	61	62	63	(16)	7	16	25	34	43	52	61	70	79
(8)	64	65	66	67	68	69	70	71	72	(17)	8	17	26	35	44	53	62	71	80
(9)	73	74	75	76	77	78	79	80	81	(18)	9	18	27	36	45	54	63	72	81

				Rep. III											Rep. IV				
(19)	1	20	12	58	77	69	34	53	45	(28)	1	11	21	31	41	51	61	71	61
(20)	10	2	21	67	59	78	43	35	54	(29)	19	2	12	49	32	42	79	62	72
(21)	19	11	3	76	68	69	52	44	36	(30)	10	20	3	40	50	33	70	80	63
(22)	28	47	39	4	23	15	61	80	72	(31)	55	65	75	4	14	24	34	44	54
(23)	37	29	48	13	5	24	70	62	81	(32)	73	56	66	22	5	15	52	35	45
(24)	46	38	30	22	14	6	79	71	63	(33)	64	74	57	13	23	6	43	53	36
(25)	55	74	66	31	50	42	7	26	18	(34)	28	38	48	58	68	78	7	17	27
(26)	64	56	75	40	32	51	16	8	27	(35)	46	29	39	76	59	69	25	8	18
(27)	73	65	57	49	41	33	25	17	9	(36)	37	47	30	67	77	60	16	26	9

				Rep. IX											Rep. X				
(73)	1	65	48	40	23	60	79	35	18	(82)	1	38	75	49	59	15	70	26	36
(74)	46	2	66	58	41	24	16	80	36	(83)	73	2	39	13	50	60	34	71	27
(75)	64	47	3	22	59	42	34	17	81	(84)	37	74	3	58	14	51	25	35	72
(76)	73	29	12	4	68	51	43	26	63	(85)	64	20	30	4	41	78	52	62	18
(77)	10	74	30	49	5	69	61	44	27	(86)	28	65	21	76	5	42	16	53	63
(78)	28	11	75	67	50	6	25	62	45	(87)	19	29	66	40	77	6	61	17	54
(79)	37	20	57	76	32	15	7	71	54	(88)	46	56	12	67	23	33	7	44	81
(80)	55	38	21	13	77	33	52	8	72	(89)	10	47	57	31	68	24	79	8	45
(81)	19	56	39	31	14	78	70	53	9	(90)	55	11	48	22	32	69	43	80	9

Plan 7 | 6 × 6 Triple Lattice

Block			Rep. I							Rep. II			
(1)	1	2	3	4	5	6	(7)	1	7	13	19	25	31
(2)	7	8	9	10	11	12	(8)	2	8	14	20	26	32
(3)	13	14	15	16	17	18	(9)	3	9	15	21	27	33
(4)	19	20	21	22	23	24	(10)	4	10	16	22	28	34
(5)	25	26	27	28	29	30	(11)	5	11	17	23	29	35
(6)	31	32	33	34	35	36	(12)	6	12	18	24	30	36

			Rep. III			
(13)	1	8	15	22	29	36
(14)	31	2	9	16	23	30
(15)	25	32	3	10	17	24
(16)	19	26	33	4	11	18
(17)	13	20	27	34	5	12
(18)	7	14	21	28	35	6

Plan 8   10 × 10 Triple Lattice

Block				Rep. I							Block				Rep. II						
(1)	1	2	3	4	5	6	7	8	9	10	(11)	1	11	21	31	41	51	61	71	81	91
(2)	11	12	13	14	15	16	17	18	19	20	(12)	2	12	22	32	42	52	62	72	82	92
(3)	21	22	23	24	25	26	27	28	29	30	(13)	3	13	23	33	43	53	63	73	83	93
(4)	31	32	33	34	35	36	37	38	39	40	(14)	4	14	24	34	44	54	64	74	84	94
(5)	41	42	43	44	45	46	47	48	49	50	(15)	5	15	25	35	45	55	65	75	85	95
(6)	51	52	53	54	55	56	57	58	59	60	(16)	6	16	26	36	46	56	66	76	86	96
(7)	61	62	63	64	65	66	67	68	69	70	(17)	7	17	27	37	47	57	67	77	87	97
(8)	71	72	73	74	75	76	77	78	79	80	(18)	8	18	28	38	48	58	68	78	88	98
(9)	81	82	83	84	85	86	87	88	89	90	(19)	9	19	29	39	49	59	69	79	89	99
(10)	91	92	93	94	95	96	97	98	99	100	(20)	10	20	30	40	50	60	70	80	90	100

Block				Rep. III						
(21)	1	12	23	34	45	56	67	78	89	100
(22)	91	2	13	24	35	46	57	68	79	90
(23)	81	92	3	14	25	36	47	58	69	80
(24)	71	82	93	4	15	26	37	48	59	70
(25)	61	72	83	94	5	16	27	38	49	60
(26)	51	62	73	84	95	6	17	28	39	50
(27)	41	52	63	74	85	96	7	18	29	40
(28)	31	42	53	64	75	86	97	8	19	30
(29)	21	32	43	54	65	76	87	98	9	20
(30)	11	22	33	44	55	66	77	88	99	10

Plan 9   12 ×12 Quadruple Lattice

Rep. I

(1)	1	2	3	4	5	6	7	8	9	10	11	12
(2)	13	14	15	16	17	18	19	20	21	22	23	24
(3)	25	26	27	28	29	30	31	32	33	34	35	36
(4)	37	38	39	40	41	42	43	44	45	46	47	48
(5)	49	50	51	52	53	54	55	56	57	58	59	60
(6)	61	62	63	64	65	66	67	68	69	70	71	72
(7)	73	74	75	76	77	78	79	80	81	82	83	84
(8)	85	86	87	88	89	90	91	92	93	94	95	96
(9)	97	98	99	100	101	102	103	104	105	106	107	108
(10)	109	110	111	112	113	114	115	116	117	118	119	120
(11)	121	122	123	124	125	126	127	128	129	130	131	132
(12)	133	134	135	136	137	138	139	140	141	142	143	144

Rep. II

(13)	1	13	25	37	49	61	73	85	97	109	121	133
(14)	2	14	26	38	50	62	74	86	98	110	122	134
(15)	3	15	27	39	51	63	75	87	99	111	123	135
(16)	4	16	28	40	52	64	76	88	100	112	124	136
(17)	5	17	29	41	53	65	77	89	101	113	125	137
(18)	6	18	30	42	54	66	78	90	102	114	126	138
(19)	7	19	31	43	55	67	79	91	103	115	127	139
(20)	8	20	32	44	56	68	80	92	104	116	128	140
(21)	9	21	33	45	57	69	81	93	105	117	129	141
(22)	10	22	34	46	58	70	94	117	106	118	130	142
(23)	11	23	35	47	59	71	95	128	107	119	131	143
(24)	12	24	36	48	60	72	96	140	108	120	132	144

Rep. III

(25)	1	14	27	40	57	70	83	96	101	114	127	140
(26)	2	13	28	39	58	69	84	95	102	113	128	139
(27)	3	16	25	38	59	72	81	94	103	116	125	138
(28)	4	15	26	37	60	71	82	93	104	115	126	137
(29)	5	18	31	44	49	62	75	88	105	118	131	144
(30)	6	17	32	43	50	61	76	87	106	117	132	143
(31)	7	20	29	42	51	64	73	86	107	120	129	142
(32)	8	19	30	41	52	63	74	85	108	119	130	141
(33)	9	22	35	48	53	66	79	92	97	110	123	136
(34)	10	21	36	47	54	65	80	91	98	109	124	135
(35)	11	24	33	46	55	68	77	90	99	112	121	134
(36)	12	23	34	45	56	67	78	89	100	111	122	133

Rep. IV

(37)	1	24	30	43	53	64	82	95	105	116	122	135
(38)	2	23	29	44	54	63	81	96	106	115	121	136
(39)	3	22	32	41	55	62	84	93	107	114	124	133
(40)	4	21	31	42	56	61	83	94	108	113	123	134
(41)	5	16	34	47	57	68	74	87	97	120	126	139
(42)	6	15	33	48	58	67	73	88	98	119	125	140
(43)	7	14	36	45	59	66	76	85	99	118	128	137
(44)	8	13	35	46	60	65	75	86	100	117	127	138
(45)	9	20	26	39	49	72	78	91	101	112	130	143
(46)	10	19	25	40	50	71	77	92	102	111	129	144
(47)	11	18	28	37	51	70	80	89	103	110	132	141
(48)	12	17	27	38	52	69	79	90	104	109	131	142

Plan 10   3 × 4 Rectangular Lattice

Block	Rep. X				Rep. Y				Rep. Z		
X_1	1	2	3		Y_1	4	7	10	Z_1	6	8	12
X_2	4	5	6		Y_2	1	8	11	Z_2	2	9	10
X_3	7	8	9		Y_3	2	5	12	Z_3	3	4	11
X_4	10	11	12		Y_4	3	6	9	Z_4	1	5	7

Plan 11   4 × 5 Rectangular Lattice

Block	Rep. X				Rep. Y				Rep. Z			
X_1	1	2	3	4	Y_1	5	9	13	17	Z_1	8	11	15	18
X_2	5	6	7	8	Y_2	1	10	14	18	Z_2	2	9	16	20
X_3	9	10	11	12	Y_3	2	6	15	19	Z_3	4	7	14	17
X_4	13	14	15	16	Y_4	3	7	11	20	Z_4	1	5	12	19
X_5	17	18	19	20	Y_5	4	8	12	16	Z_5	3	6	10	13

Plan 12   5 × 6 Rectangular Lattice

Block	Rep. X					Rep. Y					Rep. Z				
X_1	1	2	3	4	5	Y_1	6	11	16	21	26	Z_1	7	13	19	25	27
X_2	6	7	8	9	10	Y_2	1	12	17	22	27	Z_2	5	14	16	23	29
X_3	11	12	13	14	15	Y_3	2	7	18	23	28	Z_3	1	8	20	21	30
X_4	16	17	18	19	20	Y_4	3	8	13	24	29	Z_4	2	9	15	22	26
X_5	21	22	23	24	25	Y_5	4	9	14	19	30	Z_5	3	10	11	17	26
X_6	26	27	28	29	30	Y_6	5	10	15	20	25	Z_6	4	6	12	18	24

Plan 13 6 ×7 Rectangular Lattice

Block	Rep. *X*							Block	Rep.*Y*					
X_1	1	2	3	4	5	6		Y_1	7	13	19	25	31	37
X_2	7	8	9	10	11	12		Y_2	1	14	20	26	32	38
X_3	13	14	15	16	17	18		Y_3	2	8	21	27	33	39
X_4	19	20	21	22	23	24		Y_4	3	9	15	28	34	40
X_5	25	26	27	28	29	30		Y_5	4	10	16	22	36	41
X_6	31	32	33	34	35	36		Y_6	5	11	17	23	29	42
X_7	37	38	39	40	41	42		Y_7	6	12	18	24	30	36

Block	Rep. *Z*					
Z_1	12	17	22	28	33	38
Z_2	2	13	24	29	35	40
Z_3	4	9	20	25	36	42
Z_4	6	11	16	27	32	37
Z_5	1	7	18	23	34	39
Z_6	3	8	14	19	30	41
Z_7	5	10	15	21	26	31

Plan 14   7 ×8 Rectangular Lattice

Block	Rep. X								Rep. Y						
X_1	1	2	3	4	5	6	7		Y_1	8	15	22	29	36	43	50
X_2	8	9	10	11	12	13	14		Y_2	1	16	23	30	37	44	51
X_3	15	16	17	18	19	20	21		Y_3	2	9	24	31	38	45	52
X_4	22	23	24	25	26	27	28		Y_4	3	10	17	32	39	46	53
X_5	29	30	31	32	33	34	35		Y_5	4	11	18	25	40	47	54
X_6	35	36	37	38	39	41	42		Y_6	5	12	19	26	33	48	55
X_7	43	44	45	46	47	48	49		Y_7	6	13	20	27	34	41	56
X_8	50	51	52	53	54	55	56		Y_8	50	14	21	28	35	42	49

Rep. Z

Z_1	9	17	25	33	41	49	51
Z_2	7	18	26	34	8	43	53
Z_3	1	10	27	35	36	47	55
Z_4	2	11	19	29	42	44	56
Z_5	3	12	20	28	37	45	50
Z_6	4	13	21	22	30	46	52
Z_7	5	14	15	23	31	39	54
Z_8	6	8	16	24	32	40	48

Plan 15   8 ×9 Rectangular Lattice

Block				Rep. X									Rep. Y					
X_1	1	2	3	4	5	6	7	8		Y_1	9	17	25	33	41	49	57	65
X_2	9	10	11	12	13	14	15	16		Y_2	1	18	26	34	42	50	58	66
X_3	17	18	19	20	21	22	23	24		Y_3	2	10	27	35	43	51	59	67
X_4	25	26	27	28	29	30	31	32		Y_4	3	11	19	36	44	52	60	68
X_5	33	34	35	36	37	38	39	40		Y_5	4	12	20	28	45	53	61	69
X_6	41	42	43	44	45	46	47	48		Y_6	5	13	21	29	37	54	62	70
X_7	49	50	51	52	53	54	55	56		Y_7	6	14	22	30	38	46	63	71
X_8	57	58	59	60	61	62	63	64		Y_8	7	15	23	31	39	47	55	72
X_9	65	66	67	68	69	70	71	72		Y_9	8	16	24	32	40	48	56	64

Rep. Z

Z_1	16	23	30	37	45	52	59	66
Z_2	2	17	32	39	46	54	61	68
Z_3	4	11	26	33	48	55	63	70
Z_4	6	13	20	35	42	49	64	72
Z_5	8	15	22	29	44	51	58	65
Z_6	1	9	24	31	38	53	60	67
Z_7	3	10	18	25	40	47	62	69
Z_8	5	12	19	27	34	41	56	71
Z_9	7	14	21	28	36	43	50	57

Plan 16　　　　　　　　　　　9 ×10 Rectangular Lattice

Block			Rep. X								Block			Rep. Y							
X_1	1	2	3	4	5	6	7	8	9		Y_1	10	19	28	37	46	55	64	73	82
X_2	10	11	12	13	14	15	16	17	18		Y_2	1	20	29	38	47	56	65	74	83
X_3	19	20	21	22	23	24	25	26	27		Y_3	2	11	30	39	48	57	66	75	84
X_4	28	29	30	31	32	33	34	35	36		Y_4	3	12	21	40	49	58	67	76	85
X_5	37	38	39	40	41	42	43	44	45		Y_5	4	13	22	31	50	59	68	77	86
X_6	46	47	48	49	50	51	52	53	54		Y_6	5	14	23	32	41	60	69	78	87
X_7	55	56	57	58	59	60	61	62	63		Y_7	6	15	24	33	42	51	70	79	88
X_8	64	65	66	67	68	69	70	71	72		Y_8	7	16	25	34	43	52	61	80	89
X_9	73	74	75	76	77	78	79	80	81		Y_9	8	17	26	35	44	53	62	71	90
X_{10}	82	83	84	85	86	87	88	9	90		Y_{10}	9	18	27	36	45	54	63	72	81

Block				Rep. Z					
Z_1	11	21	31	41	51	61	71	81	83
Z_2	9	22	32	42	52	62	64	76	84
Z_3	1	12	33	43	53	63	68	73	87
Z_4	2	13	23	44	54	55	65	79	89
Z_5	3	14	24	34	46	56	72	75	90
Z_6	4	15	25	35	45	57	67	74	82
Z_7	5	16	26	36	37	47	66	77	85
Z_8	6	17	27	28	38	48	58	78	86
Z_9	7	18	19	29	39	49	59	69	88
Z_{10}	8	10	20	30	40	50	60	70	80

[a] The symbol λ denotes the number of times that two treatments appear in the same block.

Source: From Cochran, William G. and Cox, Gertrude M., *Experimental Designs*, 2nd ed., John Wiley & Sons, 1957. With permission.

Appendix G

Fractional Factorial Design Plans

Plan 1: 2^4 Factorial in 8 Units (1/2 Replicate)

Defining contrast: *ABCD*

Estimable two-factor interactions: $AB = CD, AC = BD, AD = BC$

(1)	Effects	d.f.
ab	Main	4
cd	2-factor	3
ace	Total	7
bce		
ade		
bde		
abcd		

Plan 2: 2^5 Factorial in 8 Units (1/4 Replicate)

Defining contrasts: *ABE, CDE, ABCD*

Main effects have two factors as aliases. The only estimable two-factors are $AC = BD$ and $AD = BC$.

(1)	Effects	d.f.
ab	Main	5
cd	2-factor	2
ace	Total	7
bce		
ade		
bde		
abcd		

Plan 3: 2^5 Factorial in 16 Units (1/2 Replicate)

Defining contrast: *ABCDE*

Blocks of 4 Units

Estimable two-factors: All except *CD*, *CE*, and *DE* (confounded with blocks).

Blocks	(1)	(2)	(3)	(4)	Effects	d.f.
	(1)	*ac*	*ae*	*ad*	Block	3
	ab	*bc*	*be*	*bd*	Main	5
	acde	*de*	*cd*	*ce*	2-factor	7
	bcde	*abde*	*abcd*	*abce*	Total	15

Blocks of 8 Units

Estimable two-factors: All except *DE*.

Combine blocks 1 and 2; and blocks 3 and 4. *DE* is confounded.

Effects	d.f.
Block	1
Main	5
2-factor	9
Total	15

Blocks of 16 Units

Estimable two-factors: All.

Combine blocks 1 and 4.

Effects	d.f.
Main	5
2-factor	10
Total	15

Plan 4: 2^6 Factorial in 8 Units (1/8 Replicate)

Defining contrast: *ACE, ADF, BCF, BDE, ABCD, ABEF, CDEF*
Main effects have two-factors as aliases. The only estimable two-factor is the
  set $AB = CD = EF$.

(1)	Effects	d.f.
acf	Main	6
ade	2-factor ($AB = CD = EF$)	1
bce	Total	7
bdf		
abcd		
abef		
cdef		

Plan 5: 2^6 Factorial in 16 Units (1/4 Replicate)

Defining contrast: *ACE, ADF, BCF, BDE, ABCD, ABEF, CDEF*

Blocks of 4 Units

Estimable two-factors: The alias sets $AC = BE$, and $AD = BF$, $AE = BC$, $AF = BD$, $CD = EF$, $CF = DE$.

Blocks	(1)	(2)	(3)	(4)	Effects	d.f.
	(1)	acd	ab	acf	Block	3
	abce	aef	ce	ade	Main	6
	abdf	bcf	df	bcd	2-factor	6
	cdef	bde	abcdef	bef	Total	15

AB, ACF, BCF confounded.

Blocks of 8 Units

Estimable two-factors: Same as in blocks of four units, plus the set $AB = CE = DF$.

Combine blocks 1 and 2; and blocks 3 and 4. *ACF* is confounded.

Effects	d.f.
Block	1
Main	6
2-factor	7
3-factor	1
Total	15

Blocks of 16 Units

Estimable two-factors: Same as in blocks of eight units

Combine blocks 1–4.

Effects	d.f.
Main	6
2-factor	7
3-factor	2
Total	15

Plan 6: 2^6 Factorial in 32 Units (1/2 Replicate)

Defining contrast: *ABCDEF*

Blocks of 4 Units

Estimable two-factors: All except *AE*, *BF*, and *CD* (confounded with blocks).

Blocks	(1)	(2)	(3)	(4)	(5)	(6)	(7)	(8)
	(1)	ab	ac	bc	ae	af	ad	bd
	abef	ef	de	df	bf	be	ce	cf
	acde	acdf	abdf	acef	cd	abcd	abcf	abce
	bcdf	bcde	bcef	abde	abcdef	cdef	bdef	adef

AE, BF, CD, ABC, ABD, ACF, ADF confounded.

Effects	d.f.
Block	7
Main	6
2-factor	12
Higher order	6
Total	31

Blocks of 8 Units

Estimable two-factors: All except *CD*.

Combine blocks 1 and 2; and blocks 3 and 4; blocks 5 and 6; and blocks 7 and 8. *CD, ABC, ABD* confounded.

Effects	d.f.
Block	3
Main	6
2-factor	14
3-factor	8
Total	31

Blocks of 16 Units

Estimable two-factors: All.

Estimable three-factors: *ABC = DEF* is lost by confounding. The others are in alias pairs, e.g., *ABD = CEF*.

Combine blocks 1–4; and blocks of 5–8. *ABC* confounded.

Effects	d.f.
Block	1
Main	6
2-factor	15
3-factor	9
Total	31

Blocks of 32 Units

Estimable two-factors: All.

Estimable three-factors: These are arranged in 10 alias pairs.

Combine blocks 1–8.

Effects	d.f.
Main	6
2-factor	15
3-factor	10
Total	31

Plan 7: 2^7 Factorial in 8 Units (1/16 Replicate)

Defining contrast: *ABG, ACE, ADF, BCF, BDE, CDG, EFG, ABCD, ABEF, ACFG, ADEG, BCEG, BDFG, CDEF, ABCDEFG.*

Main effects have two-factors as aliases. No two-factors are
estimable.

(1)	Effects	d.f.
abcd	Main	7
abef	Total	7
acfg		
adeg		
bceg		
bdfg		
cdef		

Plan 8: 2^7 Factorial in 16 Units (1/8 Replicate)

Defining contrast: *ABCD, ABEF, ACEG, ADFG, BCFG, BDEG, CDEF.*

Blocks of 4 Units

Estimable two-factors: Only the alias sets $AE = BF = CG$; $AF = BE = DG$; $AG = CE = DF$; $BG = DE = CF$.

Blocks	(1)	(2)	(3)	(4)	Effects	d.f.
	(1)	abg	acf	ade	Block	3
	efg	cdg	bdf	bce	Main	7
	abcd	abef	aceg	adfg	2-factor	4
	abcdefg	edef	bdeg	bcfg	Higher order	1
					Total	15

$AB = CD = EF,$
$AC = BD = EG,$
$AD = BC = FG$ confounded.

Plan 9: 2^7 Factorial in 16 Units (1/8 Replicate)

Defining contrast: *ABCD, ABEF, ACEG, ADFG, BCFG, BDEG, CDEF.*

Blocks of 8 Units

Estimable two-factors: Same as blocks of 4 units, plus the alias sets
  $AB = CD = EF, AC = BD = EG; AD = BC = FG.$

Blocks	(1)	(2)	Effects	d.f.
	(1)	*abg*	Block	1
	abcd	*acf*	Main	7
	abef	*ade*	2-factor	7
	aceg	*bce*	Total	15
	adfg	*bdf*		
	bcfg	*cdg*		
	bdeg	*efg*		
	cdef	*abcdefg*		

ABG confounded.

Blocks of 16 Units

Estimable two-factors: Same as in blocks of 8 units.

Combine blocks 1 and 2 of the plan for blocks of 8 units.

Effects	d.f.
Main	7
2-factor	7
Higher order	1
Total	15

Plan 10:   2⁷ Factorial in 32 Units (1/4 Replicate)

Defining contrast: *ABCDE, ABCFG, DEFG.*

Blocks of 4 Units

Estimable two-factors: *AB, AC, BC,* and *DF =EG* are lost by confounding. All other two-factors are estimable, except that *DE = FG* and *DG = EF,* so that members of these alias pairs cannot be separated.

Blocks	(1)	(2)	(3)	(4)	(5)	(6)	(7)	(8)
	(1)	*de*	*ab*	*cdg*	*ac*	*bdg*	*bc*	*adg*
	defg	*fg*	*cdf*	*cef*	*bdf*	*bef*	*adf*	*aef*
	abcdf	*abcdg*	*ceg*	*abde*	*beg*	*acde*	*aeg*	*bcfg*
	abceg	*abcef*	*abdefg*	*abfg*	*abcdefg*	*acfg*	*bcdefg*	*bcde*

AB, AC, BC, and *DF =EG, ADG, BDG, CDG* confounded.

Effects	d.f.
Block	7
Main	7
2-factor	14
Higher order	3
Total	31

Plan 11: 2^7 Factorial in 32 Units (1/4 Replicate)

Defining contrast: *ABCDE, ABCFG, DEFG.*

Blocks of 8 Units

Estimable two-factors: All except *DF* =*EG* (confounded with blocks). However, *DE* = *FG* and DG = *EF* are alias pairs which cannot be separated.

Blocks	(1)	(2)	(3)	(4)	Effects	d.f.
	(1)	*bdg*	*ab*	*de*	Effects	d.f.
	bc	*bef*	*ac*	*fg*	Block	3
	adf	*cef*	*bdf*	*adg*	Main	7
	aeg	*abfg*	*beg*	*aef*	2-factor	17
	defg	*acfg*	*cdf*	*bcde*	Higher order	4
	abcdf	*abde*	*ceg*	*bcfg*	Total	31
	abceg	*acde*	*acdefg*	*abcdg*		
	bcdefg	*cdg*	*abdefg*	*abcef*		

DF = *EG, ADE, AEF* confounded.

Blocks of 16 Units

Estimable two-factors: All except that *DE* = *FG, DG* = *EF,* and *DF* = *EG* are alias pairs.

Combine blocks 1 and 2; and blocks 3 and 4. *AEF* confounded.

Effects	d.f.
Block	1
Main	7
2-factor	18
Higher order	5
Total	31

Blocks of 32 units

Estimable two-factors: Same as in blocks of 16 units.

Combine blocks 1–4.

Effects	d.f.
Main	7
2-factor	18
3-factor	6
Total	31

Plan 12 2^7 Factorial in 64 Units (1/2 Replicate)

Defining contrast: *ABCDEFG*

Blocks of 4 Units

Estimable two-factors: All except *AB, AC, BC, EF, EG,* and *FG* (confounded with blocks).

Blocks	(1)	(2)	(3)	(4)	(5)	(6)	(7)	(8)
	(1)	*ab*	*ac*	*bc*	*ae*	*be*	*ce*	*abce*
	abcd	*cd*	*bd*	*ad*	*bcde*	*acde*	*abde*	*de*
	defg	*abdefg*	*acdefg*	*bcdefg*	*adfg*	*bdfg*	*cdfg*	*abcdfg*
	abcefg	*cefg*	*befg*	*aefg*	*bcfg*	*acfg*	*abfg*	*fg*

	(9)	(10)	(11)	(12)	(13)	(14)	(15)	(16)
	af	*bf*	*cf*	*abcf*	*ef*	*abef*	*acef*	*bcef*
	bcdf	*acdf*	*abdf*	*df*	*abcdef*	*cdef*	*bdef*	*adef*
	adeg	*bdeg*	*cdeg*	*abcdeg*	*dg*	*abdg*	*acdg*	*bcdg*
	bceg	*aceg*	*abeg*	*eg*	*abcg*	*cg*	*bg*	*ag*

AB, AC, BC, EF, EG, FG, ADE, ADF, ADG, BDE, BDF, BDG, CDE, CDF, CD confounded.

Effects	d.f.
Block	15
Main	7
2-factor	15
Higher order	26
Total	63

Plan 13: 2^7 Factorial in 64 Units (1/2 Replicate)

Defining contrast: *ABCDEFG*

Blocks of 8 Units

Estimable two-factors: All.

Estimable three-factors: All except *ABC, ADE, AFG, BDF, EG, BEG, CDG, CEF* confounded.

Blocks	(1)	(2)	(3)	(4)	(5)	(6)	(7)	(8)
	(1)	bc	ac	ab	ag	af	ae	ad
	abdg	de	df	dg	bd	be	bf	bg
	abef	fg	eg	ef	ce	cd	cg	cf
	acdf	abdf	abde	acde	abcf	abcg	abcd	abce
	aceg	abeg	abfg	acfg	adef	adeg	adfg	aefg
	bcde	acdg	bcdg	bcdf	befg	bdfg	bdeg	bdef
	bcfg	acef	bcef	bceg	cdfg	cefg	cdef	cdeg
	defg	bcdefg	acdefg	abdefg	abcdeg	abcdef	abcefg	abcdfg

Effects	d.f.
Block	7
Main	7
2-factor	21
Higher order	28
Total	63

Blocks of 16 Units

Estimable two-factors: All.

Estimable three-factors: All except *ABC, ADE, AFG* (confounded).

Combine blocks 1 and 2; and blocks 3 and 4; blocks 5 and 6; and blocks 7 and 8.

Effects	d.f.
Block	3
Main	7
2-factor	21
Higher order	32
Total	63

Blocks of 32 Units

Estimable two-factors: All.

Estimable three-factors: All except *ABC* (confounded).

Combine blocks 1–4; and blocks 5–8.

Effects	d.f.
Block	1
Main	7

2-factor	21
3-factor	34
Total	63

Blocks of 64 Units

Estimable two-factors: All.

Estimable three-factors: All.

Combine blocks 1–8.

Effects	d.f.
Main	7
2-factor	21
3-factor	35
Total	63

Plan 14: 2^8 Factorial in 16 Units (1/16 Replicate)

Defining contrast: *ABCD, ABEF, ABGH, ACEH, ACFG, ADEG, ADFH, BCEG, BCFH, BDEH, BDFG, CDEF, CDGH, EFGH, ABCDEFGH.*

Blocks of 4 Units

Estimable two-factors: Only the 4 alias sets AE = *BF* = *CH* = *DG*; *AF* = *BE* = *CG* = *DH*; *AG* = *BH* = *CF* = *DE*; *AH* = *BG* = *CE* = *DF*. Except in special circumstances, only main effects are estimable.

Blocks	(1)	(2)	(3)	(4)		
	(1)	*abef*	*adeg*	*aceh*	Effects	d.f.
	abcd	*abgh*	*adfh*	*acfg*	Block	3
	efgh	*cdef*	*bceg*	*bdeh*	Main	8
	abcdefgh	*cdgh*	*bcfh*	*bdfg*	2-factor	4
					Total	15

AB, AC, AD confounded.

Blocks of 8 Units

Estimable two-factors: As in blocks of 4 units, plus the alias sets *AC* =*BD* = *EH* = *FG*; and *AD* = *EC* = *EG* = *FH*.

Combine blocks 1 and 2; and blocks 3 and 4. *AB* confounded.

Effects	d.f.
Block	1
Main	8
2-factor	6
Total	15

Blocks of 16 Units

Estimable two-factors: As in blocks of 8 units, plus the alias sets *AB* =*CD* = *EF* = *GH.*

Combine blocks 1 and 4.

Effects	d.f.
Main	8
2-factor	7
Total	15

Plan 15: 2^8 Factorial in 32 Units (1/8 Replicate)

Defining contrast: *BCDH, BDFG, CFGH, ABCEF, ABEGH, ACDEG,*
ADEFH

Blocks of 4 Units

Estimable two-factors: All interactions of *A* and *E*. All other two-factors are lost by
confounding.

Blocks	(1)	(2)	(3)	(4)	(5)	(6)	(7)	(8)
	(1)	abd	dgh	cdf	afg	ach	bcg	bfh
	ae	bde	abcf	abgh	efg	ceh	adfh	acdg
	abcdfgh	cfgh	bcef	begh	bcdh	bdfg	defh	cdeg
	bcdefgh	acefgh	adegh	acdef	abcdeh	abdefg	abceg	abefh

BD, BF, BH, CF, DF, DH, FH, and their aliases confounded.

Effects	d.f.
Block	7
Main	8
2-factor	13
Higher order	3
Total	31

Blocks of 8 Units

Estimable two-factors: All interactions of *A* and *E*, and the alias pairs *BC =DH, BF =
DG,* and *BG = DF, BH = CD.*

Combine blocks 1 and 2; and blocks 3 and 4; blocks 5 and 6; and blocks 7 and 8. *BD,
CF, FH* confounded.

Effects	d.f.
Block	3
Main	8
2-factor	17
Higher order	3
Total	31

Blocks of 16 Units

Estimable two-factors: As in blocks of 8 units, plus the alias sets *CG =FH, BD = CH
= FG.*

Combine blocks 1 and 4; and blocks 5–8. *CF* confounded.

Effects	d.f.
Block	1
Main	8
2-factor	19
Higher order	3
Total	31

Blocks of 32 Units

Estimable two-factors: All interactions of A and E, plus the alias sets $BC = DH$, $BF = DG$, $BG = DF$, $BH = CD$, $CG = FH$, $CF = GH$, $BD = CH = FG$.

Combine blocks 1 and 8.

Effects	d.f.
Main	8
2-factor	20
Higher order	3
Total	31

Plan 16: 2^8 Factorial in 64 Units (1/4 Replicate)

Defining contrast: *ABCEG, ABDFH, CDEFGH*

Blocks of 4 Units

Estimable two-factors: All except *AF, AH, BC, BG, CG, DE,* and *FH* (confounded with blocks).

Blocks	(1)	(2)	(3)	(4)	(5)	(6)	(7)	(8)
	(1)	*adg*	*ach*	*beh*	*eg*	*fh*	*bef*	*acf*
	adefh	*abce*	*bfg*	*abdf*	*bcd*	*ade*	*abdh*	*bgh*
	bcdeg	*efgh*	*cdef*	*cdgh*	*adfgh*	*abcg*	*cdfg*	*cdeh*
	abcfgh	*bcdfh*	*abdegh*	*acefg*	*abcefh*	*bcdefgh*	*acegh*	*abdefg*

	(9)	(10)	(11)	(12)	(13)	(14)	(15)	(16)
	ab	*ce*	*afg*	*df*	*acd*	*cg*	*dh*	*agh*
	cfgh	*bdg*	*bch*	*aeh*	*abeg*	*bde*	*aef*	*bcf*
	acdeg	*acdfh*	*degh*	*bcefg*	*cefh*	*abfh*	*bcegh*	*defg*
	bdefh	*abefgh*	*abcdef*	*abcdgh*	*bdfgh*	*acdefgh*	*abcdfg*	*abcdeh*

AF, AH, BC, BG, CG, DE, FH, ACD, ADG, BEF, BEH, CDF, CEF, DFG, EFG confounded.

Effects	d.f.
Block	15
Main	8
2-factor	21
Higher order	19
Total	63

Blocks of 8 Units

Estimable two-factors: All except *BC* and *FH* (confounded with blocks).

Combine blocks 1 and 2; and blocks 3 and 4; blocks 5 and 6; blocks 7 and 8; blocks 9 and 10; blocks 11 and 12; blocks 13 and 14; and blocks 15 and 16. *BC, FH, ACD, BEF, BEH, CEF, DFG* confounded.

Effects	d.f.
Block	7
Main	8
2-factor	26
Higher order	22
Total	63

Blocks of 16 Units

Estimable two-factors: All.

Combine blocks 1–4; and blocks 5–8; blocks 9–12; and blocks 13–16. *ACD, BEF, DFG* confounded.

Effects	d.f.
Block	3
Main	8
2-factor	28
Higher order	24
Total	63

Blocks of 32 Units

Estimable two-factors: All.

Combine blocks 1–8; and blocks 9–16. *ACD* confounded.

Effects	d.f.
Block	1
Main	8
2-factor	28
Higher order	26
Total	63

Blocks of 64 Units

Estimable two-factors: All.

Combine blocks 1–16.

Effects	d.f.
Main	8
2-factor	28
Higher order	27
Total	63

Plan 17: 2^8 Factorial in 128 Units (1/2 Replicate)

Defining contrast: *ABCDEFGH*

Blocks of 16 Units

Estimable two-factors: All.

Estimable three-factors: All.

Blocks	(1)	(2)	(3)	(4)	(5)	(6)	(7)	(8)
	(1)	*acfh*	*fh*	*ac*	*af*	*ch*	*ah*	*cf*
	abcd	*bdfh*	*abcdfh*	*bd*	*bcdf*	*abdh*	*bcdh*	*abdf*
	adeg	*abefgh*	*abefgh*	*abeg*	*defg*	*begh*	*degh*	*befg*
	bceg	*cdefgh*	*bcefgh*	*cdeg*	*abcefg*	*acdegh*	*abcegh*	*acdefg*
	adfh	*ab*	*ad*	*abfh*	*dh*	*bf*	*df*	*bh*
	bcfh	*cd*	*bc*	*cdfh*	*abch*	*acdf*	*abcf*	*acdh*
	efgh	*aceg*	*eg*	*acefgh*	*aegh*	*cefg*	*aefg*	*cegh*
	abcdefgh	*bdeg*	*abcdeg*	*bdefgh*	*bcdegh*	*abdefg*	*bcdefg*	*abdegh*
	abgh	*ef*	*abfg*	*eh*	*bfgh*	*ae*	*bg*	*aefh*
	aceh	*adfg*	*acef*	*adgh*	*cefh*	*dg*	*ce*	*dfgh*
	bdeh	*bcfg*	*bdef*	*bcgh*	*abdefh*	*abcg*	*abde*	*abcfgh*
	cdgh	*abcdef*	*cdfg*	*abcdeh*	*acdfgh*	*bcde*	*acdg*	*bcdefh*
	abef	*gh*	*abeh*	*fg*	*be*	*afgh*	*befh*	*ag*
	acfg	*adeh*	*acgh*	*adef*	*cg*	*defh*	*cfgh*	*de*
	bdfg	*bceh*	*bdgh*	*bcef*	*abdg*	*abcefh*	*abdfgh*	*abce*
	cdef	*abcdgh*	*cdeh*	*abcdfg*	*acde*	*bcdfgh*	*acdefh*	*bcdg*

ABCD. ABEF. ACFG, ADEG, BCEG, BDFG, CDEF confounded.

Effects	d.f.
Block	7
Main	8
2-factor	28
3-factor	56
Higher order	28
Total	127

Blocks of 32 Units

Estimable two-factors: All.

Estimable three-factors: All.

Combine blocks 1 and 2; and blocks 3 and 4; blocks 5 and 6; blocks 7 and 8. *ABCD, ABEF, CDEF* confounded.

Effects	d.f.
Block	3
Main	8
2-factor	28
3-factor	56

Higher order   32
Total   127

Blocks of 64 Units

Estimable two-factors: All.

Estimable three-factors: All.

Combine blocks 1–4; and blocks 5–8. *ABCD* confounded.

Effects	d.f.
Block	1
Main	8
2-factor	28
3-factor	56
Higher order	34
Total	127

Blocks of 128 Units

Estimable two-factors: All.

Estimable three-factors: All.

Combine blocks 1- 8.

Effects	d.f.
Main	8
2-factor	28
3-factor	56
Higher order	35
Total	127

Blocks of 8 Units

Estimable two-factors: All except *EG* and *FH* (confounded). Start with the plan for blocks of 32 units. The first 4 rows of blocks 1 and 2 form the first block of 8 units: i.e., this block contains 1, *abcd, adeg, bceg, acfh, bdfh, abefgh, cdefgh*. Similarly, rows 5–8 of blocks 1 and 2 give the second block, rows 9–12 the third and rows 13–16 the fourth. The remaining 12 blocks are formed likewise from blocks 3 and 4; blocks 5 and 6; and blocks 7 and 8.

Effects	d.f.
Block	15
Main	8
2-factor	26
Higher order	78
Total	127

Plan 18: 3^4 Factorial in 27 Units (1/3 Replicate)

Defining contrast: 2 d.f. from *ABCD*, equivalent to putting $D = ABC$ (*Y*)

Blocks of 9 Units

Estimable two-factors: 16 of the 24 d.f. are clear. *CD* (*I*) is lost by
  confounding. Also *AB* (*J*) = *CD* (*J*); *AC* (*I*) =*BD* (*I*); *AD* (*I*) = *BC* (*I*).

Blocks	(1)	(2)	(3)
	0000	0021	0012
	0122	0110	0101
	0211	0202	0220
	1022	1010	1001
	1111	1102	1120
	1200	1221	1212
	2011	2002	2020
	2100	2121	2112
	2222	2210	2201

Effects	d.f.
Block	2
Main	8
2-factor	16
Total	26

If all interactions of *D* are negligible, the analysis may be written:

Effects	d.f.
Block	2
Main	8
AB, AC, BC	12
Error (from interactions of *D*)	4
Total	26

Plan 19:   3⁵ Factorial in 81 Units (1/3 Replicate)

Defining contrast: 2 d.f. from *ABCDE*

Blocks of 9 Units

Estimable two-factors: All except *AE* (*J*), which is confounded with blocks.

Blocks *ab*	(1) *cde*	(2) *cde*	(3) *cde*	(4) *cde*	(5) *cde*	(6) *cde*	(7) *cde*	(8) *cde*	(9) *cde*
00	000	201	102	120	021	222	111	012	210
10	122	020	221	212	110	011	200	101	002
20	211	112	010	001	202	100	022	220	121
01	110	011	212	200	101	002	221	122	020
11	202	100	001	022	220	121	010	211	112
21	021	222	120	111	012	210	102	000	201
02	220	121	022	010	211	112	001	202	100
12	012	210	111	102	000	201	120	021	222
22	101	002	200	221	122	020	212	110	011

Effects	d.f.
Block	8
Main	10
2-factor	38
Higher order	24
Total	80

Blocks of 27 Units

Estimable two-factors: All.

Combine blocks 1–3; blocks 4–6; blocks 7–9.

Effects	d.f.
Block	2
Main	10
2-factor	40
Higher order	28
Total	80

Blocks of 81 Units

Estimable two-factors: All.

Combine blocks 1–9.

Effects	d.f.
Main	10
2-factor	40
Higher order	30
Total	80

Plan 20: 4×2^4 Factorial in 32 Units (1/2 Replicate)

Defining contrast: *ABCDE*

Blocks of 8 Units

Estimable two-factors: All except *DE* (confounded with blocks).

Blocks ab	(1) cde	(2) cde	(3) cde	(4) cde
00	100	010	111	001
01	011	101	000	110
10	011	101	000	110
11	100	010	111	011
20	111	001	100	010
21	000	110	011	101
30	000	110	011	101
31	111	001	100	010

Effects	d.f.
Block	3
Main	7
2-factor	17
Higher order	4
Total	31

Blocks of 16 Units

Estimable two-factors: All.

Combine blocks 1, 2; and blocks 3, 4.

Effects	d.f.
Block	1
Main	7
2-factor	18
Higher order	5
Total	31

Blocks of 32 Units

Estimable two-factors: All.

Combine blocks 1, 2, 3, and 4.

Effects	d.f.
Main	7
2-factor	18
Higher order	6
Total	31

Source: From Cochran, William G. and Cox, Gertrude M., *Experimental Designs*, 2nd ed., John Wiley & Sons, 1957. With permission.

Appendix H

Significant Studentized Ranges for 5% and 1% Level New Multiple Range Test

Error d.f.	Protection level	ρ = Number of Means for Range Being Tested													
		2	3	4	5	6	7	8	9	10	12	14	16	18	20
1															
	.05	18.0	18.0	18.0	18.0	18.0	18.0	18.0	18.0	18.0	18.0	18.0	18.0	18.0	18.0
	.01	90.0	90.0	90.0	90.0	90.0	90.0	90.0	90.0	90.0	90.0	90.0	90.0	90.0	90.0
2															
	.05	6.09	6.09	6.09	6.09	6.09	6.09	6.09	6.09	6.09	6.09	6.09	6.09	6.09	6.09
	.01	14.0	14.0	14.0	14.0	14.0	14.0	14.0	14.0	14.0	14.0	14.0	14.0	14.0	14.0
3															
	.05	4.50	4.50	4.50	4.50	4.50	4.50	4.50	4.50	4.50	4.50	4.50	4.50	4.50	4.50
	.01	8.26	8.5	8.6	8.7	8.8	8.9	8.9	9.0	9.0	9.0	9.1	9.2	9.3	9.3
4															
	.05	3.93	4.01	4.02	4.02	4.02	4.02	4.02	4.02	4.02	4.02	4.02	4.02	4.02	4.02
	.01	6.51	6.8	6.9	7.0	7.1	7.1	7.2	7.2	7.3	7.3	7.4	7.4	7.5	7.5
5															
	.05	3.64	3.74	3.79	3.83	3.83	3.83	3.83	3.83	3.83	3.83	3.83	3.83	3.83	3.83
	.01	5.70	5.96	6.11	6.18	6.26	6.33	6.40	6.44	6.50	6.60	6.60	6.70	6.70	6.80
6															
	.05	3.46	3.58	3.64	3.68	3.68	3.68	3.68	3.68	3.68	3.68	3.68	3.68	3.68	3.68
	.01	5.24	5.51	5.65	5.73	5.81	5.88	5.95	6.00	6.00	6.10	6.20	6.20	6.30	6.30
7															
	.05	3.35	3.47	3.54	3.58	3.60	3.61	3.61	3.61	3.61	3.61	3.61	3.61	3.61	3.61
	.01	4.95	5.22	5.37	5.45	5.53	5.61	5.69	5.73	5.80	5.80	5.90	5.90	6.00	6.00
8															
	.05	3.26	3.39	3.47	3.52	3.55	3.56	3.56	3.56	3.56	3.56	3.56	3.56	3.56	3.56
	.01	4.74	5.00	5.14	5.23	5.32	5.40	5.47	5.51	5.50	5.60	5.70	5.70	5.80	5.80
9															
	.05	3.20	3.34	3.41	3.47	3.50	3.52	3.52	3.52	3.52	3.52	3.52	3.52	3.52	3.52
	.01	4.60	4.86	4.99	5.08	5.17	5.25	5.32	5.36	5.40	5.50	5.50	5.60	5.70	5.70
10															
	.05	3.15	3.30	3.37	3.43	3.46	3.47	3.47	3.47	3.47	3.47	3.47	3.47	3.47	3.48
	.01	4.48	4.73	4.88	4.96	5.06	5.13	5.20	5.24	5.28	5.36	5.42	5.48	5.54	5.55

Significant Studentized Ranges for 5% and 1% Level New Multiple Range Test (continued)

Error d.f.	Protection level	ρ = Number of Means for Range Being Tested													
		2	3	4	5	6	7	8	9	10	12	14	16	18	20
11	.05	3.11	3.27	3.35	3.39	3.43	3.44	3.45	3.46	3.46	3.46	3.46	3.46	3.47	3.48
	.01	4.39	4.63	4.77	4.86	4.94	5.01	5.06	5.12	5.15	5.24	5.28	5.34	5.38	5.39
12	.05	3.08	3.23	3.33	3.36	3.40	3.42	3.44	3.44	3.46	3.46	3.46	3.46	3.47	3.48
	.01	4.32	4.55	4.68	4.76	4.81	4.92	4.96	5.02	5.07	5.13	5.17	5.22	5.24	5.26
13	.05	3.06	3.21	3.30	3.35	3.38	3.41	3.42	3.44	3.45	3.45	3.46	3.46	3.47	3.47
	.01	4.26	4.48	4.62	4.69	4.74	4.84	4.88	4.94	4.98	5.04	5.08	5.13	5.14	5.15
14	.05	3.03	3.18	3.27	3.33	3.37	3.39	3.41	3.42	3.44	3.45	3.46	3.46	3.47	3.47
	.01	4.21	4.42	4.55	4.63	4.70	4.78	4.83	3.87	4.91	4.96	5.00	5.04	5.06	5.07
15	.05	3.01	3.16	3.25	3.31	3.36	3.38	3.40	3.42	3.43	3.44	3.45	3.46	3.47	3.47
	.01	4.17	4.37	4.50	4.58	4.64	4.72	4.77	4.81	4.84	4.90	4.94	4.97	4.99	5.00
16	.05	3.00	3.15	3.23	3.30	3.34	3.37	3.39	3.41	3.43	3.44	3.45	3.46	3.47	3.47
	.01	4.13	4.34	4.45	4.54	4.60	4.67	4.72	4.76	4.79	4.84	4.88	4.91	4.93	4.94
17	.05	2.98	3.13	3.22	3.28	3.33	3.36	3.38	3.40	3.42	3.44	3.45	3.46	3.47	3.47
	.01	4.10	4.30	4.41	4.50	4.56	4.63	4.68	4.72	4.75	4.80	4.83	4.86	4.88	4.89
18	.05	2.97	3.12	3.21	3.27	3.32	3.35	3.37	3.39	3.41	3.43	3.45	3.46	3.47	3.47
	.01	4.07	4.27	4.38	4.46	4.53	4.59	4.64	4.68	4.71	4.76	4.79	4.82	4.84	4.85
19	.05	2.96	3.11	3.19	3.26	3.31	3.35	3.37	3.39	3.41	3.43	3.44	3.46	3.47	3.47
	.01	4.05	4.24	4.35	4.43	4.50	4.56	4.61	4.64	4.67	4.72	4.76	4.79	4.81	4.82

Significant Studentized Ranges for 5% and 1% Level New Multiple Range Test (continued)

Error d.f.	Protection level	ρ = Number of Means for Range Being Tested													
		2	3	4	5	6	7	8	9	10	12	14	16	18	20
20															
	.05	2.95	3.10	3.18	3.25	3.30	3.34	3.36	3.38	3.40	3.43	3.44	3.46	3.46	3.47
	.01	4.02	4.22	4.33	4.40	4.47	4.53	4.58	4.61	4.65	4.69	4.73	3.76	4.78	4.79
22															
	.05	2.93	3.08	3.17	3.24	3.29	3.32	3.35	3.37	3.39	3.42	3.44	3.45	3.46	3.47
	.01	3.99	4.17	4.28	4.36	4.42	4.48	4.53	4.57	4.60	4.65	4.68	4.71	4.74	4.75
24															
	.05	2.92	3.07	3.15	3.22	3.28	3.31	3.34	3.37	3.38	3.41	3.44	3.45	3.46	3.47
	.01	3.96	4.14	4.24	4.33	4.39	4.44	4.49	4.53	4.57	4.62	4.64	4.67	4.70	4.72
26															
	.05	2.91	3.06	3.14	3.21	3.27	3.30	3.34	3.36	3.38	3.41	3.43	3.45	3.46	3.47
	.01	3.93	4.11	4.21	4.30	4.36	4.41	4.46	4.50	4.53	4.58	4.62	4.65	4.67	4.69
28															
	.05	2.90	3.04	3.13	3.20	3.26	3.30	3.33	3.35	3.37	3.40	3.43	3.45	3.46	3.47
	.01	3.91	4.08	4.18	4.28	4.34	4.39	4.43	4.47	4.51	4.56	4.60	4.62	4.65	4.67
30															
	.05	2.89	3.04	3.12	3.20	3.25	3.29	3.32	3.35	3.37	3.40	3.43	3.44	3.46	3.47
	.01	3.89	4.06	4.16	4.22	4.32	4.36	4.41	4.45	4.48	4.54	4.58	4.61	4.63	4.65
40															
	.05	2.86	3.01	3.10	3.17	3.22	3.27	3.30	3.33	3.35	3.39	3.42	3.44	3.46	3.47
	.01	3.82	3.99	4.10	4.17	4.21	4.30	4.34	4.37	4.41	4.46	4.51	4.54	4.57	4.59
60															
	.05	2.83	2.98	3.08	3.14	3.20	3.42	3.28	3.31	3.33	3.37	3.40	3.43	3.45	3.47
	.01	3.76	3.92	4.03	4.12	4.17	4.23	4.27	4.31	4.34	4.39	4.44	4.47	4.50	4.53
100															
	.05	2.80	2.95	3.05	3.12	3.18	3.22	3.26	3.29	3.32	3.36	3.40	3.42	3.45	3.47
	.01	3.71	3.86	3.98	4.06	4.11	4.17	4.21	4.25	4.29	4.35	4.38	4.42	4.45	4.48
∞															
	.05	2.77	2.92	3.02	3.09	3.15	3.19	3.23	3.26	3.29	3.34	3.38	3.41	3.44	3.47
	.01	364	3.80	3.90	3.98	4.04	4.09	4.14	4.17	4.20	4.26	4.31	4.34	4.38	4.41

Source: Printed with permission of the Biometric Society, North Carolina.

Appendix I

Student t Distribution

Degrees of Freedom	Upper-Tail Area									
	.4	.25	.1	.05	.025	.01	.005	.0025	.001	.0005
1	0.325	1.000	3.078	6.314	12.706	31.821	63.657	127.32	318.31	636.62
2	.289	.816	1.886	2.920	4.303	6.965	9.925	14.089	22.327	31.598
3	.277	.765	1.638	2.353	3.182	4.541	5.841	7.453	10.214	12.924
4	.271	.741	1.533	2.132	2.776	3.747	4.604	5.598	7.173	8.610
5	0.267	0.727	1.476	2.015	2.571	3.365	4.032	4.773	5.893	6.869
6	.265	.718	1.440	1.943	2.447	3.143	3.707	4.317	5.208	5.959
7	.263	.711	1.415	1.895	2.365	2.998	3.499	4.029	4.785	5.408
8	.262	.706	1.397	1.860	2.306	2.896	3.355	3.833	4.501	5.041
9	.261	.703	1.383	1.833	2.262	2.821	3.250	3.690	4.297	4.781
10	0.260	0.700	1.372	1.812	2.228	2.764	3.169	3.581	4.144	4.587
11	.260	.697	1.363	1.796	2.201	2.718	3.106	3.497	4.025	4.437
12	.259	.695	1.356	1.782	2.179	2.681	3.055	3.428	3.930	4.318
13	.259	.694	1.350	1.771	2.160	2.650	3.012	3.372	3.852	4.221
14	.258	.692	1.345	1.761	2.145	2.624	2.977	3.326	3.787	4.140
15	0.258	0.691	1.341	1.753	2.131	2.602	2.947	3.286	3.733	4.073
16	.258	.690	1.337	1.746	2.120	2.583	2.921	3.252	3.686	4.015
17	2.57	.689	1.333	1.740	2.110	2.567	2.898	3.222	3.646	3.965
I8	.257	.688	1.330	1.734	2.101	2.552	2.878	3.197	3.610	3.922
19	.257	.688	1.328	1.729	2.093	2.539	2.861	3.174	3.579	3.883
20	0.257	0.687	1.325	1.725	2.086	2.528	2.845	3.153	3.552	3.850
21	.257	.686	1.323	1.721	2.080	2.518	2.831	3.135	3.527	3.819
22	.256	.686	1.321	1.717	2.074	2.508	2.819	3.119	3.505	3.792
23	.256	.685	1.319	1.714	2.069	2.500	2.807	3.104	3.485	3.767
24	.256	.685	1.318	1.711	2.064	2.492	2.797	3.091	3.467	3.745
25	0.256	0.684	1.316	1.708	2.060	2.485	2.787	3.078	3.450	3.725
26	.256	.684	1.315	1.706	2.056	2.479	2.779	3.067	3.435	3.707
27	.256	.684	1.314	1.703	2.052	2.473	2.771	3.057	3.421	3.690
28	.256	.683	1.313	1.701	2.048	2.467	2.763	3.047	3.408	3.674
29	.256	.683	1.311	1.699	2.045	2.462	2.756	3.038	3.396	3.659
30	0.256	0.683	1.310	1.697	2.042	2.457	2.750	3.030	3.385	3.646
40	.255	.681	1.303	1.684	2.021	2.423	2.704	2.971	3.307	3.551
60	.254	.679	1.296	1.671	2.000	2.390	2.660	2.915	3.232	3.460
120	.254	.677	1.289	1.658	1.980	2.358	2.617	2.860	3.160	3.373
∞	.253	.674	1.282	1.645	1.960	2.326	2.576	2.807	3.090	3.291

Note: The table provides the values of t_α that correspond to a given upper-tail area and a specified number of degrees of freedom.

Source: Pearson, E.S. and Hartley, K.O., *Biometrika Tables for Statisticians*, Vol. I, Cambridge University Press, London, 1966. Partly derived from Table III of R.A. Fisher and F. Yates, *Statistical Tables for Biological, Agricultural and Medical Research*, Longman Group, London (previously published by Oliver & Boyd. Edinburgh, 1963). Reproduced with permission of the authors and publishers.

Appendix J

Coefficients and the Sum of Squares of Sets of Orthogonal Polynomials When There Are Equal Interval Treatments

Degree of Polynomials	Comparison	Number of Levels							Sum of Squares of the Coefficients $\left(\sum c^2\right)$
		T_1	T_2	T_3	T_4	T_5	T_6	T_7	
1	Linear	−1	+1						2
2	Linear	−1	0	+1					2
	Quadratic	+1	−2	+1					6
3	Linear	−3	−1	+1	+3				20
	Quadratic	+1	−1	−1	+1				4
	Cubic	−1	+3	−3	+1				20
4	Linear	−2	−1	0	+1	+2			10
	Quadratic	+2	−1	−2	−1	+2			14
	Cubic	−1	+2	0	−2	+1			10
	Quartic	+1	−4	+6	−4	+1			70
5	Linear	−5	−3	−1	+1	+3	+5		70
	Quadratic	+5	−1	−4	−4	−1	+5		84
	Cubic	−5	+7	+4	−4	−7	+5		180
	Quartic	+1	−3	+2	+2	−3	+1		28
	Quintic	−1	+5	−10	+10	−5	+1		252
6	Linear	−3	−2	−1	0	+1	+2	+3	28
	Quadratic	+5	0	−3	−4	−3	0	+5	84
	Cubic	−1	+1	+1	0	−1	−1	+1	6
	Quartic	+3	−7	+1	+6	+1	−7	+3	154
	Quintic	−1	+4	−5	0	+5	−4	+1	84
	Sextic	+1	−6	+15	−20	+15	−6	+1	924

Source: Reprinted by permission from *Statistical Methods*, by Snedecor, George W. and Cochran, William G., 6th ed., Iowa State University Press, Ames, 1971.

Appendix K

MINITAB

K.1 Introduction

The purpose of this appendix is to introduce the users of this text to MINITAB. MINITAB has been used extensively in performing various statistical analyses. Throughout this book, you have been introduced to the various designs in agricultural and natural science experiments. In each chapter we analyzed problems using a step approach. This was done to give you a better understanding of the nature of the data and how to analyze and interpret the results. We now turn to the use of MINITAB to perform a variety of statistical analyses. To use MINITAB, some basic skills of working around Windows are assumed.

K.2 MINITAB Basics

To start MINITAB from your program files, open MINITAB by double clicking the icon. ▧ Mtb.exe MINITAB opens up with two windows: the "Session" window on the top half of the screen and the "Data" window on the bottom half of the screen, as shown in Figure K.1.

The data window contains the worksheet where you can enter data. Each column holds one variable. The default names for variables are C_1, C_2, C_3, ..., etc., and they are automatically listed on the top row of the worksheet. Each column can be renamed as you wish. It is recommended that the names of the variables should be entered in the cells right under these default names. The worksheet can have as many columns and rows as your computer's available memory allows.

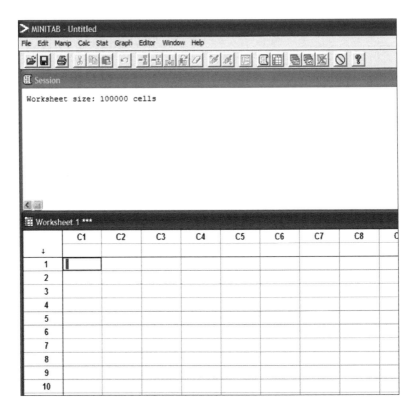

FIGURE K.1
MINITAB "Session" and "Data" windows.

K.2.1 Storing Data

To use MINITAB for any analysis, you need to enter the data either manually or by transferring them from ready files. Entering data in MINITAB is typical as in any other data analysis software. We use Example 4.2.1.2 in Chapter 4 to show how we have entered the data as shown in Figure K.2. As you note, the data for treatment is entered in column 1 for all replications. In the second column we have entered the treatments numbered from 1 to 8. Since we had five replications, the treatment values are duplicated for each of the replications as shown in Figure K.2.

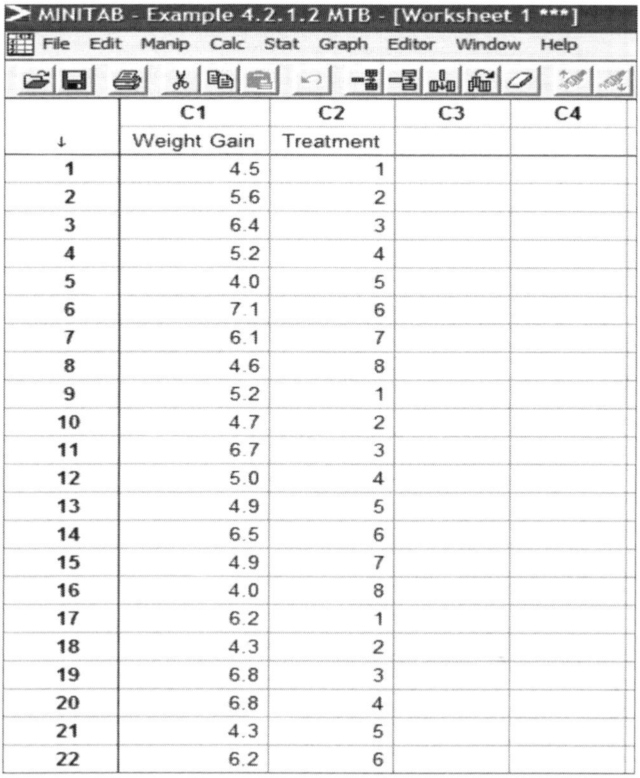

FIGURE K.2
Data entry in a MINITAB worksheet.

K.2.2 Analysis of the Data

Once you have entered or imported the data in a MINITAB worksheet you are ready to perform the desired analysis. As this problem deals with a single-factor experimental design, we will use the one-way analysis of variance.

On the menu bar, look for "Stat" button. Right click on it and you will see a drop down menu with several options for analysis. Go to ANOVA One Way. You will see a screen as shown in Figure K.3.

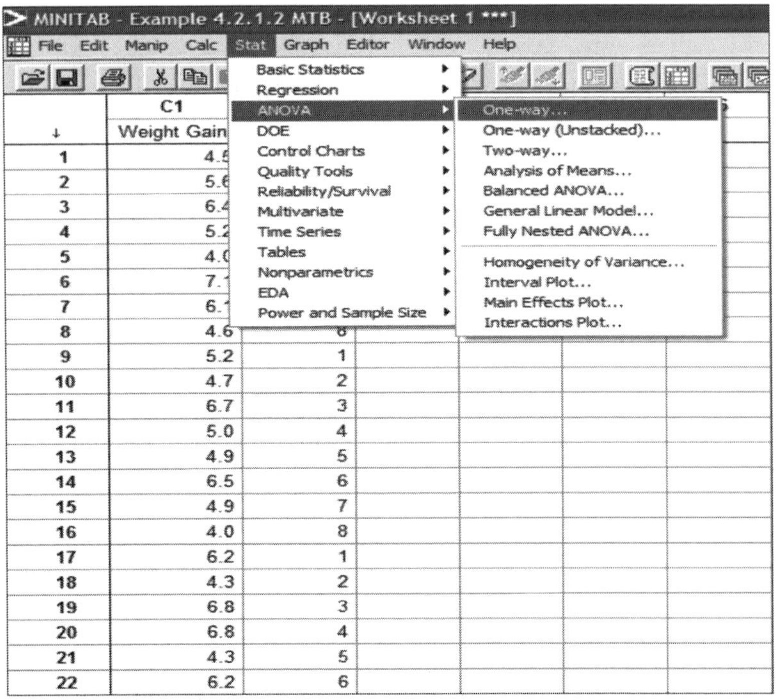

	C1				
↓	Weight Gain				
1	4.5				
2	5.6				
3	6.4				
4	5.2				
5	4.0				
6	7.1				
7	6.1				
8	4.6	6			
9	5.2	1			
10	4.7	2			
11	6.7	3			
12	5.0	4			
13	4.9	5			
14	6.5	6			
15	4.9	7			
16	4.0	8			
17	6.2	1			
18	4.3	2			
19	6.8	3			
20	6.8	4			
21	4.3	5			
22	6.2	6			

FIGURE K.3
"Stat" screen pull down menu.

Observe that a dialog box appears on the screen as shown in Figure K.4. On the left side you will see C_1 showing data (weight gain) and C_2 showing treatment. On the right side you will see two boxes one labeled "Response" and the other "Factor." By placing the cursor on C_1 and double clicking on it you will see that the data appears in the response box. Similarly, you can move your cursor to the Factor box. You will now be able to double click on the treatment, and it will show up in the box. Then click on "Ok." The session window will show the result of the analysis as shown in Figure K.5.

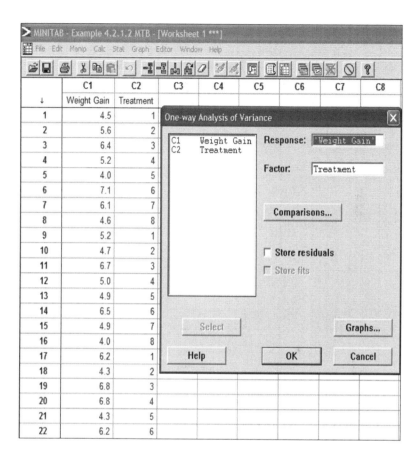

FIGURE K.4
Dialog box for analysis.

You will note that the ANOVA table in the middle of this screen provides all the information for analysis and decision making. You will recall from the discussion in various chapters that the decision process for a hypothesis test can be based on the probability value (p-value) for the given test. In this particular instance, the p-value (0.000) provides sufficient evidence that the average weight gain for at least one treatment from the others when α is 0.05 is different. In the individual 95% confidence intervals table that appears right after the ANOVA table in Figure K.5, you will note that several intervals overlap. You need to do multiple comparison tests to see the differences that exist among the treatments.

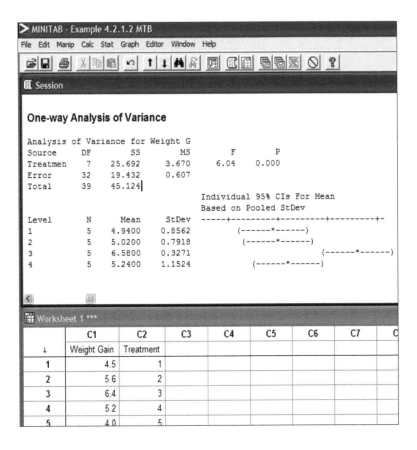

FIGURE K.5
Results of the analysis shown in the session window.

To save the data file and the analysis, click on File in the main menu and choose Save.

K.3   Randomized Complete Block Design

We have used the data for Example 4.2.2.1 to illustrate the use of MINITAB.

Step 1: Place all the grain yield data in one column as shown in Figure K.6. As we have 6 treatments, the second column shows the treatments labeled as 1, 2, 3, 4, 5, and 6. Similarly, we have labeled the replications as 1, 2, and 3. The replications are placed in the third column as shown in Figure K.6.

	C1	C2	C3	C4	C5	C6
↓	Grain Yield	Treatments	Replication			
1	147.0	1	1			
2	159.4	2	1			
3	158.9	3	1			
4	173.6	4	1			
5	158.4	5	1			
6	157.1	6	1			
7	130.1	1	2			
8	167.3	2	2			
9	166.2	3	2			
10	170.8	4	2			
11	169.3	5	2			
12	148.8	6	2			
13	142.2	1	3			
14	150.5	2	3			
15	159.1	3	3			
16	162.5	4	3			
17	160.2	5	3			
18	139.0	6	3			

FIGURE K.6
Data for Example 4.2.2.1.

Step 2: On the menu bar, look for Stat button. Right click on it and you will see a drop down menu with several options for analysis. Go to ANOVA Two Way.

Step 3: First select ANOVA, and then select Two Way on the right screen as shown in Figure K.7.

Step 4: Observe that a dialog box appears on the screen as shown in Figure K.8. On the left side you will see C_1 showing the grain yield data, C_2 showing treatment, and C_3 the replications. On the right side you will see three boxes one labeled "Response" and the other two are Row Factor and Column Factor, respectively. By placing the cursor on C_1 and double clicking on it you will see that "grain yield" appears in the response box. Similarly, you can move your cursor to the Row Factor and Column Factor boxes. You will now be able to double click on the treatment, and it will show up in the "factor box"; then click on "Ok."

Step 5: Use the information found in Figure K.9 to make statistical decision.

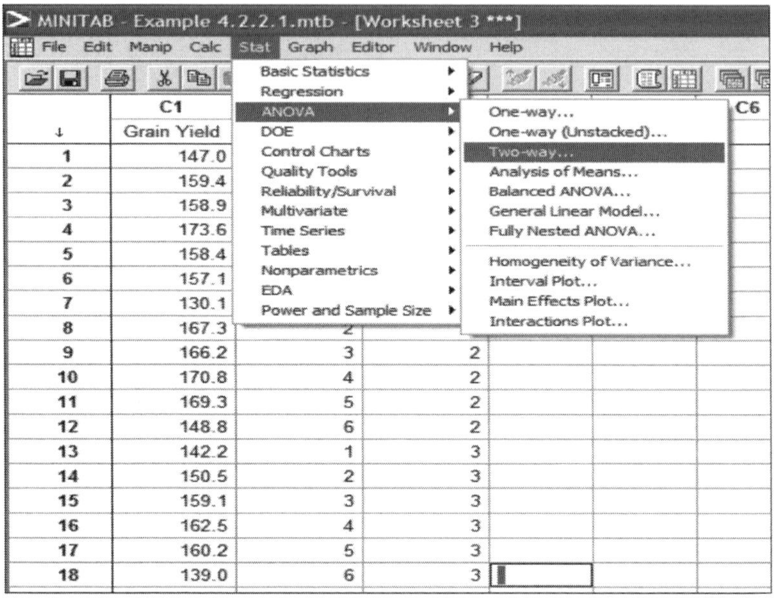

FIGURE K.7
Choice of analysis menu.

K.4 Regression Analysis

Steps for using MINITAB in regression analysis are:

Step 1: Open the open MINITAB by double clicking the icon.
Mtb.exe

Step 2: Enter the data manually to C1, C2, C3, ..., etc., or cut and paste from other sources. We have used the data from Example 9.2.2 as shown in Figure K.10.

Step 3: Select Stat ⇒ Regression ⇒ Regression from the menu as shown in Figure K.11.

Step 4: You will note that two dialog boxes appear on the screen as shown in Figure K.12. Place the cursor over C_2 and double click on it. You will note that "Yield" will appear in the Response box. Similarly, place the cursor on C_1 and double click on it. You should see the "predictor" variable appear in the box. In the case of multiple regression you can select all the predictor variables to be placed in this box.

Step 5: The results of the analysis are shown in Figure K.13.

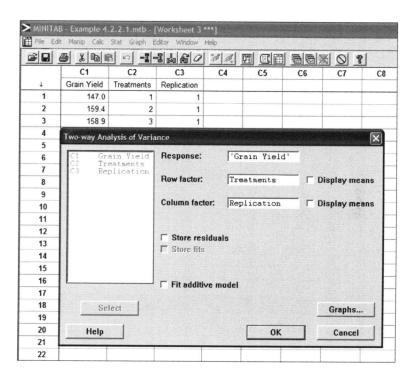

FIGURE K.8
Dialog box for two-way analysis.

As shown in Figure K.13, the top section provides the estimated regression equation. The table following the equation provides the t-statistics and the p-value for the estimated regression equation. The standard error of estimate and the coefficient of determination (R^2) and the adjusted coefficient of determination are given. The analysis of variance table provides the F-test and the p-value for this example. In addition to the analysis of variance table, the unusual observation as well as the Durbin–Watson statistic is provided. For the interpretation of the Durbin–Watson statistic, please refer to the section on serial and autocorrelation in Chapter 9.

To use the more elaborate design models in the MINITAB, the reader is referred to the following books:

Meyer, R. and Krueger, D. (2005). *A Minitab Guide to Statistics.* New Jersey: Prentice Hall.
Wakefield, D. and McLaughlin, K. (2004). *An Introduction to Data Analysis Using Minitab for Windows.* New Jersey: Prentice Hall.

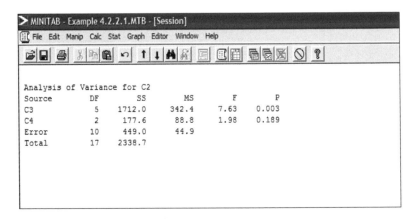

```
> MINITAB - Example 4.2.2.1.MTB - [Session]
 File  Edit  Manip  Calc  Stat  Graph  Editor  Window  Help


Analysis of Variance for C2
Source      DF       SS       MS       F       P
C3           5    1712.0    342.4    7.63    0.003
C4           2     177.6     88.8    1.98    0.189
Error       10     449.0     44.9
Total       17    2338.7
```

FIGURE K.9
Session window showing the results.

	C1	C2	C3	C4	C5	C6	C7
↓	Yield	Fertilizer					
1	50	5					
2	57	10					
3	60	12					
4	62	18					
5	63	25					
6	65	30					
7	68	36					
8	70	40					
9	69	45					
10	66	48					

> MINITAB - Example.9.2.2.mtb - [Worksheet 1 ***]

FIGURE K.10
Data input cells.

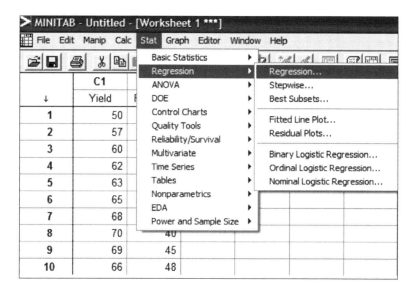

FIGURE K.11
Choice of statistical analyses.

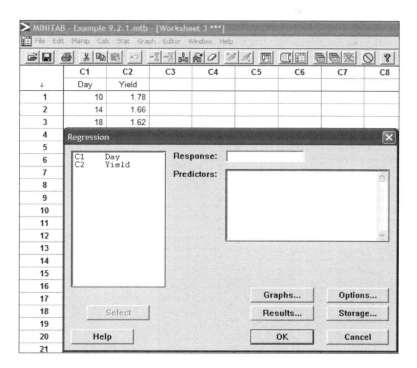

FIGURE K.12
Dialog box with options for graphs, results, option, and storage.

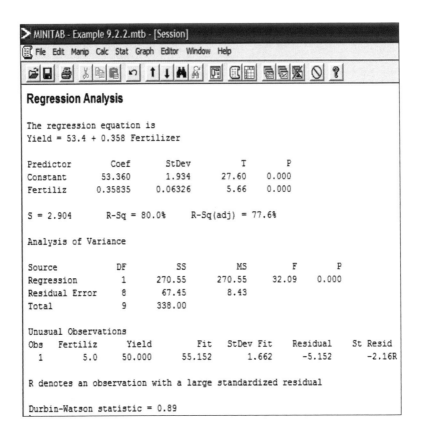

FIGURE K.13
Regression analysis.

Index

A

Accidents, *see* Missing data
Accuracy, *see also* Efficiency
  completely randomized design, 52
  covariance analysis, 321
  factorial experiments, 120
  incomplete block designs, 84
  randomized complete block design,
    64–65
Additivity assumption, 14–16, 19, 29
Agricultural research, 1–11
Aken´Ova, Ntare and, studies, 9
Aliases, 186–199, *187–190, 193*
Alphabet notation, 222
Analysis of variance (ANOVA)
  arc sine transformation, 33–34
  assumptions, 14–19
  balanced lattice designs, 87–94,
    89, 93
  completely randomized design, 51,
    56, *56*
  data transformation, 27
  fractional replication, 189–199,
    192–193, 198
  genotypic differences, 9
  Greco-Latin square, 48, *49*
  homogenous variance assumption,
    18
  Latin square designs, 76–80, *77, 80*
  logarithmic transformation, 30–31
  partially balanced lattice designs,
    without repetition, 95–102,
    96, 101
  partially balanced lattice designs,
    with repetition, 102–111, *103,
    105, 109*
  randomized complete block design,
    67–71, *69*
  split-plot designs, 137–143, *138, 141*

split-split-plot design, 163–172, *164,
    166, 168, 171*
  square-root transformation, 28
  strip-plot design, *145–146,* 145–151,
    150
  strip-split-plot design, 173–185,
    176–177, 179–181, 184
Anderson and McLean studies, 3, 24
Anderson studies, 142
Angular transformation, 32–34, *33–34*
ANOVA, *see* Analysis of variance
    (ANOVA)
Anscombe and Tukey studies, 16
Arc sine transformation, 32–34, *33–34,
    353–355*
Assumptions
  additivity, 14–16
  angular transformation, 32–34, *33–34*
  ANOVA, 14–19
  arc sine transformation, 32–34, *33–34*
  basics, 13
  data transformation, 27–34, *29–34*
  failure detection, 19–27
  homogenous variance, 18–19
  independent errors, 17–18
  logarithmic transformation, 28–32,
    30–32
  normality, 16–17
  regression and correlation analysis,
    307
  square-root transformation, 28, *29*
  violations, 14–19
Autocorrelation, 307–308, *308*

B

Balanced lattice designs
  ANOVA, 87–94, *89, 93*
  basic plans, *371–375*
  basics, 51, 85, *86–87, 89, 93*

427

Printed and bound by CPI Group (UK) Ltd, Croydon, CR0 4YY

24/10/2024

01778277-0015